ROADSIDE GEOLOGY
of MASSACHUSETTS

James W. Skehan

Mountain Press Publishing Company
2001

Third Printing, September 2006

Cover image from U.S. Geological Survey
EROS Data Center, August 24, 2000

Photos and maps © 2001 by James W. Skehan
unless otherwise credited

ROADSIDE GEOLOGY SERIES
Roadside Geology is a registered trademark of
Mountain Press Publishing Company.

Library of Congress Cataloging-in-Publication Data

Skehan, James William, 1923–
    Roadside geology of Massachusetts / James W. Skehan.
        p. cm. — (Roadside geology series)
    ISBN 0-87842-429-6
        1. Geology—Massachusetts—Guidebooks. 2. Massachusetts—
    Guidebooks. I. Title. II. Series.

QE123 .S54 2001
557.44—dc21
                                                                    00-052720

PRINTED IN THE UNITED STATES OF AMERICA

Mountain Press Publishing Company
P.O. Box 2399 • Missoula, Montana 59806
(406) 728-1900

I dedicate this volume to my first mentors,
James W. and Mary Effie Coffey Skehan, and to
many others who have mentored and inspired me along the way—
especially my professors Thomas J. Quigley, S.J.; Michael J. Ahern,
S.J.; Daniel Linehan, S.J.; Marland P. Billings; James B. Thompson Jr.;
Lincoln R. Page; Nicholas Rast; and Brian Sturt—all delightful
companions and skilled observers in the field.

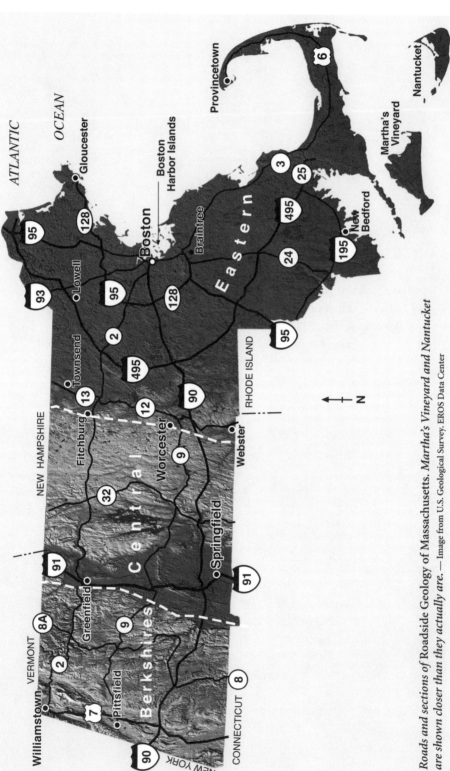

Roads and sections of Roadside Geology of Massachusetts. Martha's Vineyard and Nantucket are shown closer than they actually are. — Image from U.S. Geological Survey, EROS Data Center

CONTENTS

PREFACE   *ix*

ACKNOWLEDGMENTS   *x*

INTRODUCTION   *1*
Three Supercontinents and an Ocean   *8*
Life in Rocks   *17*
Glaciation   *18*

EASTERN SEABOARD   *33*
Merrimack Terrane   *35*
Nashoba Terrane   *37*
Avalon Terrane   *43*
Coal Basins   *54*
Meguma Terrane   *60*
Jurassic Time   *60*
Glacial Geology   *61*

Roadguides to the Eastern Seaboard   *67*
Boston Harbor Islands and Peninsulas   *67*
The Islands: Martha's Vineyard and Nantucket   *77*
U.S. 6: Cape Cod   *82*
Interstate 90 (Mass Pike): Boston—Auburn   *96*
Interstate 93: Methuen—Medford   *108*
Interstate 95: Canton—Attleboro   *112*
Interstate 95: Salisbury—Peabody   *120*
Interstate 95/Massachusetts 128: Peabody—Braintree   *127*
Interstate 195: Wareham—Seekonk   *133*
Interstate 495: Salisbury—Chelmsford   *142*
Interstate 495: Chelmsford—Westborough   *151*
Interstate 495: Westborough—Norton   *158*

Interstate 495: Norton—Cape Cod Canal   *163*

Massachusetts 2: Cambridge—Leominster   *171*

Massachusetts 3 (The Southeast Expressway):
    Quincy—Sagamore Bridge   *183*

Massachusetts 12 and 13: Webster—Townsend   *193*

Massachusetts 24: Randolph—Fall River   *204*

Massachusetts 128: Gloucester—Peabody   *210*

CENTRAL LOWLAND AND BRONSON HILL UPLAND   *219*

Colliding Continents   *219*

Merrimack Terrane   *222*

Mesozoic Rift Basins   *225*

Glacial Geology   *228*

Roadguides to the Central Lowland and Bronson Hill Upland   *235*

Interstate 90 (Mass Pike): Auburn—Springfield Area   *235*

Interstate 91: Connecticut River Valley   *242*

Massachusetts 2: Fitchburg—Connecticut River   *258*

Massachusetts 9: Worcester—Amherst   *265*

Massachusetts 32: South Monson—Royalston   *272*

THE BERKSHIRES   *283*

Laurentian Terranes   *285*

Glacial Geology   *295*

Roadguides to the Berkshires   *299*

Interstate 90 (Mass Pike): Westfield—West Stockbridge   *299*

U.S. 7: Sheffield—Williamstown   *310*

Massachusetts 2 (The Mohawk Trail):
    Greenfield—Williamstown   *321*

Massachusetts 9: Northampton—Pittsfield   *332*

A Circuitous Route through the Berkshire Hills:
    Sandisfield—Heath   *340*

GLOSSARY   *351*

SELECTED READING   *360*

INDEX   *367*

MAP SYMBOLS

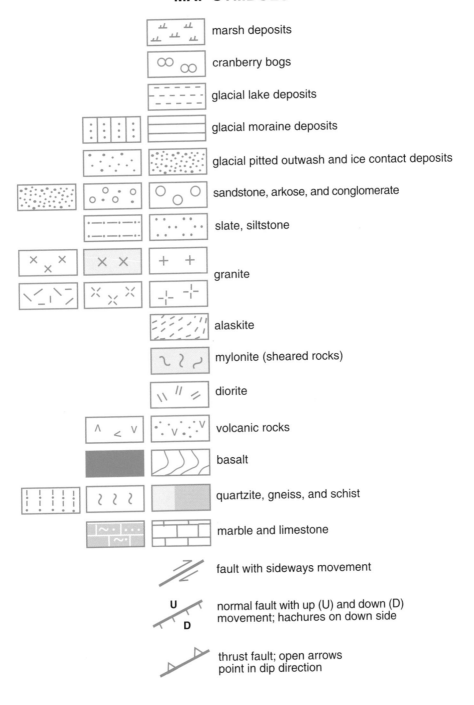

marsh deposits

cranberry bogs

glacial lake deposits

glacial moraine deposits

glacial pitted outwash and ice contact deposits

sandstone, arkose, and conglomerate

slate, siltstone

granite

alaskite

mylonite (sheared rocks)

diorite

volcanic rocks

basalt

quartzite, gneiss, and schist

marble and limestone

fault with sideways movement

normal fault with up (U) and down (D) movement; hachures on down side

thrust fault; open arrows point in dip direction

PREFACE

I divided the book into three sections—the Eastern Seaboard, the Central Lowland and Bronson Hill Upland, and the Berkshires—based on geology and geography. The Eastern Seaboard section discusses Gondwana terranes, the growing continental margin, the formation of coal basins, and the glacially shaped coastline. The Central Lowland and Bronson Hill Upland section discusses collisions of island arcs and the formation of the large rift basin that would later hold Glacial Lake Hitchcock and still later the Connecticut River. The mountainous Berkshire region contains some of the earliest history of the North American continent and the ice age glaciers' passage over these hills. An in-depth introduction to each section discusses the regional geology.

Throughout this book I have tried to describe the geology using terms that are as close to everyday English as possible. The geological words that I have retained are packages full of meaning. Once you unwrap a word or phrase, such as *plate tectonics,* and picture the concept or process, you can understand many interesting ideas. I have also included a glossary with snippets of geologic jargon. Use it as a quick reference to help lead you to the bigger ideas. The glossary contains some words you ought to know to understand basic ideas or concepts.

For readers who want to use this book to maximum advantage in the field, I suggest several additional resources: the colorful *Bedrock Geologic Map of Massachusetts* (scale 1:250,000) together with its companion sheet of rock formations and cross sections published by the U.S. Geological Survey; the U.S. Geological Survey topographic quadrangle maps (scale 1:24,000) and double quadrangle maps (scale 1:25,000); and the U.S. Geological Survey bedrock and glacial geology quadrangle maps. Regional atlases that contain street maps for townships will help readers locate specific sites.

ACKNOWLEDGMENTS

I gratefully acknowledge the unparalleled contribution to this volume by Christine C. McDonald Bronchuk, my research assistant, who carried full responsibility for the production of illustrations and whose organizational skills and meticulous attention to detail kept track of all aspects of the manuscript production. She coordinated with Frances Ahearn, to whom I owe an enormous debt of gratitude for her wizardry in transforming early drafts of illustrations into finished computerized geologic maps and other illustrations. Scott Barlow was also very helpful during a critical stage in the digitizing of photographs, computer generated diagrams, maps, and overlays on photographs illustrating geological structures.

Photographers, credited by name in the text, generously contributed their time and skill to record geological features in the field. Those photographs and other illustrations and diagrams that lack attribution are by the author. The entire text was read by Mary Havreluk and Donna K. Chambers, who, having no previous geological background, identified an initial list of terms for the glossary that were required to clarify the geological story for the interested general reader.

Geologists who contributed constructive reviews of early drafts of manuscript chapters include Edward S. Belt, Robert F. Boutilier, David P. Dethier, Arthur Goldstein, Lindley S. Hanson, J. Christopher Hepburn, Paul Karabinos, Carl Koteff, William A. Newman, Robert N. Oldale, John D. Peper, and Margaret D. Thompson. Several other geologists provided helpful discussions of both bedrock and glacial geology. I am especially indebted to Bob Oldale of the U.S. Geological Survey, for whatever success he may have had in helping to educate me in glacial geology. However, I take full responsibility for any errors of fact or interpretation of either the bedrock or glacial geology.

Without the generous financial support and encouragement of the following this volume would never have come to fruition: the Boston College Jesuit Community and especially Joseph A. Appleyard, S.J., Francis R. Herrmann, S.J., and James M. Collins, S.J.; John E. Ebel, Director of Weston

Observatory; Vincent J. Murphy, Weston Geophysical Inc.; Randolph J. Martin III; and Glen Dash of the Dash Foundation. It is impossible to properly thank all those who contributed to the production of this book in intangible but important ways, but I would especially like to thank Patricia C. Tassia, my longtime assistant. Tracy S. Downing; Linda W. Katz; staff members at Weston Observatory; and Patricia Donovan, librarian at Weston Observatory Library. I am very grateful to Jennifer Carey, Kim Ericsson, and Kathleen Ort of Mountain Press who were very helpful in the editorial process and who, together with David Alt, greatly improved all aspects of the manuscript.

ERA	PERIOD	EPOCH million years ago	IMPORTANT GEOLOGIC EVENTS IN MASSACHUSETTS
CENOZOIC	QUATERNARY	Pleistocene — 1.6 —	Wisconsinan ice age begins 80,000 years ago and covers Massachusetts between 25,000 and 15,000 years ago. Marine sediments deposited during Sangamon interglacial stage. Remnant till of Illinoian ice sheet deposited 140,000 years ago.
CENOZOIC	TERTIARY	Pliocene — 5 — Miocene — 24 — Oligocene — 36 — Eocene — 58 — Paleocene — 65 —	Deposition of glauconitic sands, coarse sands, and gravel.
MESOZOIC	CRETACEOUS	 — 145 —	Intrusive rhyolite in northeastern Massachusetts. Deposition of variegated clays, silts, and lignite coal at Gay Head. Marine sediments of coastal plain deposited far inland.
MESOZOIC	JURASSIC	 — 208 —	Rift volcanism initiates opening of Atlantic Ocean and breakup of Pangaea. Rift basins open in the Connecticut Valley region. Basalt flows and dikes, including Medford dike. Deposition of fossiliferous redbeds. Dinosaurs leave tracks.
MESOZOIC	TRIASSIC	 — 245 —	Deposition of coarse clastic sediments.
PALEOZOIC	PERMIAN	 — 286 —	Final assembly of Pangaean supercontinent during the Alleghanian orogeny, 275 to 250 million years ago.
PALEOZOIC	PENNSYLVANIAN	— 320 —	Narragansett Basin and other coal basins form in Avalon terrane.
PALEOZOIC	MISSISSIPPIAN	— 360 —	Rapid uplift of Nashoba terrane.
PALEOZOIC	DEVONIAN	 — 417 —	Continued sedimentation. Acadian mountain building event—Merrimack, Nashoba, and Avalon microcontinents collide with Laurentia and its associated volcanic island chains. Collision produces extensive plutonism and dome uplift. Rift plutonism and volcanism in Avalon terrane.
PALEOZOIC	SILURIAN	 — 443 —	Initial stage of Acadian mountain building event. Sedimentary rocks deposited unconformably on Bronson Hill volcanic belt of Laurentia. Edge of Avalon terrane sinks beneath Nashoba terrane in subduction zone, generating more Burlington mylonite. Volcanic and plutonic activity begun in Ordovician time continues to build Nashoba and Merrimack terranes.
PALEOZOIC	ORDOVICIAN	 — 495 —	Shelburne Falls and possibly Bronson Hill volcanic island chains, which formed along margin of Laurentia, collide with continent in the Taconic mountain building event. Rifts open in Avalon and produce alkaline plutonic activity.
PALEOZOIC	CAMBRIAN	 — 545 —	Fossiliferous continental shelf sediments—Stockbridge marble and Cheshire quartzite—deposited on Laurentian margin. Trilobite-bearing sediments deposited on margins of Avalon.
PRECAMBRIAN	PROTEROZOIC EON	 — 2,500 —	Avalon and associated microcontinents separate from Gondwana 550 million years ago. Boston rift basin forms in Avalon about 570 million years ago. Major faulting and shearing along margin of Gondwana forms the Burlington mylonite. Magmas from the Avalon volcanic chain intruded the mylonite, forming the Dedham and Milford granites. Rodinia supercontinent completely assembled by 750 million years ago, then breaks up, giving rise to Gondwana supercontinent. Grenvillian mountain building event affects Grenville gneisses on eastern margin of Laurentia, 1.2 to 1.1 billion years ago.
PRECAMBRIAN	ARCHEAN EON		

Geologic timescale. —Geologic times from Palmer, 1983; Tucker and McKerrow, 1995; Bowring and others, 1993

Introduction

From Mount Greylock in the Berkshires to Boston Harbor Islands National Recreation Area, Massachusetts has a wealth of geologic wonders. Sparkling, quartz-rich sandy beaches, tidal estuaries, and rocky headlands ornament the coast, while granite, schist, and basalt crisscross the interior. For such a small state the geology is remarkably diverse. You can find rocks and minerals from almost every geologic setting imaginable. They all ended up in Massachusetts because continents kept colliding together here, compacting rocks into tight bands. Pioneer geologists at the colleges and universities of Massachusetts and the U.S. Geological Survey advanced the science of geology by studying this complex array of rocks, fossils, and glacial deposits.

Human settlement of Massachusetts intertwines with the geology. Native Americans chiseled tools from volcanic rocks at the Blue Hills. A bountiful spring emerging from glacial gravels in Beacon Hill enticed the Puritans to settle at Boston. Workers in Quincy built the first commercial railroad in North America with blocks of local granite. Saugus claims the restored ironworks at Hammersmith as the "Birthplace of the American Iron and Steel industry," dating to 1643. Poor-quality coal from 300-million-year-old swamps heated some settlements and stoked industrial fires.

Geologic Time

The oldest rocks in Massachusetts, the Grenville gneisses, are over 1 billion years old. By contrast, the most recent continental ice sheet dumped its last load of rock in Massachusetts a mere 15,000 to 14,000 years ago. A lot can happen in 1 billion years—supercontinents can collide and then rift apart several times—and mountains the size of the Swiss Alps can form and then erode away. Scientists devised the geologic timescale to help sort out these momentous events.

The geologic timescale of eons, eras, periods, and epochs records the relative stacking order of rock formations. Boundaries between periods reflect changes in the fossils present or other important events recorded in the rocks. For example, the boundary between Precambrian time and Cambrian time marks the first widespread presence of marine creatures with hard parts, or skeletons. Radiometric age dating of radioactive minerals in igneous rocks determines the time of the boundaries in the geologic timescale.

Radioactive elements in minerals are geologic clocks. A parent isotope, an atom of a chemical element with a specific mass, decays radioactively to become a daughter isotope. By establishing the ratio of parent isotopes to daughter isotopes and measuring the rate of decay of the parent, geologists determine the length of time since the parent mineral crystallized. Such radiometric age determinations are now remarkably accurate.

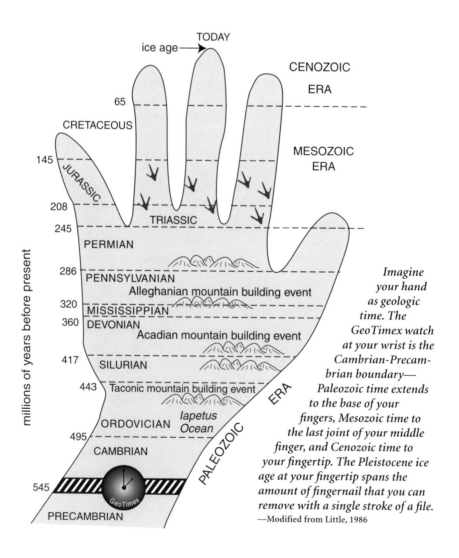

Imagine your hand as geologic time. The GeoTimex watch at your wrist is the Cambrian-Precambrian boundary— Paleozoic time extends to the base of your fingers, Mesozoic time to the last joint of your middle finger, and Cenozoic time to your fingertip. The Pleistocene ice age at your fingertip spans the amount of fingernail that you can remove with a single stroke of a file.
—Modified from Little, 1986

Rock Cycle

The rock cycle, proposed by James Hutton two centuries ago, explains how rocks form through interacting geologic processes. For example, rocks in uplifted mountains weather and erode into fragments. Water and gravity may transport and deposit these rock fragments on the ocean shore as beach sands. Upon burial beneath other sediments, the sand grains cement together, or lithify, into sandstone. If the sandstone is buried deep in the earth's crust, heat and pressure transform the sandstone into quartzite, a metamorphic rock.

Deeply buried rocks may melt into magma, which upon cooling becomes igneous rock. The kind of igneous rock that forms depends on the chemical composition of the rocks that are melted and the speed with which the rocks cooled. The rocks may then be uplifted into mountains, and a new cycle begins. Massachusetts contains many plutons of igneous rocks—age dating of these rocks reveals a long and dramatic history of continental collisions.

Uniformitarianism

James Hutton proposed one of the most fundamental ideas in geology: processes operating on and below the present surface of the earth have

Rock cycle. —Modified from Press and Siever, 1994

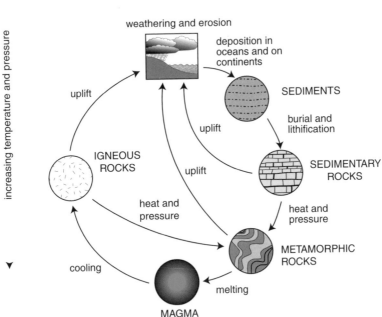

done so more or less consistently throughout history. This principle is called uniformitarianism—some people mistake it for a religion! This principle will help you develop a feel for how geologic history unfolds. Processes that are actively building mountains today, such as those uplifting the Andes of South America or the Himalayas of Asia, are the same processes that formed mountains long ago. Mountain building events, volcanic eruptions, earthquakes, and other geological processes that we observe today are also responsible for the ancient geological features that we observe along the roads of Massachusetts.

Metamorphism

Heat and pressure—from continental collisions, magmatic intrusions, and burial deep within the earth—metamorphosed many of the rocks in Massachusetts. Metamorphism changes the minerals and texture of a rock. The types and sizes of minerals in the rocks tell us how hot the rocks got and how much pressure they endured. The alignment of the minerals tells geologists from what direction the forces were applied and how many times the rocks were deformed. The chemistry of the rock tells us what the original rock may have been. Geologists learned much of the geologic history of supercontinents, plate tectonics, and mountain building events through careful analysis of metamorphic rocks.

The grade of metamorphism, from low to high, describes the intensity of the pressure and temperature of metamorphism. For example, if a shale changes to a slate the metamorphism is low grade; the conversion of shale to a schist that contains the minerals garnet and sillimanite is moderate to high grade. Rocks in direct contact with a magmatic intrusion may undergo high-grade metamorphism, while rocks farther away experience lower grades. If the pressures and/or temperatures are really high, the rock may shear, flow, or melt, producing mylonites, migmatites, and igneous rocks.

Mylonites and Migmatites

It's unfortunate that the words *mylonite* and *migmatite* are so similar in look and sound—we may find it hard to keep them straight. They are both what we might call yo-yo rocks. These rocks of the continental crust have been jammed deep within the earth, deformed, and then brought up to the surface again.

A mylonite is a streaky or banded rock that forms as rock masses shear past each other during metamorphism. A migmatite forms as a mixture of solid rock and magma that is either injected into the rock from elsewhere or melted directly out of the rock. As you might expect, mylonites and migmatites grade into one another. Migmatitic mylonites are present in the uplands around Boston.

Thin pegmatites of Dedham granite crosscut the vertical foliation of mylonite. A folded dike, an offshoot of the Dedham granite, cuts across the foliation. At junction of Massachusetts 30 and Rice Road in Wayland.

Rocks shear in fault zones so deep and hot that some minerals, such as quartz, may lose their original structure, recrystallize, and flow ductilely. Other minerals such as feldspar and biotite may break brittlely. The shapes of deformed quartz and feldspar indicate the direction of relative movement of the sheared rock layers when they slipped past one another like cards in a deck.

Trying to discover what the original rock was before it was sheared into mylonite is a bit like trying to reconstruct an ear of corn from cornmeal. Even so, it is possible to tell whether the original rock was pale or dark. And with luck, it may be possible to find fragments of the original rock that somehow escaped the shearing.

Faults and Thrusts

Compressional and extensional forces in the earth break rocks. Rocks near the surface break brittley. Rocks at depth break ductilely. The breaks are called faults. Normal faults often bound rift basins, such as the Connecticut Valley, that form under extensional forces. As the earth's crust pulls apart, basins form when blocks drop down along normal faults. A strike-slip fault, such as the San Andreas fault in California, moves parallel to the strike, or surface trend, of the fault. A dip-slip fault moves down the dip, or angle, of the fault plane. The Clinton-Newbury fault zone in Massachusetts, which forms the boundary between the Nashoba and Merrimack terranes, has moved obliquely because it is partly a strike-slip and partly a dip-slip fault. Rocks on one side of the Clinton-Newbury fault slid a considerable distance parallel to the fault relative to rocks on the other side.

The forces of some mountain building events in Massachusetts pushed and lifted huge sections of rock over others, forming thrust sheets—large masses of rock that moved a long distance up a low-angle fault. This fault mechanism often places older rocks on top of younger rocks. In intense metamorphism and thrusting, rocks may shear and flow.

Plate Tectonics

In the 1960s, geologists developed a theory that the outer crust of the earth consists of great slabs of rock that slide around the globe. These plates, which are relatively rigid, are up to 60 miles thick. When they collide with one another, they give rise to mountain chains, such as the Berkshire Hills and Taconic Range of western Massachusetts. The plate tectonics theory has revolutionized the way geologists interpret the rocks of Massachusetts— an area that has undergone at least three plate collisions in the last 500 million years.

At collisional boundaries, the heavier plate, commonly composed of oceanic crust, sinks beneath the lighter continental crust in a trench at a subduction zone. The trench consumes large masses of rock. Heat and pressure soften and ultimately melt the down-going slab, transforming it into magma that rises buoyantly from the subduction zone along fractures and faults. Rising magma heats and metamorphoses continental rocks, forms magma chambers at depth in the earth, and erupts as lava, often explosively, where fractures reach the earth's surface. A chain of volcanoes typically forms above the sinking plate. Today, an oceanic plate is sinking beneath the coast of the Pacific Northwest, forming the active volcanoes of the Cascade Range.

In other parts of the earth, plates rift, or spread, apart. Continued rifting culminates in the formation of an ocean basin. The present Atlantic Ocean began to open up around 200 million years ago. It continues to grow as new crust is added at the mid-Atlantic rift, visible above water in Iceland. Deep fractures in the earth's crust carry basaltic lavas to the surface along

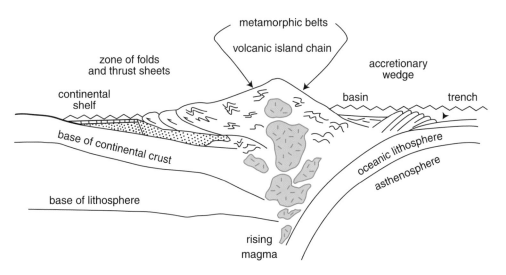

Oceanic crust sinking beneath the edge of a continent. —Modified from Stanley, 1993

the rift. The Connecticut Valley is an ancient rift valley that formed at the same time the Atlantic basin rifted open. Basalt flows dating from this time occupy parts of the valley.

Igneous and Volcanic Rocks

Magma generated by the movement of continental and oceanic plates either cools slowly at depth, crystallizing into igneous rocks, or it explodes or oozes onto the surface, crystallizing quickly into volcanic rocks. The slow cooling of igneous rocks permits minerals to grow into crystals visible to the naked eye. Granite, granodiorite, diorite, and gabbro are all igneous rocks, cooled at depth but differing in chemistry.

Volcanic rocks cool quickly on the surface of the earth. Not all of the minerals have time to grow, so some minerals in volcanic rocks are not visible to the naked eye. A rock that cools so quickly that almost no crystals form is essentially glass, or obsidian. Rhyolite, andesite, and basalt are volcanic rocks, all cooled quickly but from magmas of different chemistries. Rhyolite is the extrusive chemical equivalent of granite, andesite of diorite, and basalt of gabbro.

Light-colored rocks such as granite have more feldspar and quartz— minerals with silica, sodium, and potassium—than dark-colored rocks. Light-colored rocks are called *felsic,* a mnemonic adjective from *fel*dspar and *si*lica that describes the general chemistry. It's an odd term but used often in geology. Its counterpart is the term *mafic,* a mnemonic word derived from *ma*gnesium and *ferric,* which means "containing iron." Dark-colored rocks like diorite and gabbro contain minerals rich in magnesium and iron such as amphibole and pyroxene. A rock composed almost entirely of mafic minerals is called *ultramafic.*

THREE SUPERCONTINENTS AND AN OCEAN

Three supercontinents—Rodinia, Laurentia, and Pangaea—left their marks in the small state of Massachusetts. Cratons—the nuclei of continents—and microcontinents, pieces of which we recognize in Massachusetts, began colliding a little over 1 billion years ago in Precambrian time to form the supercontinent Rodinia. The resulting, nearly worldwide, mountain building event, the Grenville, deformed and metamorphosed the rocks of the Berkshire highlands. Hot magma generated in the collision formed plutons that intruded older rocks.

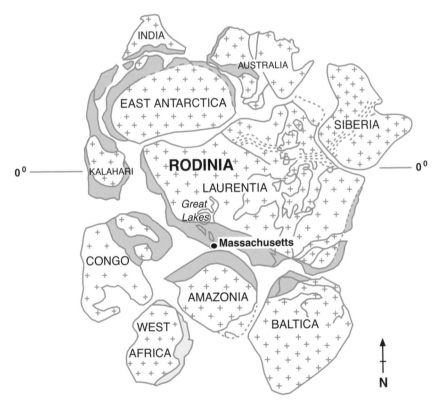

The assembly of the late Precambrian supercontinent, Rodinia, by about 750 million years ago. —Skehan, 1997, modified from Hoffman, 1991

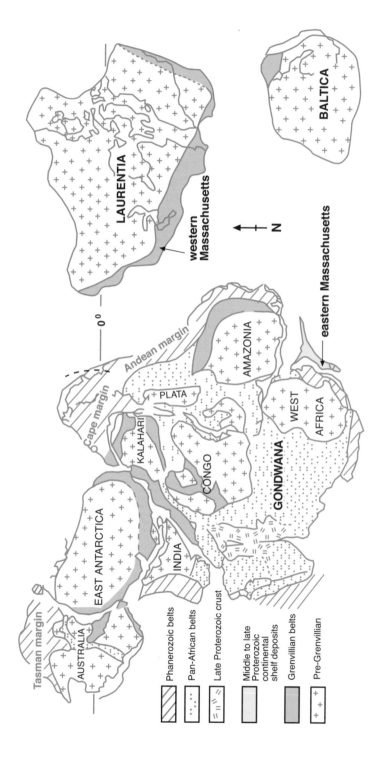

The assembly of the late Precambrian supercontinent, Gondwana, about 550 million years ago, just after the breakout of Laurentia from the Rodinian supercontinent. —Skehan, 1997, modified from Hoffman, 1991

Continent-size pieces assembled to form Rodinia by about 750 million years ago. Rodinia broke apart after that time, and the Laurentian craton, the nucleus of North America, separated from its former neighbors. Some of the other pieces rotated and reassembled in the Southern Hemisphere into a new supercontinent, Gondwana, by about 550 million years ago. Gondwana included land that is now modern Africa, South America, India, Australia, and Antarctica.

Laurentia, named for a prominent mountain range in the Canadian Shield, consists in part of rocks that are more than 2.5 billion years old. Since it broke away from Rodinia it has increased in size—notably, several volcanic island chains, similar to the volcanic island of Japan, and at least two microcontinents have collided with and welded onto eastern Laurentia's margin. The Merrimack, Nashoba, Avalon, and Meguma terranes that make up the eastern half of Massachusetts were probably microcontinents or island chains that broke off Gondwana and collided with each other on their trip toward Laurentia between 550 and 370 million years ago.

The Iapetus Ocean, the large Paleozoic sea between Laurentia and Gondwana, began to open up when Gondwana and Laurentia split apart about 500 to 550 million years ago. It grew so large that migration across it was difficult, and organisms evolved independently on either side, creating divergent plant and animal faunas. The ocean was finally swallowed up when Gondwana reunited with Laurentia in Devonian time to form Pangaea, the third supercontinent.

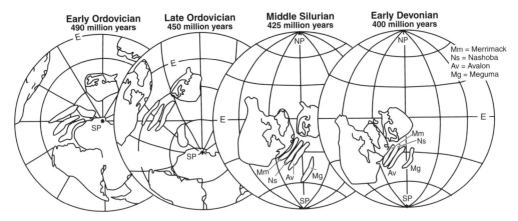

The Merrimack (Mm), Nashoba (Ns), Avalon (Av), and Meguma (Mg) terranes that fringed the Gondwanan supercontinent broke away possibly about 500 million years ago, in early Ordovician time. Equator (E); North Pole (NP); South Pole (SP).
—Modified from Torsvik and others, 1992; Meissner and others, 1994

The early stages of the assembly of Pangaea include the Acadian mountain building event that marked the collision of Laurentia with the Gondwanan terranes between 425 and 370 million years ago. The final stage involved the collision of the Meguma terrane with the Avalon and other Gondwanan terranes during the Alleghanian mountain building event 300 to 250 million years ago. Pangaea was fully assembled by the end of Permian time, about 245 million years ago.

Pangaea began to rift apart about 200 million years ago. The breakup of Pangaea gave rise in Mesozoic time to the modern continents of North America and Europe, and the early formation of the Atlantic Ocean. Early in the opening of the North Atlantic, many Mesozoic basins that extend from Nova Scotia to the Carolinas—including the Hartford Basin in the Connecticut Valley—began to form. Numerous basalt dikes represent evidence for rifting not only in the Connecticut Valley, but elsewhere in Massachusetts and in the Appalachians. These basins pulled apart through Jurassic time, and the Atlantic Ocean continues to widen by rifting to this day.

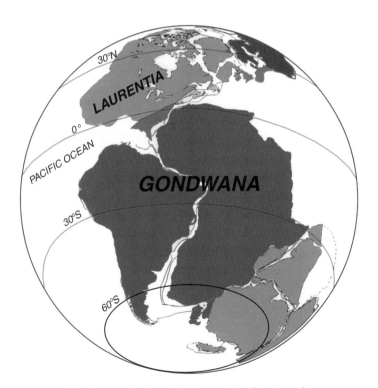

Gondwana docks with Laurentia, forming the Pangaean supercontinent. —Modified from Dalziel, 1997

Major Mountain Building Events

Major mountain building events occur when plates of the earth's crust collide. The rocks of Massachusetts record at least five such events.

Grenville Mountain Building Event. The earliest mountain building event to affect the rocks of Massachusetts is the Grenville. It took place 1.2 to 1.1 billion years ago when many small continents or cratons collided to form the supercontinent Rodinia. Rocks of Grenvillian age are present in the Berkshires.

Avalonian Mountain Building Event. The Avalonian mountain building event occurred when the margin of the Gondwanan supercontinent was intensely sheared, possibly by strike-slip faulting, in late Precambrian time. Magmas of the Avalon volcanic chain intruded the sheared zone over an offshore subduction zone.

Taconic Mountain Building Event. The Taconic mountain building event, the third major collision to affect rocks in Massachusetts, thrust up and deformed the continental shelf, slope, and rise of Laurentia. Rocks near the Connecticut River valley and to the west show evidence of this collision. Based on the ages of rocks involved in the deformation and metamorphism, we know that this collision took place in Ordovician time and volcanic island activity was associated with it. Geologists continue to debate which of two volcanic chains collided with Laurentia and the exact timing of these events.

Acadian Mountain Building Event. The fourth major collision in Massachusetts, the Acadian mountain building event, occurred when Avalon and the other amalgamated Gondwanan microcontinents smashed into Laurentia in middle Devonian time. This collision, recorded in the extensive plutonic and high-grade metamorphic rocks of central Massachusetts, firmly welded these blocks of land together. The zone of intense deformation, several tens of miles wide, extends north to western Maine and south to Long Island Sound. The mountain building event appears to have swept from the southeast to the northwest across the rocks of New England.

A date of 395 million years for the formation of migmatite in the Nashoba terrane possibly represents a deformational episode involving the Gondwanan microcontinents as they approached Laurentia just before the Acadian mountain building event. Detailed studies in Maine bracket the event between 423 and 384 million years ago. Age dates obtained on metamorphic rocks in Massachusetts fall within a similar age range, from about 425 to 370 million years ago.

Alleghanian Mountain Building Event. As late as the last quarter of the twentieth century, a number of geologists thought that the Alleghanian mountain building event was a relatively insignificant tectonic event of Pennsylvanian or Permian time in southern New England. Sedimentary rocks of the coal basins in Massachusetts, however, were folded, metamorphosed, and thrust-faulted in late Pennsylvanian time. It appears that yet another block of land—the Meguma microcontinent—collided with the eastern shore of Laurentia at the same time the ancestral African continent collided with southern New England in the final assembly of Pangaea.

Age dates on igneous and metamorphic rocks attributed to the Alleghanian event span an enormous range, from 354 to 250 million years. Geologists now realize this collision renewed compression in the Acadian collision zone. Radiometric dates on some metamorphic minerals in the Merrimack and Nashoba terranes give dates within this range. The Alleghanian mountain building event gave rise to Rhode Island's 275-million-year-old Narragansett Pier granite and 250-million-year-old Westerly granite of the Avalon terrane. The Alleghanian collision deformed the rocks of Massachusetts as far west as the Bronson Hill volcanic belt. This collision produced the Massabesic uplift in north-central Massachusetts, created a core of Devonian and Pennsylvanian granite in the Wachusett belt of the Merrimack terrane, and emplaced a granite of similar age in the Glastonbury dome of central Massachusetts and Connecticut. It appears that the Alleghanian mountain building event may have been one of the more widespread and complex events in southern New England, involving renewed compression and overthrusting.

Laurentian Terranes

The rocks of the Berkshire Hills once formed the ancient margin of Laurentia, the early North American continent. Laurentian terranes include the Grenville gneisses of Precambrian age that formed the basement of the continent, continental shelf deposits along the shore of the Iapetus Ocean, and the remains of two offshore volcanic island chains—the Shelburne Falls volcanic chain in the Berkshire foothills and the Bronson Hill volcanic chain, which lie just east of the present-day Connecticut Valley. The Taconic mountain building event, which resulted from the collision of the continent with these volcanic island chains, thrusted, pushed, and metamorphosed the collection of Precambrian to early Paleozoic rocks between 485 to 440 million years ago in Ordovician time. Geologists divide the Laurentian terranes of Massachusetts into six structural and stratigraphic divisions.

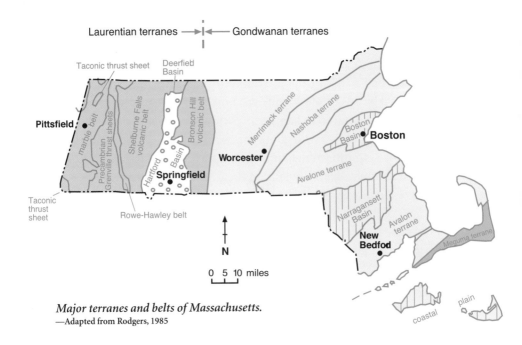

Major terranes and belts of Massachusetts.
—Adapted from Rodgers, 1985

Gondwanan Terranes

The four terranes that now make up central and eastern Massachusetts—the Merrimack, Nashoba, Avalon, and Meguma—are large blocks of
the earth's crust separated from one another by major faults. Most geologists agree that at least the Avalon and Meguma terranes were once part
of the Gondwanan supercontinent. I believe all four were derived from
Gondwana.

We know that Avalon was part of Gondwana because distinctive fossils,
trilobites of Cambrian time, exist in Avalon terrane in Massachusetts and in
Gondwanan rocks in Africa. Avalon must have broken away from Gondwana
during the end of Precambrian time, about 550 million years ago, or at the
very latest by Ordovician time, 490 million years ago. Fossils of the Meguma
terrane are also related to those of the northwest African margin.

The Avalon terrane originated as a volcanic island chain near the south
pole along the fringe of the Gondwanan supercontinent. Late Precambrian
quartzites of eastern Massachusetts may have originally been deposited as
beach sands on the continental shelf of Gondwana. These deposits were later
caught up in the Avalonian mountain building event, the deformation that
produced the Avalonian island chain.

The Nashoba and Merrimack terranes, which lie west of the Avalon
and Meguma terranes, probably formed seaward of them near Gondwana

during Cambrian, Ordovician, and Silurian time. The older rocks that form the foundation upon which the Cambrian through Silurian strata were deposited may have broken away from Gondwana and headed for Laurentia just ahead of the Avalon terrane. During Cambrian time, the Avalon chain of volcanic rocks began its journey northward from the south pole, along with its companion terranes. They collided with one another from time to time along the way. Granite plutons within each terrane record those collisions.

Laurentian and Gondwanan Terranes of Massachusetts

Laurentian terranes (from west to east)
1. Taconic thrust sheet
2. Vermont-Stockbridge marble belt
3. Berkshire Massif
4. Hoosac thrust sheet and Rowe-Hawley belt
5. Shelburne Falls volcanic belt

—Connecticut Valley belt of Mesozoic time—

6. Bronson Hill volcanic belt

Gondwanan terranes (from west to east)
1. Merrimack terrane
   Ware belt
   Gardner belt
   Wachusett Mountain belt
   Southbridge belt
   Nashua belt
   Rockingham belt

   —Clinton-Newbury fault zone—

2. Nashoba terrane

   —Bloody Bluff fault zone—

3. Avalon terrane
   Dedham granite
   Boston Basin
   Milford granite (Rhode Island batholith)
   Narragansett Basin
   Fall River batholith

   —Nauset fault—

4. Meguma terrane

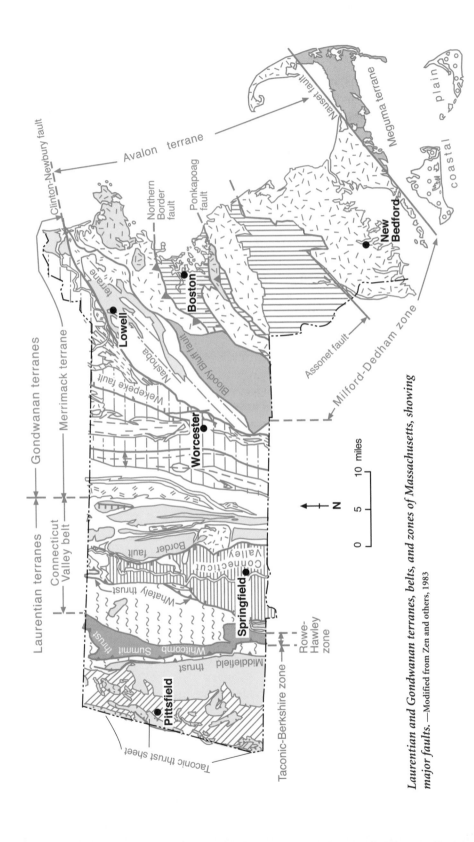

Laurentian and Gondwanan terranes, belts, and zones of Massachusetts, showing major faults. —Modified from Zen and others, 1983

LIFE IN ROCKS

As sediments lithify into sedimentary rocks, the remains of plants and animals are preserved as fossils in the rock record, though later metamorphism may destroy them. Although the abundant, shelly marine fauna of brachiopods, corals, and bryozoa found elsewhere in New England must have also existed in Massachusetts, continental collisions transformed sedimentary rocks into schists and marbles. Intense deformation favors neither the preservation of fossils nor the motivation of paleontologists to expend the time and laborious effort to find any remaining fragments of fossils. But, despite the plethora of metamorphic rocks in Massachusetts, many rocks contain well-preserved fossils. For example, giant trilobites of middle Cambrian age occur in rocks near Boston, and ostracodes of Silurian to Devonian time occur in sedimentary rocks in northeastern Massachusetts.

The Norfolk and Narragansett Basins of southeastern Massachusetts and Rhode Island comprise over 1,000 square miles and at least 10,000 feet of fossiliferous coal-bearing strata. These beds contain about three hundred species of fossil plants and animals. Fossil tracks suggest that the first animal visitors to the shores of Massachusetts—long before the *Mayflower*—may have been amphibians, emerging onto beaches in search of sustenance at least by Pennsylvanian time.

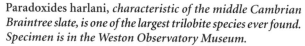

Paradoxides harlani, *characteristic of the middle Cambrian Braintree slate, is one of the largest trilobite species ever found. Specimen is in the Weston Observatory Museum.*

The red strata of Mesozoic time in the Connecticut Valley yield a rich harvest of dinosaur footprints, the largest collection of which are in the Amherst College Pratt Museum. The museum also contains abundant fossil fish that record lake environments in the valley. Fossiliferous outcrops of Cretaceous, Miocene, and Pleistocene age occur in Martha's Vineyard, Nantucket, and the Boston area.

GLACIATION

It is little wonder that the early settlers thought they saw evidence of Noah's flood in New England. Huge boulders perch on the tops of hills, extensive deposits of sand and gravel cover the land, and exposed bedrock is swept clean and polished. The surficial glacial geology dictated where settlers could farm and build—it fashioned much of the history of the early settlement and growth of the commonwealth.

Louis Agassiz, a pioneer geologist of the nineteenth century, was influential in establishing the theory of continental glaciation. Agassiz grew up in Switzerland and saw bedrock smoothed and polished by the movement of modern-day ice—small glaciers trailing out of the Alps. He published his theory that a great ice age was responsible for surface deposits in *Études sur les Glaciers* in 1840 and presented more evidence for his theory in *Systeme Glaciere* in 1847. He noted striations and grooves that marked the direction of movement of pebble-studded tongues of ice as they slid down valley floors. "Every terminal moraine," he wrote, "is the retreating footprint of some glacier."

Agassiz found additional evidence of glaciation upon arriving in North America in autumn 1846. In his book *Geological Sketches* Agassiz wrote that when his steamer docked in Halifax, Nova Scotia, he "sprang on shore and started at a brisk pace for the heights above the landing. On the first undisturbed ground, after leaving the town, I was met by the familiar signs, the polished surfaces, the furrows and scratches, the line-engraving of the glacier, so well known in the Old World; and I became convinced . . . that here also this great agent had been at work, although it was only after a long residence in America . . . that I fully understood the universality of its action."

Louis Agassiz became professor of geology at Harvard University in Cambridge in 1848. His ideas, previously published in French, were slow to be accepted widely in North America. To many, including other scientists, it seemed incredible that great sheets of ice had covered the northern regions of North America, as well as northern Europe, and that glaciers had deposited boulders, sand, and gravel across the landscape. As time went on, how-

Louis Agassiz   *Reverend Edward Hitchcock*

ever, geologists accumulated overwhelming evidence in support of this novel idea. Continental glaciation became an accepted explanation for many of the surficial erosional and depositional features of the New England landscape.

Agassiz was not the only person to suspect glaciation in North America. Reverend Edward Hitchcock, professor of geology and president of Amherst College, wrote the *Final Report on the Geology of Massachusetts*. As it went to press in 1841, he added a "Postscript on 'Glacio-aqueous action'" to the two-volume report. Hitchcock had recently become aware of Agassiz's book *Études sur les Glaciers* and papers by European geologists William Buckland and Charles Lyell on the same topic. He recognized the importance and relevance of these works to the surficial geology of Massachusetts. In the main body of the report he had described a number of features around the state whose origin he attributed, somewhat hesitatingly, to "diluvial action," but in the postscript he adds a new model of scientific thought:

> By the labours of these distinguished men, Agassiz, Buckland, and Lyell, the whole subject of diluvium, [running water] has been made to assume an aspect so new and interesting, that I am unwilling my Report should go out of my hands unaccompanied by a brief view of the facts and inferences concerning it . . . an outline of the glacial theory, and its application to this country. . . . To conclude: the theory of glacial action has imparted a fresh and a lively interest to the diluvial phenomena of this country. It certainly explains most of those phenomena in a satisfactory manner. It seems to me, however,

that the term Glacio-aqueous action more accurately expresses this agency than the term glacial action: for the effects referable to water are scarcely less than those produced by ice.

And Hitchcock was right: glacio-aqueous action more accurately expresses the complete glacial mechansim. Geologists now recognize the extensive distribution of glacial lakes and the mighty meltwater rivers that transported debris from the glacier to the sites of deposition.

Ice Ages

Large continent-size glaciers covered the earth at various times between middle Precambrian and Pleistocene time. Around 1.6 million years ago, the beginning of Pleistocene time, the earth began to experience the most recent episodes of glaciation. During the past 1 million years, continental glaciers spread southward across New England, fully advancing and retreating as many as four times. The continental glacier was nearly 1 to 2 miles thick near its center, or area of maximum snow accumulation, near Hudson Bay.

The latest glacial episode, represented by the Wisconsinan ice sheet, left its mark on Massachusetts. Evidence of an earlier episode, the Illinoian ice sheet, was largely erased or buried in Massachusetts by the Wisconsinan sheet. The Wisconsinan ice sheet extended as far south in New England as Nantucket, Martha's Vineyard, and Long Island. It began its advance about 80,000 years ago.

Maximum extent of the continental ice sheets that covered North America during the Pleistocene glaciation. Directions of ice flow were perpendicular to curved lines. —Modified from Oldale, 1992

This great ice sheet blanketed all of New England by about 23,000 to 22,000 years ago. As it moved over southern New England, the southern margin fanned out into lobes—broad, bulbous, pancake-shaped margins— bulging southward. Geologists identify five lobes in and near Massachusetts from west to east: the Hudson Valley lobe, the Connecticut Valley–Worcester Plateau lobe, the Narragansett Bay–Buzzards Bay lobe, the Cape Cod Bay lobe, and the South Channel lobe.

Each lobe sloped toward the other, giving rise to a dimple where they met, which is also where melting water tended to concentrate. Terminal moraines outline the southern edges of these lobes. The half-moon shape of Cape Cod marks the southernmost extent of the Cape Cod Bay lobe. The northern end of Buzzards Bay is the site of the dimple between the Buzzards Bay lobe and the Cape Cod Bay lobe.

In 1925, William C. Alden of the U.S. Geological Survey estimated the thickness of glacial ice over southern New England. He assumed that the tops of the high peaks, such as Wachusett Mountain in central Massachusetts and Mount Monadnock in New Hampshire, were glaciated. Using the distance of 90 miles from the terminal moraine on Block Island, Rhode Island, to Wachusett Mountain at an elevation of 2,006 feet, and 118 miles to Mount Monadnock at an elevation of 3,165 feet, Alden calculated that the minimum slope of the glacier's surface had to be 27 feet per mile to rise over the peak crests. Assuming that relative elevations are the same as today, the thickness of the ice over the summit of Wachusett Mountain would have been slightly more than 400 feet, and the thickness of ice over Worcester about 1,500 feet. Other calculations based on slightly different assumptions come up with a minimum thickness of the ice sheet at Worcester of just under 2,000 feet.

Glacial erosion stripped off a substantial amount of rock material from the underlying bedrock. The preglacial, weathered bedrock was probably similar to the reddish, clay-rich, decomposed rocks that you can see today in the unglaciated central and southern Appalachian Mountains. Many of the stones and matrix of the Wisconsinan glacial debris are unweathered, indicating that earlier advances of the continental glacier, including the Illinoian ice sheet, probably scraped off more deeply weathered material. The interval between the latest ice sheet advances was not long enough to allow much weathering of the newly exposed bedrock.

By about 21,000 years ago the glacier front began to recede northward. Geologists determined the time from radiocarbon dates of organic materials overlying end moraines on Long Island. The melting glacier deposited these end moraines between 21,000 and 19,500 years ago, after it built the Ronkonkoma terminal moraine. Dates from postglacial organic materials on Martha's Vineyard indicate that the ice melted from it earlier than 15,300 years ago. The Cape Cod Bay lobe stood at Boston between 15,000 and 16,000 years ago and receded into southwest Maine by 14,500 years ago.

As the glacier receded, huge volumes of water poured onto the land. Glacial lakes formed in depressions and basins throughout Massachusetts. The largest, Glacial Lake Hitchcock, filled the Connecticut River valley from northern Vermont and New Hampshire to Connecticut. Many of the small lakes and wetlands that dot the present landscape are remnants of these glacial lakes.

The weight of the continental ice sheet had depressed the underlying crust. After the ice melted from the region and the lakes drained, the earth's crust rebounded at the rate of about 4.74 feet per mile, with an upward tilt to the northwest. The uplift triggered downcutting by rivers, exposing glacial lakebeds, particularly in the Connecticut River valley.

Glacial Landforms and Deposits

As it moved over the landscape, the glacier scraped up and transported weathered rock, vegetation, and soils. The glacier and meltwater shaped the landscape and deposited the debris as glacial landforms—moraines, drumlins, kames, outwash sands and gravels, eskers, delta plains, kettles, and lakebeds.

Glacial geologists typically lump glacial features into two broad groups: ice-contact deposits and fluvial, or meltwater, deposits. Features that form directly in contact with the ice, including moraines, eskers, deltas, kettles, and drumlins, are called ice-contact features. Meltwater deposits include all glacial deposits that running or standing water have deposited or reworked.

Sediments deposited by water are generally stratified—they have visible layers, or beds. Depositional environments are often complex and, thus, the layers are difficult to decipher. Geologists look for subtle changes in sediment size and composition and for sedimentary structures. Because these sediments are generally soft, exposures along riverbanks or in gravel and clay pits often slump, hiding the layers and structures from view.

Moraines. A moraine is a prominent linear mound of unsorted rock debris that builds at the toe of a glacier. The classic, simple model of moraine formation describes a process of debris embedded in the ice falling to the ground as the edge of the glacier melts. A large pile of debris accumulates in one place if the glacier continues to move forward at the same rate that it melts backward. The debris continues to pile up as long as the edge of ice remains stationary.

Geologists also recognize a more complex mechanism at work in the formation of moraines. Careful studies of the Sandwich moraine on Cape Cod show that the ice has literally shoved the material into place. Occasionally, the melting ice lobe surged forward and pushed sheets of stratified outwash sediments in front of it. The series of sheets, thrust one atop another in the Sandwich moraine, slope down in the

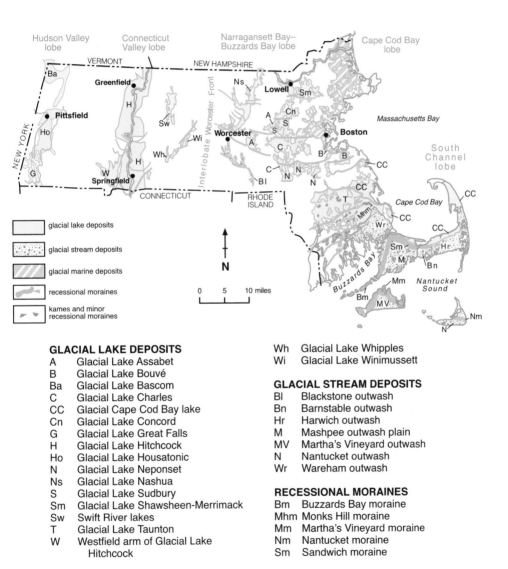

GLACIAL LAKE DEPOSITS

A	Glacial Lake Assabet
B	Glacial Lake Bouvé
Ba	Glacial Lake Bascom
C	Glacial Lake Charles
CC	Glacial Cape Cod Bay lake
Cn	Glacial Lake Concord
G	Glacial Lake Great Falls
H	Glacial Lake Hitchcock
Ho	Glacial Lake Housatonic
N	Glacial Lake Neponset
Ns	Glacial Lake Nashua
S	Glacial Lake Sudbury
Sm	Glacial Lake Shawsheen-Merrimack
Sw	Swift River lakes
T	Glacial Lake Taunton
W	Westfield arm of Glacial Lake Hitchcock
Wh	Glacial Lake Whipples
Wi	Glacial Lake Winimussett

GLACIAL STREAM DEPOSITS

Bl	Blackstone outwash
Bn	Barnstable outwash
Hr	Harwich outwash
M	Mashpee outwash plain
MV	Martha's Vineyard outwash
N	Nantucket outwash
Wr	Wareham outwash

RECESSIONAL MORAINES

Bm	Buzzards Bay moraine
Mhm	Monks Hill moraine
Mm	Martha's Vineyard moraine
Nm	Nantucket moraine
Sm	Sandwich moraine

End moraines, kames, and stratified glacial deposits of late Wisconsinan age in Massachusetts. —After Stone and Peper, 1982

opposite direction that the ice moved forward. Eventually the ice rode over the thrusted sheets of sediments and deposited a layer of basal till. When the ice melted back, new outwash was deposited behind the moraine.

Geologists also discovered that the debris picked up and carried along by the ice sheet actually migrates within the ice. It floats up to the ice surface along shear planes near the front of the active, or live, ice. Live ice is the principal source of debris for moraines and outwash. The shear zone acts like a conveyor belt carrying debris from the base of the glacier to the place of deposition at the stagnant end of the glacier.

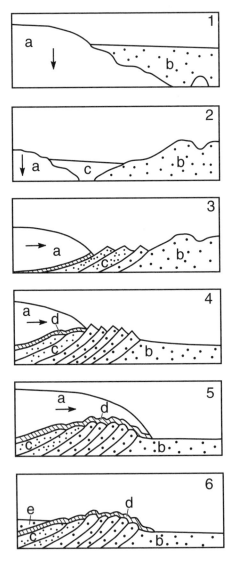

The formation of a coastal end moraine. In stages 1 and 2, stagnant ice (a) wastes away, and meltwater deposits outwash (b and c) at the melting edge of the ice. In stages 3 through 6, ice advancing in the direction of the arrow thrusts frozen stratified outwash up into sheets. It deposits basal till (d) onto the outwash. As the ice retreats, new outwash (e) is deposited behind the moraine.
—Oldale and O'Hara, 1984

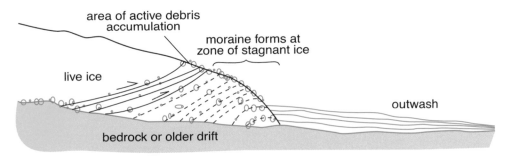

Cross section of glacier showing shear zones. —After Koteff and Pessl, 1981

A terminal moraine marks the farthest position of an ice sheet from its source of snow. The prominent ridge of the Ronkonkoma terminal moraine on Long Island marks the southernmost margin in New England of the most recent ice sheet. The Block Island, Martha's Vineyard, and Nantucket terminal moraines mark the southernmost extent along the coast of Rhode Island and Massachusetts. An end moraine is any other moraine that grows at the toe of a glacier but is not the farthest one from the center of snow accumulation. Most moraines are end moraines.

Ground moraine is glacial debris deposited as the glacier recedes. As a glacier melts back, it may not remain in one place long enough to deposit a large mound of debris, though it still continues to drop sediments. It thus creates a generally flat to hummocky ground moraine of unsorted clay, sand, gravel, cobbles, and boulders.

Till. The unsorted glacial debris that comprises till is deposited directly by a glacier; it is not reworked by meltwater. Till consists of a mixture of clay, silt, sand, gravel, and boulders of various sizes and shapes, compacted by the weight of overlying ice. Ground moraine is a particular kind of till consisting of glacial debris deposited by receding ice. Ground moraine was only briefly covered by thin ice, so it is not so dense or compact as some till.

A thin mantle of till, generally about 15 feet thick, covers a large part of Massachusetts, and geologists generally consider it ground moraine in the absence of any evidence to the contrary.

Eskers. Eskers—long, sinuous or meandering ridges of debris—resemble stream channels on a map. The steep, narrow ridges can be 30 or 35 feet high. In cross section, an esker resembles a single MacDonald's arch with the steep slopes dipping 30 to 35 degrees. Esker sediments range in size from silt and sand to pebbles, cobbles, and boulders up to 18 inches in diameter.

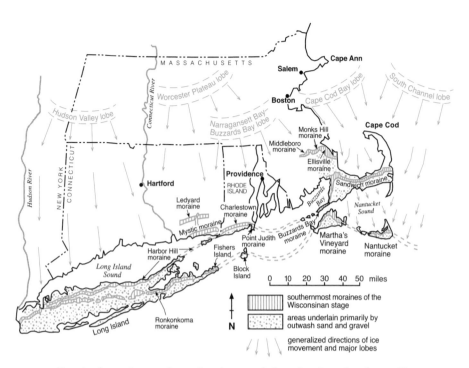

Terminal moraines and associated outwash deposits of sand and gravel in southern New England. —Modified from Mather, 1952; Koteff, 1974; and Larson, 1982

Streams of glacial meltwater deposit eskers as they flow under hydraulic pressure in tunnels through the ice or flow in open channels—ice crevasses. As the confining ice melts, the gravel slumps and forms ridges. Many eskers form in the lowest part of valleys. Because the channels generally form within a few tens of feet of, or on, the ground surface, the underlying topography controls their location. An exceptional esker in Gleasondale, however, cuts across the crest of a drumlin. Here the ice channel must have been at least 100 feet above the base of the ice.

Geologist Carl Koteff, who once described a glacier's conveyor belt mechanism as a "dirt machine," now believes that esker tubes contribute substantially to the volume of outwash. He believes that the conveyor belt mechanism—the upward shearing at the interface between the live ice and stagnant ice—still supplies considerable "dirt" to moraines, but the esker tube contributes substantially to the flow of outwash. The confining ice walls of the tube are constantly collapsing inward when melting, dragging debris from the bottom of the ice sheet into the tube. Meltwater under great hydraulic pressure carries the debris beyond the ice margin to form outwash.

Drumlins. Throughout Massachusetts, the thin mantle of till and ground moraine thickens into characteristic streamline landforms. Drumlins, or elliptical hills, form when moving ice meets enough resistance along its base that it is easier for the ice to ride up and over the glacial material than to pick up and transport it. The passage of ice over the top of the debris streamlines it. The crest of a drumlin arcs from end to end like the cross section of a football from tip to tip. The long axis of the drumlins in Massachusetts point in a southerly direction—the direction the ice was traveling.

When I was a graduate student at Harvard, a punster professor asked the question, "What is a drumlin?" to which a freshman responded, "It's a hill of a lot of till." The professor gave full credit for the answer because he liked the pun, but the answer did not describe the three-dimensional shape.

Drumlins consist of till and may contain glacial debris from more than one ice sheet. Most drumlins in Massachusetts are probably composed of Illinoian till with a thin layer of Wisconsinan till over the top. Drumlins often rest on bedrock, and because in many cases later stratified deposits partially bury them, you cannot see their full height and thickness at the surface, though they may be as much as 100 feet tall.

In Massachusetts, drumlins cluster abundantly in some areas and are nearly absent from others. In central Massachusetts near Worcester and Brookfield, geologists have mapped 275 drumlins. Drumlins are very sparse in the mountainous terrain of the Berkshire Highlands composed of gneisses. In the Bronson Hill Upland where schists are abundant, the ice compacted, streamlined, and wedged clay-rich and mica-rich sediments against the north side of rock buttresses. These clay-rich sediments are often as compact as cement, and it may have been more difficult for the ice sheet to pick up and transport them than to override and streamline them into drumlins.

Outwash. When glacial ice melts, its runoff may form a series of rivers or braided streams. These rivers carry rock fragments from the end of the glacial conveyor belt and esker tubes, spreading layers of sand and gravel over a broad area, and forming an outwash plain. You can see examples of miniature outwash plains along the base of bare sand and gravel slopes, perhaps along a highway, after a heavy rain.

Meltwater from the glacier may distribute an apron of sediment downstream from the ice sheet, with the sediment grading with distance from coarse gravel to fine sand and clay. In broad areas free of ice blocks, the outwash plains retain a form that is readily identifiable. Such features are widespread over large parts of Massachusetts. Much of Raynham and the western part of Taunton are built on the gently dipping surface of an outwash plain.

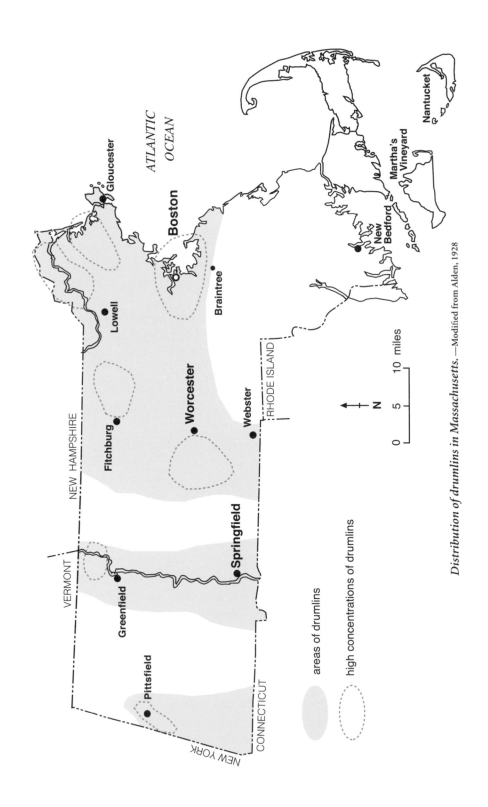

Distribution of drumlins in Massachusetts. —Modified from Alden, 1928

NEW HAMPSHIRE

VERMONT

NEW YORK

CONNECTICUT

RHODE ISLAND

ATLANTIC OCEAN

Boston

Gloucester

Lowell

Braintree

Fitchburg

Worcester

Webster

Greenfield

Springfield

Pittsfield

New Bedford

Martha's Vineyard

Nantucket

N

0   5   10 miles

areas of drumlins

high concentrations of drumlins

Kettle Lakes. Stagnant blocks of ice downstream from the glacier margin may become partly or wholly buried by accumulating sand and gravel. As the ice blocks melt, the unsupported glacial material collapses and forms craterlike depressions, or kettles. Walden Pond is a kettle formed in a delta plain. The Mashpee pitted plain of Cape Cod, a broad sand and gravel delta plain, is pockmarked with numerous kettles, kettle ponds, and lakes. The ice sheet lavishly ornamented Massachusetts with such features. A kettle contains water if the groundwater level is higher than the bottom of the kettle or if the base beneath the kettle is not porous and hence traps rainwater.

Formation of a kettle. A block of ice breaks off the glacier and sits on outwash (A). Outwash builds up around the ice (B), and when it melts, the sediment collapses into a kettle (C). —Modified from Alden, 1925

Kames. A kame, a low mound or short ridge, forms where glacial meltwater deposits material in a depression bordered by ice. Both water and ice contribute to its structure. Because it is deposited by water, a kame has stratified beds, but the beds deposited directly on or against the ice collapse into a mound when the ice melts.

The shape of the deposits determine whether geologists call it simply a kame or a kame field, kame terrace, kame plain, or kame delta. A kame terrace consists of stream or lake sediments deposited between the ice and a valley wall. A kame plain is an outwash plain bounded by ice. A kame field is a group of closely spaced kames. A kame delta forms as ice deposits sediment into a standing body of water. If the glaciofluvial deposits do not have a distinguishing form, geologists map them as undifferentiated stratified sands and gravels.

Kames may contain lenses of flowtill, which is identical in composition to till but occurs as layers up to 4 feet thick that lie on top of deposits laid down in water. The flowtill was first deposited on the ice surface as till, then flowed as muddy debris toward lower areas as the ice melted. Geologists have recognized flowtill in many areas since Joseph Hartshorn first described it near Taunton.

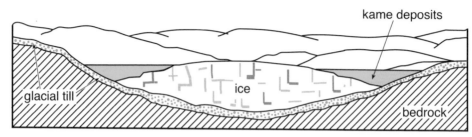

Cross section of a valley glacier with characteristic kame deposits between the valley wall and the crest of the glacier. —Alden, 1925

Glacial Lakes

As the glaciers melted, barriers such as moraines, ice, and bedrock impounded the meltwater into glacial lakes. The depth, shape, and size of the lakes changed with time as the level of the dams or spillways changed. Ice dams melted or became bouyant enough to float, and dams of unconsolidated sediments washed away. Buried bedrock ridges exposed by erosion stood firm after overlying sediments were stripped away.

The distance between the melting glacier and lake it feeds depends on the difference in elevation between the spillway and the base of the glacier. If the spillway is higher than the base of the glacier, the lake will butt up against the glacier. If the spillway is lower, the meltwater will flow downhill until it reaches the elevation of the spillway. Spillways overflow if water enters the lake faster than it evaporates.

Deltas. Perhaps the most common sand and gravel deposit in New England is a delta laid down in a glacial lake. Water deposits sand and gravel in front of melting ice, giving rise to broad, flat, sandy plains. These deposits are especially abundant and widespread in Cape Cod, eastern Massachusetts, and in the Connecticut Valley. All such gentle and easily dug deposits are prized sites for cemeteries, sand and gravel pits, and subdivisions. They usually contain potable groundwater aquifers.

Instead of nearly horizontal layers of sand and gravel, such as in the outwash plain, delta deposits form slanted layers because they were deposited on the sloping surface of a lake basin—either at the edge of the glacier or at some distance downstream. The sloping layers are called foreset beds and tilt down in the direction that the meltwater flows into the lake. As the package of delta foreset beds builds out into the lake, gently dipping beds of glacial outwash are deposited on top of the more steeply dipping foreset beds. Because of their position, these beds are called topset beds.

Lake Bottom Deposits. Sediments at the bottom of a glacial lake often have couplets, or varves, of dark clay alternating with a light silty layer, a seasonal phenomenon. The coarser-grained sediments—silts and sands—are deposited in summer when warm weather triggers melting of sediment-laden ice. This layer grades up into a layer of finer sediments—clay—deposited during winter when the surface water freezes and fine particles and organic matter settle. You can imagine that if you counted every varve in the lake bottom deposits, you would know the age of the lake. Ernst Antevs did just this for Glacial Lake Hitchcock and estimated that it existed for about 4,000 years.

Blowing Sand

Strong winds blew along the edge of the North American ice sheet during the melting and recession of the glacier. Extensive outwash plains,

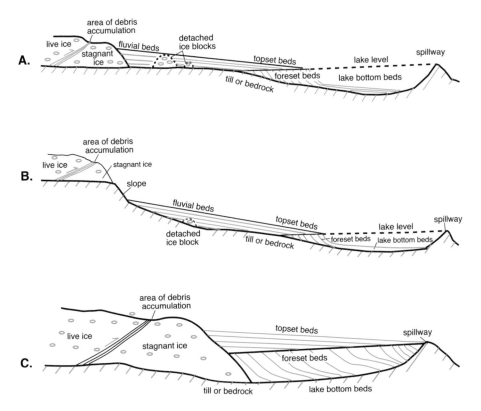

Cross sections of topset and foreset beds deposited near and in a glacial lake at the edge of a melting glacier. Spillway (A) is at same elevation as the base of the melting glacier. Spillway (B) is lower than the base of the melting glacier. Spillway (C) is higher than the base of the melting glacier. —After Koteff, 1974

lake bottom deposits, and other water-laid deposits provided ample sand and silt for the wind to transport. Wind-blown, or eolian, deposits of 1 to 2 feet in thickness blanket parts of the Connecticut Valley and southeastern Massachusetts, including Cape Cod. The eolian deposits of Massachusetts are fine to medium, well-sorted sand. They are not called loess, a common windblown sediment, because they contain very few silt-size particles. Because these deposits are near the surface weathering zone, they are commonly oxidized to a light brown to yellow or reddish brown. Severe frost heaving during the retreat of the ice sheet disturbed the eolian layer so that these wind-blown sediments are mixed with, and commonly resemble, the glacial deposits below.

The blowing sand blasted and polished numerous boulders, cobbles, and pebbles that were churned onto the surface by frost action. You can recognize sandblasted rocks, or ventifacts, by the delicate pitting of the polished, greasy surfaces. These rounded to flat sandblasted surfaces are commonly faceted. The wind usually blows from a predominant direction, but frost action may have rolled the stones over, baring another side to the wind and creating a facet. The blowing sand also produced flutes, or rounded grooves, on bedrock surfaces.

Life after Glaciation

Tundra flora developed after the ice sheet withdrew from the region. A dominantly coniferous forest followed, later succeeded by a mixed coniferous-deciduous forest. Ice age animals were living in Massachusetts shortly after the ice melted from the area.

Archeologists have found and examined a number of mid- to early Archaic hearth sites on Cape Cod. These sites have yielded 135 stone tools and several thousand chips that represent the time period between 9,500 and 7,000 years before present. Mastadon teeth that give an age date of about 11,000 years ago have been found in sediments dredged off the coast of Marblehead on the North Shore.

Eastern Seaboard

Eastern Massachusetts is a collage of bedrock hills made even hillier by countless drumlins. Glacial till, lake deposits, and outwash blanket much of the bedrock, but boulders carved from bedrock—large glacial erratics—litter the surface almost everywhere. Plymouth Rock, perhaps the most famous erratic in New England, is a granitic boulder sitting in a kame field.

The plentiful harbors attracted sailors to the coastal region of Massachusetts. Sandy offshore bars and barrier islands help protect the rockbound coast. Behind the barrier beaches lies a great natural resource, the marshland habitat of birds and animals.

Among the many natural resources, colonists discovered an abundance of colorful granites such as the Quincy, Milford, Chelmsford, Fitchburg, and Hardwick. Skilled quarrymen arrived from Europe to work these granites. Sand and gravel left by the Pleistocene ice sheets fueled a concrete and

View northeast from Bass Rocks, Gloucester, across Good Harbor Beach to Brier Neck, a bedrock promontory of Cape Ann granite.

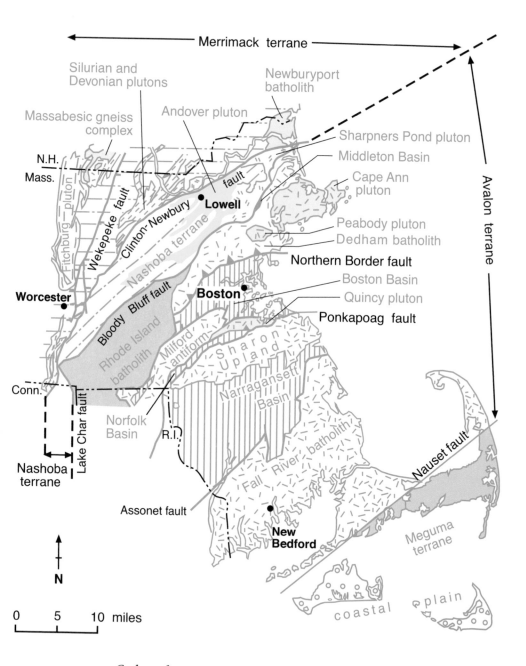

Geology of eastern Massachusetts. —Modified from Zen and others, 1983

construction industry, and factories soon lined the coastal rivers—the Charles, Merrimack, Neponset, Taunton, Nashua, Miller, Blackstone, and Quaboag.

As the population increased, colleges sprung up to educate the citizenry. Harvard University was the first, and the greater Boston area now boasts sixty-five institutions of higher learning. And certainly, geologists at these schools have had plenty of complex rocks to study.

Three terranes of the supercontinent Gondwana—the Merrimack, Nashoba, and Avalon—dominate the bedrock of eastern Massachusetts. A fourth, the Meguma terrane, is buried beneath the southern edge of Cape Cod. These terranes, most likely volcanic island chains that developed along the fringe of Gondwana near the south pole, broke away and moved north about 550 million years ago. They collided with each other, perhaps several times, during the 180-million-year-long trip. Though the Pilgrims made a lengthy trip to Massachusetts, the rocks of eastern Massachusetts traveled even farther. The terranes finally welded firmly to Laurentia in the Acadian mountain building event about 370 million years ago in Devonian time.

MERRIMACK TERRANE

The Merrimack terrane, primarily in central Massachusetts, extends to the coast across northeastern Massachusetts. The Clinton-Newbury fault zone forms its eastern border. The western margin of the Merrimack terrane is probably the major collision zone between Gondwana on the east and the Bronson Hill belt of Laurentia on the west.

Nashua Belt

The Silurian and Devonian sedimentary rocks of the Nashua belt are mainly devoid of igneous intrusions. The low-grade metasedimentary rocks contain well-developed sedimentary structures. The Oakdale formation—beds of interlayered calcareous siltstones, slate, and lesser amounts of quartzite and marble—was probably deposited in submarine fans at the base of canyons on the ocean floor. Ocean currents laden with suspended particles flowed down the continental slope and submarine canyons, contributing to the deposits.

The Worcester formation contains carbonaceous slate and phyllite. These beds are older, and possibly separated by an unconformity from, the fossiliferous coal-bearing beds and age-dated conglomerate at and near the Worcester coal mine.

Rockingham Belt

The most easterly belt of the Merrimack terrane, the Rockingham belt, occupies a wedge whose southernmost point is just north of Clinton in north-central Massachusetts. The belt, an extension of the Rockingham anticlinorium in southern Maine and southeastern New Hampshire, consists mainly of the Kittery quartzite of Ordovician to Silurian age and the Eliot and Berwick formations of Silurian age. The metamorphic grade is generally low in the eastern part of the belt and rises to garnet metamorphic zone in the west. Two distinctive plutons, the Ayer and Chelmsford granites, intruded the Rockingham belt in Silurian time.

Many brittle faults cut the Rockingham and Nashua belts during adjustments of the earth's crust near the end of Paleozoic time and into Mesozoic time. The Clinton-Newbury fault zone defines the eastern boundary of the Rockingham belt.

Ayer and Chelmsford Granites. The Ayer granite contains biotite and large crystals of feldspar 2 or more inches long. The crystals commonly align parallel to the gneissic foliation. The granite crystallized in early Silurian time, about 430 million years ago.

The Chelmsford granite, quarried for over a hundred years for curbstones and architectural purposes, is an attractive banded and foliated rock. Radiometric age dating using zircon gives an age date of 430 million years, the time the magma crystallized. Another type of age date that measures the time since metamorphism gives a date of 373 million years, in Devonian time.

The Silurian age for the formation of the granites and the Devonian age for their deformation suggests that the Nashoba terrane plunged in a subduction zone beneath the Merrimack terrane in Silurian time, generating magma that then cooled to form the granite. The widespread Acadian mountain building event later deformed these rocks in Devonian time.

Berwick and Eliot Formations. The Berwick formation of Silurian age consists of thin- to thick-bedded calcareous phyllites and schists that originated as limy sandstone, siltstone, and shale. When such rocks are metamorphosed, they often develop interesting, colorful lime silicate minerals. Minerals in the Berwick formation include tremolite, actinolite, hornblende, epidote, calcium-bearing garnet, and diopside, depending on the grade, or intensity, of metamorphism. The Dracut diorite of Silurian time intrudes the Berwick formation.

The Eliot formation, also of Silurian age, resembles the Oakdale formation of the Nashua belt to the west. It is phyllite to limy phyllite, a rock of low metamorphic grade.

NASHOBA TERRANE

The Nashoba terrane, a narrow belt sandwiched between the Merrimack and Avalon terranes, is bounded on the west by the Clinton-Newbury fault zone and on the east by the Bloody Bluff fault zone. The Nashoba terrane extends from the Atlantic coastline in northeastern Massachusetts south to Chester, Connecticut, near Long Island Sound.

The Nashoba terrane began to form in early Paleozoic time in the ocean between Avalon and Laurentia. Its strata, dipping steeply to the west, consist mainly of metamorphic volcanic rocks in the east and metasedimentary rocks in the west. Plutonic rocks of Ordovician, Silurian, and Devonian time intrude the metamorphic rocks. Heat and pressure generated by repeated collisions between the terranes during the trip from Gondwana formed and deformed these plutonic and metamorphic rocks. The Avalon terrane ran into and began subducting beneath the Nashoba terrane in Ordovician and Silurian time.

Clinton-Newbury Fault Zone

The Nashoba terrane's western limit, where it abuts its neighbor, the Merrimack terrane, is the broad, west-dipping Clinton-Newbury fault zone. The Clinton-Newbury fault probably originated as a subduction zone—where the edge of the Nashoba microcontinent sank beneath the edge of the Merrimack terrane. Mylonite and ductile faults dominate the fault zone's early, prolonged history, some 450 to 370 million years ago from Ordovician to Devonian time. The fault is no longer active.

I first encountered the fault zone during the construction of the deep bedrock Cosgrove Tunnel (formerly the Wachusett-Marlboro Tunnel), which brings water from Wachusett Reservoir southeast beneath a drainage divide to Southborough. A complex series of closely spaced faults cuts the 1.5-mile-broad fault zone.

By about Permian time the Clinton-Newbury fault zone had become a right-lateral fault. It consisted of several branches and was probably active in several episodes. During powerful fault movements, it bit off an enormous chunk of the Merrimack terrane and carried it to the southwest to a place that is as yet unknown. As this fault developed, the northwest block of the Merrimack terrane moved northeast and the Nashoba block moved to the southwest and filled the area of the missing Merrimack terrane.

In northern Massachusetts, the Merrimack terrane extends for 80 miles in an east-west direction from the Atlantic Ocean at Salisbury to Athol in central Massachusetts. In southern Massaachusetts, the Merrimack terrane is only 23 miles wide. The northeast-trending Rockingham belt near the New Hampshire line is nearly 40 miles broad, but near Clinton it is a mere

1.5 miles wide. Although different in origin from the San Andreas fault in California, the Clinton-Newbury fault zone must have resembled it, sliding large blocks of land great distances. Perhaps geologists will identify the missing block of Merrimack rock in some other region bordering the Atlantic Ocean.

The east-northeast-trending Clinton-Newbury fault zone abruptly cuts off a number of northeast-trending Silurian and Devonian formations. The Berwick formation, the Newbury volcanic rocks, and the Dracut diorite in northeastern Massachusetts end abruptly at the Clinton-Newbury fault zone. Small fragments of the Oakdale and Worcester formations of the Nashua belt and the Ayer granite that originated as part of the western Rockingham belt are present in isolated fault blocks. The Pennsylvanian coal beds in Worcester are also present in small fault blocks.

Massabesic and Nashoba Uplifts

The Nashoba terrane and the Massabesic gneiss complex of the Merrimack terrane share many similarities in their evolution. The Massabesic gneiss at the northeast end of the Fitchburg pluton is bounded on the west and east by the Campbell Hill and the Wekepeke normal faults, respectively, and is essentially an uplifted fault block. The Massabesic gneiss moved upward relative to the Oakdale sediments to the east of the Wekepeke fault. Massive intrusions of relatively light, felsic igneous rocks of Devonian age, and later a 275-million-year-old granite, rendered the area buoyant.

The Nashoba and Massabesic structures rose rapidly about 354 to 325 million years ago in Mississippian time in response to jostling between the fault-bounded Avalon and Merrimack terranes during the onset of the Alleghanian collision. The two areas are probably modified flower structures—high-angle faults resembling a bouquet in cross section.

Rocks of the Nashoba Terrane

The color and degree of deformation distinguish major rocks of the Nashoba terrane. When identifying rocks, look for the dominant color and grain size of feldspars in the granites, the intensity of folding in layered rocks, and the development of coarse muscovite crystals in schists and the foliation in plutonic and mylonitic rocks.

Mylonitic and migmatitic rocks occur near faults in the Nashoba terrane. From the chemistry of these rocks, we know that many of the highly metamorphosed schists and gneisses of the Nashoba terrane were formerly stratified sedimentary and volcanic rocks. The Nashoba terrane also has a number of diorite plutons—an igneous rock intermediate between granite and gabbro. Diorite contains considerable plagioclase and hornblende.

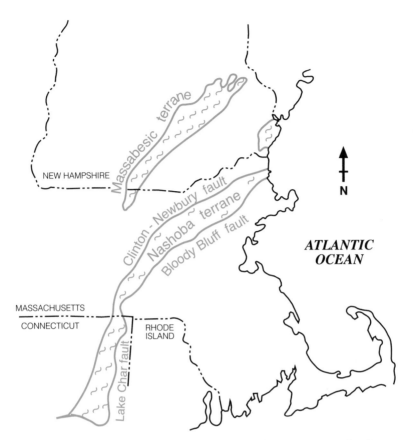

The Massabesic and Nashoba terranes in southeastern New England.
—Rast and Skehan, 1993

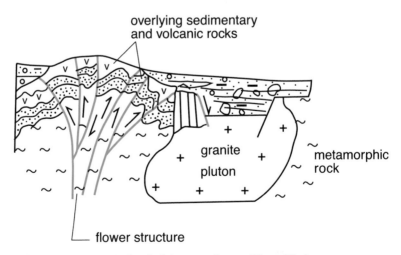

Flower structure. Similar faulting served to rapidly uplift the Nashoba and Massabesic terranes. —Rast and Skehan, 1993

Tadmuck Brook Schist. The Tadmuck Brook schist, a mylonitic rock sheared to a fare-thee-well, is present along the western margin of the Nashoba terrane. Closely associated with plate subduction and deformation along the Clinton-Newbury fault zone, it has undergone several episodes of metamorphism, at least one of which may have happened in Silurian time. This dark gray rock with lenses of quartz looks deceptively like a slate—a low-grade metamorphic rock—but it is much more deformed and complex than a slate. The Tadmuck Brook schist contains crystals of gray sillimanite

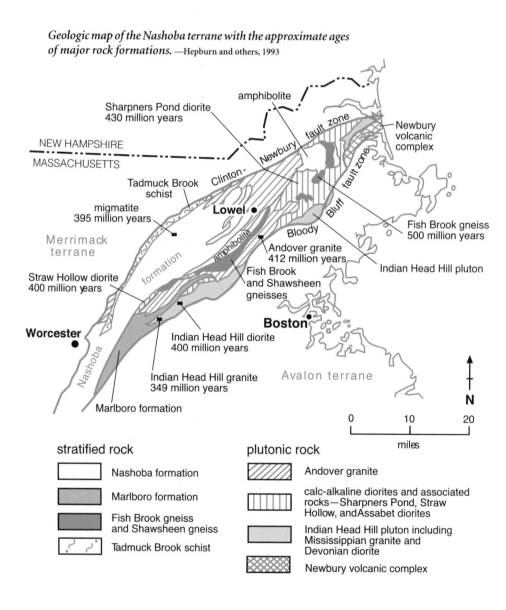

Geologic map of the Nashoba terrane with the approximate ages of major rock formations. —Hepburn and others, 1993

grown on crystals of pink andalusite as much as 18 inches long. Tiny flakes of mica give the rock a silky sheen.

Andalusite and sillimanite have the same chemical composition, but they crystallize under different conditions of temperature and pressure. Andalusite forms at medium temperatures and pressures. It changes to sillimanite as the temperature and/or pressure increases. The andalusite crystals of the Tadmuck Brook schist probably grew while the rocks sank deep beneath the Merrimack terrane into regions of medium temperatures and pressures. Then, as the rocks heated up and/or the pressure increased, sillimanite began to form.

Nashoba Formation. The Nashoba formation consists of alternating layers of gneiss and schist. The biotite and hornblende gneiss commonly has large crystals of feldspar. The light to dark gray, shimmering schist contains muscovite, biotite, garnet, and sillimanite. Smaller amounts of quartzite, marble, and amphibolite, a metamorphic rock composed primarily of amphibole, are interlayered with the schist and gneiss. These high-grade gneisses and schists make up the western half of the Nashoba terrane. The presence of abundant muscovite, biotite, and sillimanite indicates that the original sediment was rich in alumina, and in some places must have approximated a low-grade bauxite ore, a clay deposit that is the principal commercial source of aluminum.

These highly aluminous sedimentary rocks were subjected to conditions of metamorphism as high as the sillimanite zone. Subsequently, low-grade metamorphic conditions retrograded the minerals—the crystal structure and chemistry readjusted to lower pressure and temperature conditions. Large muscovite crystals, one-fourth to one-half inch long, were probably once sillimanite or andalusite crystals.

Near the base of the schists, a thick unit of finely foliated amphibolite, the Boxford member of the Nashoba formation, interfingers with the Fish Brook gneiss of late Cambrian age. For this reason, we know the Nashoba formation is probably late Cambrian to Ordovician in age.

Fish Brook and Shawsheen Gneisses. The Fish Brook gneiss stretches about 40 miles from Georgetown to Hudson and is about 1 mile broad. A pearly white to gray rock with biotite, quartz, and plagioclase, it probably originated as volcanic rock. From radiometric age dating, we know the igneous rock crystallized 500 million years ago and was metamorphosed about 425 million years ago in late Silurian time. Below the Fish Brook gneiss lies a micaceous rock containing sillimanite, the Shawsheen gneiss. Near its eastern margin, it is sulfidic and weathers rusty to yellow. The Shawsheen gneiss may be Cambrian or late Precambrian in age because it lies below and thus may be older than the Fish Brook gneiss.

Marlboro Formation. The Marlboro strata of early to middle Cambrian age, about 540 to 515 million years old, are mainly dark green to black amphibole schists and dark gneisses interlayered with light-colored quartz and feldspar gneiss. The Marlboro formation originated as light and dark volcanic rocks. Distinctive, thin, contorted layers of quartzite are either an unforgettable raspberry sherbet color, due to thousands of microscopically small garnets, or pistachio green, due to thousands of small epidote crystals. The garnet formed from manganese-rich cherts, and the epidote formed from limy and clayey cherts.

Andover Granite. The Andover granite is a light, commonly pink granite with muscovite and garnet. It is associated with pink, very coarse-grained microcline feldspar pegmatites and younger, gray to pink, fine-grained granite and aplite—a granitic rock with a granular texture. The Andover granite formed from the melting of deeply buried sedimentary rock. The magma of the Andover granite pervasively intruded all of the metasedimentary rocks of the Nashoba block except the Tadmuck Brook schist. The oldest phase of the Andover granite may have crystallized as early as 450 million years ago in late Ordovician time, and a younger phase crystallized in early Devonian time, about 412 million years ago. A collision between the Nashoba and Avalon terranes while they were en route from Gondwana may have generated the magma of the Andover granite.

Indian Head Hill Granite and Diorite. The Indian Head Hill granite, an orangish pink rock with biotite, intrudes the Marlboro formation, Andover granite, Fish Brook gneiss, and the eastern part of the Nashoba formation, all in the southeastern part of the Nashoba block near Sudbury. The Indian Head Hill granite is 349 million years old—Mississippian time—and the closely associated Indian Head Hill diorite is 400 million years old—early Devonian time. A granite similar to the Indian Head Hill granite occurs nearly 40 miles to the northeast, also along the Bloody Bluff fault. It intrudes the Sharpners Pond diorite, which has a Silurian age of 430 million years ago, consistent with the age of the Indian Head Hill diorite.

Sharpners Pond, Straw Hollow, and Assabet Diorites. These three diorites all have calc-alkaline chemistry, which suggests they probably formed in or near a subduction zone. The Sharpners Pond diorite, a dark igneous rock composed mainly of biotite mica and black hornblende, is the largest pluton in the Nashoba terrane. It intrudes the Boxford member of the Nashoba formation as well as older phases of the Andover granite. The Straw Hollow and Assabet diorites are gray, medium-grained, and foliated. They intrude the uppermost member of the Nashoba formation.

AVALON TERRANE

The Avalon terrane originated as a late Precambrian chain of volcanic and plutonic islands along the edge of Gondwana. The Avalon terrane broke away from Gondwana about 550 million years ago and welded to Laurentia about 370 million years ago. Later, the Pangaean supercontinent split apart when the Atlantic Ocean opened in Jurassic time, about 200 million years ago, a process that continues today. Part of the Avalon terrane is now in southern Great Britain and in Belgium.

Age dates of 610 and 595 million years for granites of the Avalon terrane and 600 million years for associated volcanic rocks are generally consistent on both sides of the Atlantic Ocean. The Avalon terrane extends from Belgium through the southern British Isles, Ireland, the Avalon Peninsula of Newfoundland, coastal Maritime Provinces of Canada, and coastal Maine to Long Island Sound.

The Bloody Bluff fault zone separates the eastern margin of the Nashoba terrane from the western margin of the Avalon terrane. The west-dipping Bloody Bluff fault is mainly a brittle deformation feature much newer than the Burlington mylonite zone, but it follows the location of the older subduction zone. The Bloody Bluff fault probably formed toward the end of Paleozoic time. It is named for its location near Bloody Bluff in the Minute Man National Historical Park in Lexington.

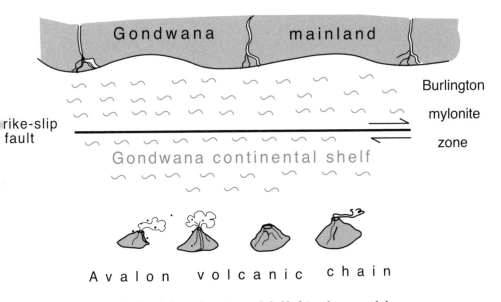

Faulted and sheared continental shelf of Gondwana and the Avalon volcanic island chain, about 600 million years ago.

Burlington Mylonite Zone

The 1- to 5-mile-wide Burlington mylonite zone formed in several stages. First, mylonite formed in Precambrian time as the edge of Gondwana was intensely sheared, possibly by strike-slip faulting. Magma of the Avalon volcanic island chain intruded the mylonite zone. Later, mylonite formed during the subduction of the Avalon terrane beneath the Nashoba terrane in Silurian time. The Burlington mylonite zone is really an umbrella term that encompasses a group of mylonites that represent a range of ages, compositions, and textures.

It is hard to determine in many cases what the original rocks were in the zone. Mylonite is a sheared rock that hardly resembles the original rock. Lenses of intensely sheared but recognizable quartzite, schist, basalt, and gabbro are preserved within the mylonite and are probably remnants of rocks that were ground up in the shearing process. In addition to the dominantly light-colored rocks of the Burlington mylonite, black to dark green mylonitic amphibolite—metamorphosed basalts—are present in the westernmost part of the Avalon block near the Bloody Bluff fault zone. These may be chunks of the Marlboro formation from the adjacent Nashoba terrane that got mixed in during brittle faulting in Mesozoic time.

The first stage in the mylonitization of the Burlington rocks was nearly complete by the time the magma of the Dedham granite intruded. The plutonic and volcanic rocks of the Avalon terrane of southeastern New England are between 625 and 589 million years old. The Dedham granite

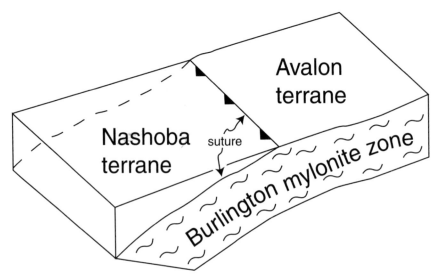

The Burlington mylonite zone first formed in Precambrian time but sheared again as the Avalon terrane was dragged beneath the Nashoba terrane in Silurian time.

yields ages of about 620 million years for plutons southwest of Boston and 610 million years for those north of Boston. I will use the younger date throughout this book for Avalonian granites where there is not a more reliable date.

At the edge of the Dedham granite batholith, magma intruded along the foliation of the Burlington mylonite. The rock, a spectacular migmatitic mylonite, looks like a layered gneiss. Thin granitic layers alternate with sheared layers of recrystallized quartz, broken feldspar, and

Geologic events in the Avalon terrane.

AGE (millions of years ago)	GEOLOGIC EVENTS	MAJOR ROCK UNITS
Triassic — 245 —	Middleton Basin: rift basin along Bloody Bluff fault zone	
Permian — 286 —		Dighton conglomerate Rhode Island fm.
Pennsylvanian Mississippian	Coal basin formation	Sachuest conglomerate Pondville conglomerate
— 360 —	Bloody Bluff fault zone	Wamsutta formation
Devonian		Peabody granite
— 417 —	Alkalic plutonism	Franklin granite Quincy granite
Silurian	Burlington mylonite zone: shearing between Avalon and Nashoba terranes	Cape Ann granite
— 443 —		
Ordovician		
— 495 —		Braintree slate
Cambrian		Weymouth formation
	Sedimentation	Cambridge slate
— 545 —	Boston Basin: rifting, volcanism, and sedimentation	Roxbury conglomerate Mattapan, Lynn, and Brighton volcanic rocks
Late Proterozoic	Calc-alkaline plutonism	Westwood granite Milford granite Dedham granite
	Burlington mylonite zone forms from shearing on margin of Gondwana; formation of Avalon island chain	Fall River granite
	Continental shelf sedimentation	Westboro formation

mafic minerals. Unaided by a microscope, a geologist might interpret it as a granitic gneiss older than the Westboro quartzite.

Folded and migmatitic mylonites are widespread and well exposed throughout the Avalonian uplands surrounding Boston.

Precambrian Rocks

The extensive Precambrian plutonic and volcanic rocks that form the Avalon microcontinent all crystallized while it was still attached to Gondwana. The Boston Basin originated as a rift basin within the Avalon island chain near the end of Precambrian time. Some rocks of Cambrian time were deposited in the basin and around the fringe of the volcanoes just after the island chain began its trip north to Laurentia.

Westboro Formation. The oldest rocks of the Avalon terrane are quartzites and shaly rocks of the Westboro formation and intrusive gabbros. The quartzites were once continental shelf sediments deposited on the edge of Gondwana during Precambrian time. We do not know their exact age, but they are clearly older than the ductile shearing that occurred when the Avalon island chain was forming in late Precambrian time.

Dedham and Westwood Granites. The rocks of the Avalon volcanoes occupy southeastern Massachusetts and underlie much of the Narragansett Basin as well as the smaller Woonsocket and Norfolk Basins. As the Avalon volcanic island chain formed on the margin of Gondwana, the continental shelf broke up and foundered. The volcano's magma, which crystallized into the 610-million-year-old maroon to pink Dedham granite, intruded blocks of continental shelf made up of Westboro quartzite and Burlington mylonite. The 599-million-year-old light pink to white Westwood granite is a relatively shallow granite that intruded the associated pile of Mattapan volcanic rocks and diorite. The emplacement of this granite marked the close of Avalonian magmatic activity in the Boston Basin. The Westwood granite has relatively fewer dark minerals and is finer grained than the Dedham granite—otherwise the two granites are difficult to distinguish.

Milford Granite. The Milford granite, also called the Rhode Island batholith, is about 610 million years old and is equivalent to the Dedham granite. The beautiful Milford granite, a sparkling pale pink and white rock, weathers to salmon or buff. Quartz and feldspar are crystallized into sugary aggregates, and black biotite clusters are sprinkled throughout. "Milford pink" is the commercial name of the architectural stone used for the Boston Public Library in Copley Square and the elegant columns of Pennsylvania Railroad Station in New York City.

Dedham granite with dark, biotite-rich segregations. Fine-grained granitic dikes cut both the granite and the segregations. On Oak Street in Natick, 0.8 miles south of Massachusetts 30.

Though the Milford granite resembles the Dedham granite, geologists mapped them separately because of the Milford granite's well-developed foliation and moderate granulation of quartz and feldspar. On a map, the batholith displays a crudely arched pattern between Natick and Rhode Island that is called the Milford antiform. Gray to buff, glassy quartzites and dark schists of the Westboro formation east, north, and west of the core of the batholith contribute to the antiformal shape. This antiform contains several phases of emplacement of Milford granite and granodiorite as well as gabbro plutons.

Parts of the Milford granite have undergone at least one, and in places, two episodes of mylonitization. Mylonitic Milford granite is present in the east-trending Nobscot mylonite zone, some 500 feet wide, extending through Southboro and Framingham.

Fall River Batholith and Associated Granites. The Fall River batholith, an entire complex of igneous rocks in southeastern Massachusetts, extends west from the coast along Cape Cod Bay to the eastern margin of the Narragansett Basin along Mount Hope Bay. The most abundant rock is the

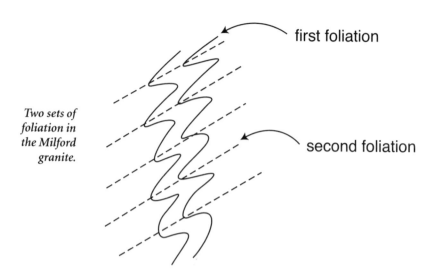

Two sets of foliation in the Milford granite.

first foliation

second foliation

Fall River granite, a light gray, medium-grained granite with only a small amount of dark minerals such as biotite and hornblende. Many such light-colored granites in the southern part of the Avalon terrane have age dates that cluster right around 600 million years, and they typically are less deformed than the 610-million-year-old Dedham granite farther north.

The boomerang-shaped batholith intruded a great variety of metamorphic rocks. Curved strips of these older Precambrian rocks are preserved as roof pendants—so-called because they ornament the top of the Fall River batholith. These rocks were part of the sedimentary sequence of the Gondwanan continental shelf.

Gabbros. Gabbros, the intrusive chemical equivalent of basalts, are typical of volcanic island chains, such as the Avalon. Gabbros crystallize slowly at depth and typically have large, visible minerals. Gabbros of different ages look amazingly alike, long puzzling geologists. In Massachusetts, some are of Precambrian age and others, radiometrically dated, give ages of 444, 427, and 375 million years, from late Ordivician to middle Devonian time. The rocks are often a mixture of coarse-grained gabbro and finer-grained diorite, sometimes called gabbro-diorites.

Boston Basin

The Boston Basin originated as a faulted rift basin within the Avalon volcanic island chain. Geophysical data suggests that the Boston Basin may extend tens of miles farther east under the Atlantic Ocean. The basin filled with volcanic and sedimentary rocks, the latter deposited by rivers, including meltwater rivers from glaciers that formed at high elevations in the

Avalon volcanic islands. These strata are younger than most of the Avalon rock formations that surround the Boston Basin.

Faults bound the basin. The west-dipping Northern Border thrust fault, manifested on the land surface as a rocky slope between the Cambridge lowlands and the highlands to the northwest along Massachusetts 60, forms the northern boundary of the Boston Basin. The Blue Hills and Ponkapoag faults, the latter crossing I-93 southeast of the Blue Hills, form the southern boundary. North-dipping thrust faults, which are offset by a series of steep normal faults, may form the western margin. The Mount Hope, Neponset, and Blue Hills faults cut through the southern part of the basin.

Along the faulted southwest margin, coarse-grained sediments of the basin sit atop the eroded surface of Westwood granite and Mattapan volcanic rocks. The sediments are younger than the 599-million-year-old Westwood granite but older than the 540-million-year-old, early Cambrian fossiliferous sedimentary rocks near Weymouth that also rest directly on an eroded surface of Avalonian granite.

The Precambrian rocks of the Boston Basin consist of the Dedham and Westwood granites; Mattapan, Brighton, and Lynn volcanic rocks; the Roxbury conglomerate; and the Cambridge slate.

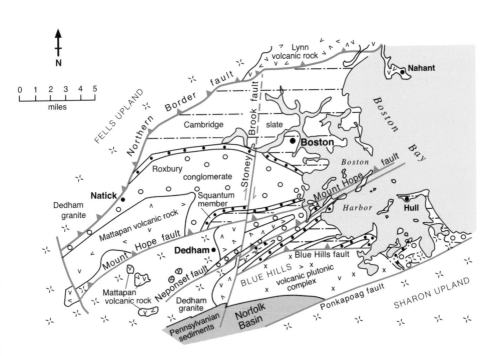

Major formations and structures of the Boston Basin. —Modified from Billings, 1979

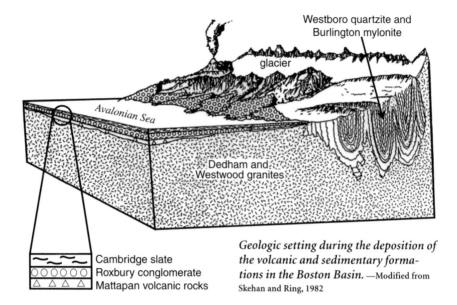

Westboro quartzite and
Burlington mylonite

glacier

Avalonian Sea

Dedham and
Westwood granites

Cambridge slate
Roxbury conglomerate
Mattapan volcanic rocks

Geologic setting during the deposition of the volcanic and sedimentary formations in the Boston Basin. —Modified from Skehan and Ring, 1982

Mattapan, Lynn, and Brighton Volcanic Rocks. The Mattapan and Lynn volcanic rocks erupted about 596 million years ago from the Avalon volcanoes. These rocks include fine-grained rhyolites, dark to light green lavas, and ashflows that include crystal tuffs, breccias, and mudflows. The Mattapan volcanic rocks are present southwest of Boston and the Lynn volcanic rocks north of Boston. Native Americans fashioned the rhyolites into tools and arrowheads.

The Brighton volcanic rocks, also from the Avalon volcanoes, consist of highly altered basaltic lavas and ashfalls that probably coated the landscape as the Boston Basin formed. They are typically below the Boston Basin sediments or interlayered near their base. Both the Brighton volcanic rocks and the associated 589-million-year-old quartz diorite dikes that intrude the Westwood granite southwest of Boston have been hydrothermally altered.

Roxbury Conglomerate. The nearly 2,000-foot-thick Roxbury formation consists of conglomerate, shale, sandstone, quartzite, arkose, and altered basaltic volcanic rocks. The conglomerate is often called puddingstone. Oliver Wendell Holmes (1809–1894) described the appearance of the pebbles in the fine-grained matrix as "plums in a pudding." His poetic theory of the origin of the puddingstone is as delightful as it is imaginative.

In Holmes's 1830 poem "The Dorchester Giant," the giant's unruly children:

... flung [the pudding] over to Roxbury Hills,
They flung it over the plain

> And all over Milton and Dorchester too
> Great lumps of pudding the giants threw . . .

The Dorchester member of the Roxbury formation is a 1,550-foot-thick sequence that ranges in grain size from conglomerate to slate and in color from pink to red, gray, white, and green.

The Squantum member is a 400-foot-thick sequence of poorly sorted sediment. A grayish, purplish, or greenish matrix of sand to clay encloses partly rounded to angular rocks ranging from pebbles to large boulders. The rock fragments consist of granite, quartzite, and altered basalt. Nearly a century ago, geologists interpreted the sediment as a glacial till. Ever since, they have debated the glacial origin of the Squantum member. Geologists have not found any glacially scratched pebbles in the sediments, but rhythmically bedded siltstones above the sediment look like varves and bolster the hypothesis of a glacial origin. It was deposited while Avalon sat near the south pole, possibly during the vast episode of glaciation that Paul Hoffman of Harvard University calls the time of the "Snowball Planet."

Cambridge Slate. The Cambridge slate is a dark green, rhythmically layered rock. Light gray silt to fine sand layers alternate with dark layers of clay to fine silt. The slate underlies the northern part of the Boston Basin. It is known principally from borings and tunnels north of Boston but is well represented in outcrops south of Boston and on some of the islands in Boston Harbor. In parts of the Nut Island–Deer Island tunnel under Boston Harbor, the pinkish Cambridge slate resembles the rhythmically bedded siltstones of the Squantum member of the Roxbury conglomerate.

Cambrian Trilobites and Other Fossils

Rocks of early to middle Cambrian age contain the first fossil records of animals with hard parts such as shells or chitinous carapaces. Rocks older than Cambrian age contain a relatively sparse collection of life-forms, most of which were soft-bodied animals superficially resembling the jellyfish. Age dates obtained by Robert Tucker of Washington University and Samuel Bowring of the Massachusetts Institute of Technology place the beginning of Cambrian time at 545 or 544 million years ago. By the beginning of Cambrian time, all but one animal phyla were represented in the rock record.

The early Cambrian rocks of the Weymouth formation of Boston's South Shore contain shells of gastropods and other mollusks and brachiopods. Trilobites appear in the Braintree slate approximately 510 million years ago in middle Cambrian time. Trilobites were usually small animals about the size of a silver dollar, but some trilobites in eastern Massachusetts, such as *Paradoxides harlani,* reached a length of about 1 foot. Until the concept of plate tectonics was proposed in 1967, it was a great puzzle as to why these trilobites near Boston were so different from those of the same age in northwestern Vermont. The giant trilobite *Paradoxides* was common to Gondwanan terranes and could only have reached Massachusetts by way of plate tectonic movement.

Blue Hills Volcano

The Blue Hills form an east-west-trending ridge that rises from about 160 feet elevation at the base to 600 feet at the top of Big Blue. Well-exposed granitic and volcanic rocks comprise the resistant ridge. The Blue Hills volcano erupted about 440 million years ago at the beginning of Silurian time. The Quincy granite crystallized from the volcano's magma chamber, and the Blue Hills quartz porphyry erupted from the volcano. The alkaline, or potassium- and sodium-rich, chemistry of these rocks indicate they formed in a rifting environment. The magma intruded a cover of Cambrian sedimentary rocks and Precambrian Mattapan volcanic rocks. When the magma chamber erupted, it released a lot of gassy lava and the volcano collapsed, opening a broad crater, or caldera.

The Quincy granite is a beautiful architectural stone highly prized for its deep greenish color and massive character. Geologists call a rock *massive* if it has no directional grain. This feature makes the rock hard to cut into building stone because it does not naturally split in certain directions. Some of the Quincy granite is pink because it was altered by hot water solutions in the upper part of the volcano's magma chamber.

The Blue Hills quartz porphyry, a rhyolitic rock, erupted from the Blue Hills volcano. It shares a similar chemistry with the Quincy granite because it is the granite's extrusive volcanic equivalent. The granite and the porphyry contain riebeckite—an amphibole—and other soda-rich minerals.

The quartz porphyry was exposed and weathered before the boulder-rich Giant conglomerate of the Norfolk Basin was deposited during early Pennsylvanian time. The ancient weathering zone contains rounded, maroon and green residuals of quartz porphyry. The residual coloration is from iron in the volcanic rock. The old erosion surface was approximately horizontal until crustal movements compressed the Boston Basin against the Blue Hills, and they in turn rammed against the Norfolk Basin during the Alleghanian mountain building event in Permian time.

Rift Granites

Alkalic granites—that is, granites rich in potassium and sodium—originate from rifting, or extensional tectonics. Geologists call these rocks rift granites. Two bodies of rift granites occur in northeastern Massachusetts.

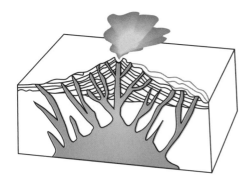

Molten magma erupts, releasing clouds of incandescent ash.

Continued eruption of lava and ash flows partially empties the magma chamber.

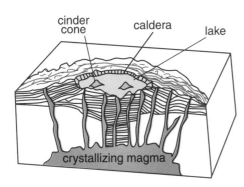

A broad caldera opens as the volcano collapses into the emptying magma chamber. The magma cools, though volcanic gas emissions may continue and hot springs may form. Small volcanic cones may develop, and a lake may flood the caldera.

Evolution of the Blue Hills volcano and its caldera.

The Cape Ann granite crystallized about 450 million years ago, and the greenish gray Peabody granite crystallized about 370 million years ago. These two granites intruded older dark gray to black plutonic rocks—gabbros.

The Franklin pluton southwest of Boston resembles the Quincy granite in its chemistry and mineralogy, but it is younger, dating from late Silurian time, about 417 million years ago. This narrow, 25-mile-long pluton probably represents a rift in the Avalon terrane and separates the sheared Milford granite to the west from the less-deformed Dedham granite and Fall River batholith.

COAL BASINS

Four rift basins formed in the Avalon terrane about 315 million years ago and filled with sediments during Pennsylvanian time. Three basins—the Narrangansett, Norfolk, and Woonsocket—are in southeastern Massachusetts. The Narrangansett and Woonsocket Basins extend into Rhode Island, and the small North Scituate Basin is entirely in Rhode Island.

The Narragansett Basin, by far the largest coal basin in the region, occupies almost 1,000 square miles, much of it low-lying, swampy forests. A thick sequence of nonmarine sedimentary rocks—conglomerate, sandstone, siltstone, shale, and coal—underlies these extensive lowlands. The basin, famous during the industrial revolution of New England for its coal, is well known today as a natural laboratory for the study of almost every aspect of geology, and especially structural geology and metamorphism of coal.

The Norfolk Basin is a narrow, 2.5-mile-wide strip of land between the Precambrian crystalline rocks of the Boston Basin and the Sharon Upland. Although the Norfolk Basin is mainly a synclinal structure, an arch in its center exposes Wamsutta basalt and rhyolite volcanic flows older than the

Before the Alleghanian mountain building event

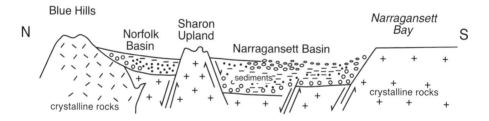

After the Alleghanian mountain building event

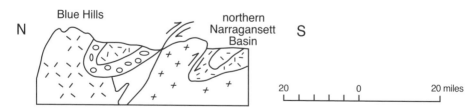

Fault-block basins in Precambrian granites. The basins filled with sediments before the tectonic collision that caused the Alleghanian mountain building event. —Modified from Skehan and others, 1986; Skehan, 1983

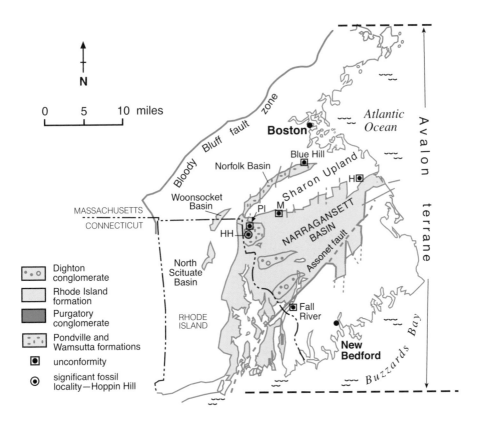

Narragansett, Norfolk, Woonsocket, and North Scituate Basins in the Avalon terrane. Significant floral locations and unconformities between Pennsylvanian and late Precambrian rocks are noted. Hanover (H); Hoppin Hill (HH); Foolish Hill, Mansfield (M); Masslite Plainville Quarry (Pl). —Modified from Skehan and others, 1986

coal sediments. Terrestrial sediments of Pennsylvanian age fill the basin and are in turn covered with Pleistocene glacial deposits.

Many coal mining towns in Massachusetts have the same names as towns in coal mining districts of Great Britain, for example, Mansfield and Norton. The first mining activity in the Narragansett Basin began with Leonard's coal mine in Mansfield in 1736. In Massachusetts and Rhode Island about forty-two prospects and mines dot the margins of the basin. Coal was mined for steam engines and domestic use. Several mines in Portsmouth, Rhode Island, produced an estimated 1.1 million tons of coal, primarily used for copper smelting, in the nineteenth and twentieth centuries.

Cooked Coal

The coals of the basins were baked from enormous peat deposits formed from abundant vegetation growing in great freshwater swamps and marshlands in humid climates. Peat is forming today in the bayous of the lush Mississippi River delta. Heat and pressure from deep burial transforms—metamorphoses—peat into coal, and increased heat and pressure transforms bituminous coal into anthracite or meta-anthracite.

The rocks of the Narragansett Basin range from essentially unmetamorphosed in the northern region to high-grade sillimanite zones of metamorphism in the southwestern Narragansett Bay area. Some meta-anthracite deposits of the Narragansett Basin are up to 40 feet thick. Experts believe that meta-anthracite deposits are generally compacted forty times their original thickness—these coal seams would originally have been an astounding 1,600-foot-thick bed of peat. To achieve such thick peat deposits, the basin must have been actively subsiding, perhaps coupled with uplift in the surrounding mountains.

The Narragansett Basin was intensely folded and faulted in Permian time in the Alleghanian mountain building event when the Meguma microcontinent pushed into the Avalon terrane. That collision buckled the basin into a crescent convex to the northwest. Meanwhile, Cambrian and Precambrian rocks wedged above and below the Avalon block were shoved onto the basin while the lower part of the Meguma microcontinent was dragged beneath the Avalon terrane. The heat and pressure converted the organic matter to coal. We know this happened in early Permian time because the Narragansett Pier granite intruded the basin sediments in southern Rhode Island about 275 million years ago.

During the collision, the sediments in the Narragansett Basin were heated through a whole spectrum of temperatures and pressures. Its northern part, as well as the Norfolk Basin to the north, were hardly metamorphosed. Temperatures were higher in Rhode Island, where a 20-mile-thick stack of thrust sheets was piled on top of the basin. The high temperatures drove off most of the combustible gases, converting the original coal to meta-anthracite. Its low volatile content makes meta-anthracite very hard to ignite, but once started, it burns with an extremely hot flame.

In a lengthy ditty entitled *Meditations on Coal*, William Cullen Bryant spoke jocosely in unflattering terms of the coal's powers of ignition:

In the conflagration at the end of the world
The last thing to burn will be Rhode Island coal.

Fossils

The fine-grained sedimentary rocks of the Narragansett Basin have yielded three hundred species of plant fossils—a greater number than from any other coal basin in the world. These plant fossils are useful in determining the relative age of the rocks. Swamp mud buried the plant remains, thus protecting them from oxidation and erosion. Over time, heat and pressure turned the swamp mud to stone, and the organic plant remains became carbon residue. The plant impressions are visible on bedding surfaces. Boston College's Weston Observatory exhibits a 12-ton block of carbonaceous shale with abundant fossils.

The rocks also reveal fossils of animals, including a number of insect wings, a spider, more than a dozen species of cockroaches, a tube worm, and a gastropod track. Geologists have also found the footprints of six species of amphibians as well as a possible amphibian skin and burrow.

Pennsylvanian-age plant and leaf fossils of the Narragansett Basin from Masslite Quarry in Plainville.

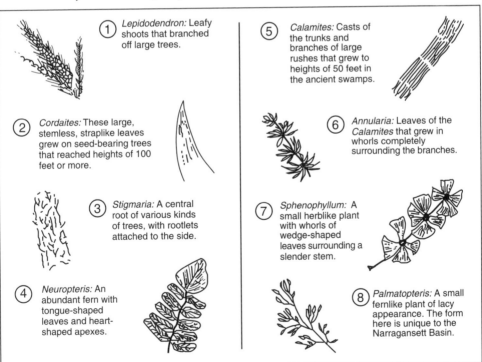

1 *Lepidodendron:* Leafy shoots that branched off large trees.

2 *Cordaites:* These large, stemless, straplike leaves grew on seed-bearing trees that reached heights of 100 feet or more.

3 *Stigmaria:* A central root of various kinds of trees, with rootlets attached to the side.

4 *Neuropteris:* An abundant fern with tongue-shaped leaves and heart-shaped apexes.

5 *Calamites:* Casts of the trunks and branches of large rushes that grew to heights of 50 feet in the ancient swamps.

6 *Annularia:* Leaves of the *Calamites* that grew in whorls completely surrounding the branches.

7 *Sphenophyllum:* A small herblike plant with whorls of wedge-shaped leaves surrounding a slender stem.

8 *Palmatopteris:* A small fernlike plant of lacy appearance. The form here is unique to the Narragansett Basin.

Rocks of the Narragansett and Norfolk Basins

The sediments of the Narragansett Basin are divided into three major formations. The distinctive smoky quartz–bearing Sachuest conglomerate of Pennsylvanian age rests on an erosion surface of the Fall River granite. The coal-bearing Rhode Island formation, in the middle, makes up the bulk of the Pennsylvanian strata. The Dighton conglomerate tops the sequence. The Pondville conglomerate and Wamsutta redbeds are the major formations in the Norfolk Basin. These two formations also occur in the northern part of the Narragansett Basin. They are exposed in hills that rise above the glacial lake deposits and outwash sequences.

Sachuest Conglomerate. The Sachuest conglomerate, named for Sachuest Point in Rhode Island, forms the base of the Pennsylvanian coal basin sequence of the southeastern part of Narragansett Basin. The conglomerate, which consists of very small pebbles, has white spots of highly weathered feldspar crystals set in a jet-black carbonaceous shale and smoky quartz matrix. The quartz and feldspar were derived from the weathering of late

Stratigraphic relationships for the Norfolk Basin and the northern and southern parts of the Narragansett Basin. —Modified from Skehan and others, 1986

ROCK STRATIGRAPHIC UNITS			Geologic Time Period	millions of years before present
NORFOLK BASIN	NORTH NARRAGANSETT BASIN	SOUTH NARRAGANSETT BASIN		
	Dighton conglomerate			286
	Rhode Island formation	Rhode Island formation.	LATE PENNSYLVANIAN	292
				294
		Purgatory conglomerate		296
	Rhode Island formation			299
Wamsutta redbeds	Wamsutta redbeds	Sachuest conglomerate	MIDDLE PENNSYLVANIAN	302
Pondville	Pondville			306
conglomerate			EARLY PENNSYLVANIAN	315
				320

Precambrian granite. In Fall River, the pebbles of the conglomerate are less weathered, and the carbonaceous shale and smoky quartz content is essentially absent due to changes in the sediment supply.

Rhode Island Formation. The Rhode Island formation consists of gray sandstone and green siltstone with lesser amounts of black shale, pebble conglomerate, and coal. Quartz forms the major component of the sandstone and conglomerate. These sediments commonly contain sedimentary structures such as graded bedding, crossbedding, channel scours, and channel fill. The entire sequence is at least 10,000 feet thick. Some estimates range up to 20,000 feet, but in those estimates parts of the formation may have been measured more than once because of large-scale deformation.

Dighton Conglomerate. The gray Dighton conglomerate consists mainly of rounded quartzite fragments, ranging in size from pebbles to cobbles and even boulders. The conglomerate contains some granite and slate fragments and lenses of sand. In the southeastern part of the basin, deep burial with elevated temperatures and pressures deformed the Dighton cobbles by pressure solution. When these cobbles were intensely squeezed in a northwest-southeast direction, the quartz dissolved in water and migrated to the northeast and southwest ends of the cobbles, where the quartz precipitated as fibrous crystals. The middle parts of the cobbles are now slender and the northeast and southwest ends feature "goatees" of quartz.

Wamsutta Redbeds. The Wamsutta redbeds—early Pennsylvanian age or possibly older—consist of coarse- to fine-grained, clastic, graded sediments including reddish conglomerates, maroon siltstones, and deep maroon slates. Magnificent sedimentary depositional structures such as crossbedding and graded bedding are well preserved in large-scale folds.

Pondville Conglomerate. The Pondville conglomerate contains a variety of rock types, including quartzite and granite. It was probably an alluvial fan deposit laid down near the head of a canyon, and ranges in thickness from about 5,300 to 7,600 feet. The oldest known fossil plants from New England, about 315 to 300 million years old, are present in the only known graphitic black slate in the Norfolk Basin, an upper member of the Pondville conglomerate.

The Giant conglomerate, deposited some 315 to 300 million years ago, is a distinctive unit within the Pondville conglomerate. Most of the boulders in the Giant conglomerate along the south margin of the Blue Hills are from the Blue Hills quartz porphyry. Many of the boulders have a rusty rind, evidence of long exposure on the old weathering surface of the Blue Hills volcano.

MEGUMA TERRANE

The northeast-trending Nauset fault slices through Cape Cod and forms the boundary between the Avalon and Meguma terranes. Many rocks in the Avalon terrane have a northeast-trending map pattern and foliation that suggests that these bands were deformed by a collision with a microcontinent moving up from the southeast, most likely the Meguma microcontinent. Part of the Meguma terrane probably sank beneath the Avalonian plate in a subduction zone, represented on the surface by the northwest-dipping Nauset fault. Another part of the Meguma terrane may have been thrust onto the Avalon terrane.

Paul Schenk of Dalhousie University named the Meguma terrane of Nova Scotia after a region in Morocco and suggested it was probably derived from West Africa. It forms the Scotian shelf south of Nova Scotia and the Bay of Fundy, and probably underlies a large part of the Bay of Maine. Rocks of the Meguma terrane are not exposed in Massachusetts, but the terrane is well known from exposures along the coast of southern Nova Scotia. Cambrian rocks are exposed near Jamestown, Rhode Island, and they probably are part of the Meguma terrane.

The initial collision of the Meguma terrane with the amalgamated Merrimack-Nashoba-Avalon terranes took place during the Acadian mountain building event. This collision formed migmatite; the Straw Hollow diorite in the Nashoba block, which is about 400 million years old; and the Indian Head Hill granite pluton, which is 349 million years old. Alleghanian movements in Permian time renewed westward and later eastward thrusting and produced uplifts, first in the Avalon, then in the Nashoba, and later in the Merrimack terrane.

JURASSIC TIME

As the Atlantic Ocean began rifting open in Jurassic time, about 200 million years ago, basaltic magma squirted up along rift faults, erupting as lava in some areas. The 300-foot-wide Medford dike north of Boston at 190 million years of age and the Jurassic basalts of the Connecticut Valley are the earliest evidence of the newly forming Atlantic Ocean. One to two thousand basalt dikes of Jurassic age, many northeast-trending, are scattered throughout the Boston area.

The Middleton Basin, along the western faulted margin of the Avalon block, formed in Triassic and Jurassic time. Red conglomerate, arkose, and red micaceous shale containing plant fossils of late Triassic to early Jurassic age fill the basin. Brittle faults fractured the Triassic rocks of the Middleton Basin in Jurassic time.

GLACIAL GEOLOGY

The most recent ice sheet, the Wisconsinan, deposited most of the glacial debris in eastern Massachusetts. At Sankaty Head, Nantucket, we see deposits from the Illinoian sheet and interglacial marine beds deposited in seawater between 140,000 and 120,000 years ago during the Sangamon age, an interglacial stage between the Wisconsinan and Illinoian ice sheets.

By about 23,000 years ago, the Wisconsinan ice sheet had advanced south across New England and had slowed as its front moved into the more temperate climate of Martha's Vineyard and Nantucket. At that time, the ice was still moving forward but its front was stationary because it was melting as fast as it was advancing. It probably melted clear of southeastern Massachusetts by about 15,000 to 14,000 years ago.

Ice Lobes

The glacial ice sheet developed several fanlike lobes along its southern margin. Where the shrinking front of the lobes paused in their retreat, the melting ice dumped sediment, making looping ridges of glacial till—moraines that precisely record temporary positions of the ice front. Terminal moraines outline the southernmost position of the lobes. Three lobes spread across eastern Massachusetts. From east to west, they are the South Channel lobe, the Cape Cod Bay lobe, and the Narragansett Bay–Buzzards Bay lobe.

Each lobe sloped toward the other, giving rise to a dimple where they met. The western margin of the Cape Cod Bay lobe abutted the eastern part of Narragansett Bay–Buzzards Bay lobe at the north end of Buzzards Bay. Martha's Vineyard is an angular moraine deposited where the Cape Cod Bay and Buzzards Bay lobes separated.

The three lobes melted at different rates. Initially, the Narragansett Bay-Buzzards Bay lobe melted first, followed by the Cape Cod Bay lobe, and finally the South Channel lobe. Geologists determined this by studying differences in the relative age of deposits. For instance, meltwater streams flowed west from the South Channel lobe and deposited sediment into Glacial Lake Cape Cod—a lake that formed after the Cape Cod Bay lobe receded northward. By 18,000 years ago, marine life-forms lived in the waters of Cape Cod Bay and the Bay of Maine. Later, when the ice fronts were near Boston, the offshore lobes melted back faster than the onshore ones.

Sea level was more than 300 feet lower during the ice age because water was tied up in the ice sheet. Evidence of the glacial margin east of Cape Cod lies beneath today's higher sea. Offshore investigations reveal that the South Channel lobe was the largest glacial lobe in southeastern New England. Based on tills recovered from core samples, we know the ice front of the South Channel lobe extended at least 60 miles southeast of Nantucket.

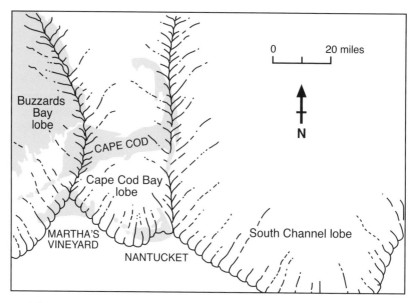

The maximum extent of the lobes over Cape Cod and the Islands during the last continental glaciation. —Modified from Oldale, 1992

Deposits of the South Channel lobe are exposed on the land surface only on the outer part of Cape Cod. No above-water deposits are present on the islands to the south.

If your intuition tells you that a glacial lobe formed the half moon of the Cape Cod shoreline, you are right on track. Most of the sediments on the central part of Cape Cod came from the Cape Cod Bay lobe. This lobe deposited part of the terminal moraine on the eastern end of Martha's Vineyard, and three recessional moraines—the Nantucket, Sandwich, and Ellisville moraines—track its meltback from southeastern Massachusetts.

The Narragansett Bay–Buzzards Bay lobe deposited the Martha's Vineyard moraine, the Block Island moraine, and the Buzzards Bay–Point Judith moraine. The melting of this lobe produced a magnificent array of sand and gravel deposits over the eastern third of mainland Massachusetts and a bit of western Cape Cod and Martha's Vineyard.

Pitted Outwash Plains

Whenever the glacier paused in its long retreat, torrents of meltwater dumped enormous loads of sand and gravel across the lowland between the ice and the ridge of the moraine. Kame hills and kettle ponds punctuate the sand and gravel plains. Areas pockmarked with kettles are called pitted plains. Most of the kettles mark places where large chunks of ice were buried in the outwash, then melted. They left depressions that were deep enough to penetrate the water table, forming modern ponds and lakes. The kames are deposits of outwash sediment that were laid down along the front of the melting glacier, then slumped into small hills as the ice that supported them finally melted.

The great Wareham pitted plain developed between the Buzzards Bay moraine and the ice sheet receding to the north. The Mashpee pitted plain formed in the interlobate area between the Cape Cod Bay and Buzzards Bay lobes.

Wind picked up sand and silt in the outwash plains and blew it across southeastern Massachusetts, including Cape Cod, to a depth of several feet. The deposits are commonly oxidized pale brown, reddish brown, and yellow and contain windblasted stones. Windblown deposits of this region are mostly fine to medium sand. They formed a thin cover over the bleak landscape that emerged from beneath the melting ice. Even with the addition of sand and silt, southeastern Massachusetts is poor farmland.

Glacial Lakes

As the continental glacier melted, it produced a succession of glacial lakes in eastern Massachusetts. Meltwater ponded between the melting ice front and any natural dam, such as a moraine, on the recently deglaciated terrain. One large lake, Glacial Lake Cape Cod, formed at the southern edge of the Cape Cod Bay lobe and north of the Sandwich moraine. Most of the lake deposits now lie beneath the waters of southern Cape Cod Bay. The Sandwich moraine and outwash impounded the lake on its south side, while the South Channel lobe prevented the lake waters from escaping to the east.

A succession of about nine lakes formed south of the Narragansett Bay-Buzzards Bay lobe as it melted. These glacial lakes were numerous, spectacular, and different from those of other parts of Massachusetts. Many small ponds and lakes formed in small depressions, and large lakes developed in a few favorable places, particularly in the major drainage basins of eastern Massachusetts. The principal glacial lakes from south to north are: Glacial Lakes Taunton, Neponset, Bouvé, Charles, Assabet, Sudbury, Concord, Nashua, and Shawsheen-Merrimack.

As the Buzzards Bay portion of the lobe receded northward, the rising meltwaters grew into Glacial Lake Taunton. Two stages of this lake covered about a total of 500 square miles. Lake waters deposited sediments as far north as Brockton, almost as far east as Kingston, and in the Jones River lowland, north of the Middleborough moraine. The sediments consist of varved clays, bedded silts and sands, and kame deltas laid down during the high-water stages of the lake.

These high-water deposits suggest to geologists that Glacial Lake Taunton's water level was about 55 to 65 feet above present-day sea level, and thus even higher than the glacial sea level. So why didn't the water just flow out to sea via the Jones River? The Cape Cod Bay lobe formed an ice dam at least as far south as the Jones River valley and blocked the outlet. Another outlet at Fall River controlled the water level of Glacial Lake Taunton until the ice melted north and freed the Jones River outlet, now

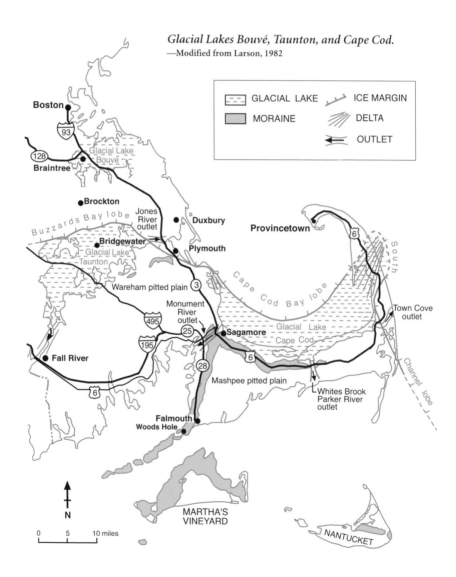

Glacial Lakes Bouvé, Taunton, and Cape Cod.
—Modified from Larson, 1982

marked by an eroded spillway between the Jones River lowland and the Jones River valley in Kingston.

Glacial Lake Bouvé was well north of, and formed later than, Glacial Lake Taunton. The lake was named for the accomplished geologist Thomas T. Bouvé, one-time president of the Boston Society of Natural History. Meltwater flowing from the Narragansett Bay–Buzzards Bay lobe that covered the Blue Hills and Boston formed the lake. The southern shore of the lake was a ridge of bedrock and till just south of Liberty Plains in South Hingham. The Blue Hills at West Quincy formed its western shore, and ice dammed its north and east sides.

A series of northward-expanding glacial lakes occupied the Nashua Valley, the most prominent drainage area in Massachusetts east of the Connecticut Valley. Glacial Lake Nashua grew to a length of about 35 miles as the ice sheet receded to the north. This lake dominates the glacial history of the Nashua River valley; it is one of the most studied glacial lakes in the country and is integral to the geological knowledge of central New England. It is discussed in detail in the roadguide for Massachusetts 12 and Massachusetts 13.

Sea Level Changes

Two counteracting processes affected sea level as the glaciers melted: sea level rose as the amount of water in the sea increased, and the land began to rise after the weight of the ice was removed. Despite this confounding relationship, geologists estimate that sea level rose rapidly but sporadically beginning about 14,500 years ago at an approximate rate of 50 feet per 1,000 years. The continent rebounded between 10,500 and 7,500 years ago after the melting ice unloaded the crust. About 10,000 years ago the rate of sea-level rise gradually decreased, and it slowed to a crawl about 7,000 years ago. Sea level probably reached its present stand about 3,500 years ago, though it is still not constant. In the last 2,000 years, sea level has risen about 6 feet. The story of relative rise of sea level with respect to rebounding land is complex and not yet fully understood.

Marine waters once covered parts of the Boston Basin and other low-lying areas, including parts of Cape Ann. Seawater encroached on the coast as the Cape Cod Bay ice lobe receded north. North of Boston, glacial deposits interfinger with fossiliferous marine sediments.

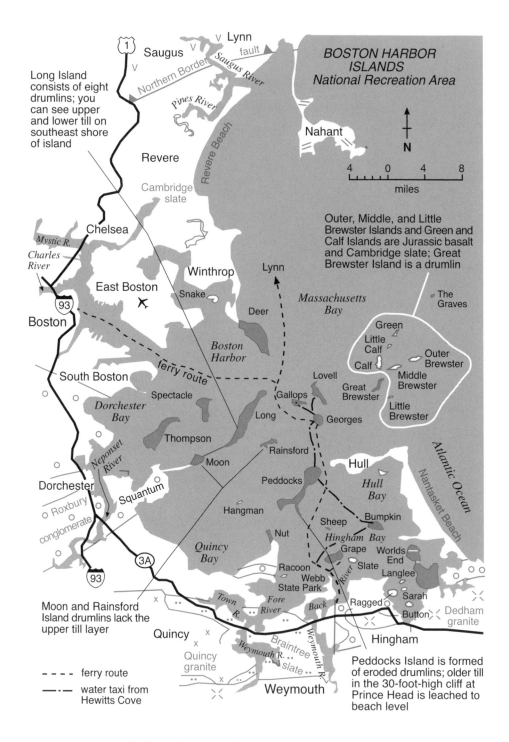

Long Island consists of eight drumlins; you can see upper and lower till on southeast shore of island

1 Saugus

V Lynn

V Saugus River fault

Northern Border

Pines River

BOSTON HARBOR ISLANDS National Recreation Area

Nahant

N

| 4 | 0 | 4 | 8 |

miles

Revere

Revere Beach

Cambridge slate

Chelsea

Mystic R.

Charles River

93

Boston

Winthrop

East Boston ✈

Snake

Deer

Lynn

Massachusetts Bay

Outer, Middle, and Little Brewster Islands and Green and Calf Islands are Jurassic basalt and Cambridge slate; Great Brewster Island is a drumlin

The Graves

Green

Little Calf

Calf

Outer Brewster

Middle Brewster

Little Brewster

Boston Harbor

ferry route

South Boston

Spectacle

Dorchester Bay

Thompson

Neponset River

Dorchester

o Roxbury

conglomerate

Squantum

Moon

Long

Gallops

Lovell

Great Brewster

Georges

Rainsford

Hull

Atlantic Ocean

Namasket Beach

Peddocks

Hull Bay

Hangman

Nut

Sheep

Bumpkin

Hingham Bay

Grape

Worlds End

Langlee

Quincy Bay

Racoon

Webb State Park

Slate

Sarah

Button

Ragged

Dedham granite

3A

93

Moon and Rainsford Island drumlins lack the upper till layer

x

Town R.

Fore River

Back River

Quincy

x

Quincy granite

Weymouth R.

Braintree slate

Weymouth R.

Weymouth

Hingham

Peddocks Island is formed of eroded drumlins; older till in the 30-foot-high cliff at Prince Head is leached to beach level

- - - - ferry route

—·— water taxi from Hewitts Cove

Geology of Boston Harbor. —Modified from Zen and others, 1983; Massachusetts Department of Environmental Management, 1998

ROADGUIDES TO THE EASTERN SEABOARD

Boston Harbor Islands
and Peninsulas

Though neither the Pilgrims nor Puritans came directly to Boston Harbor, it is one of the great old ports of the East Coast. Boston Harbor occupies about 50 square miles and is somewhat protected from the ravages of the open sea. The Nantasket Beach peninsula in Hull, an ungainly arm of land at the south end of the harbor, reaches protectively northwestward, and Winthrop Head and Deer Island, a pronglike peninsula to the north of the harbor, stretches southeastward. The Long Island causeway points its bony finger northeastward from Quincy across the middle of the harbor, adding more protection to the inner harbor.

Boston Harbor, sprinkled with thirty-eight islands with a combined 1,200 acres of land, has 180 miles of shoreline, most of which is in the public domain. Though many islands remain in their natural state, humans have added breakwaters, causeways, tunnels, and fill. The former Apple, Bird, and Governor's Islands, for example, are now part of the underpinning of Logan International Airport. The Boston Harbor Islands National Recreation Area was established in 1996 for conservation, recreation, and educational purposes.

Two major shipping lanes weave through the obstacle course of islands in Boston Harbor. President Roads skirts to the south of Deer Island and connects to both the North and South Channels. Nantasket Roads passes south of Georges Island and north of Nantasket Peninsula and Peddocks Island.

The islands of Boston Harbor consist of two distinct groups—the inner islands formed of drumlins, and outer islands formed of bedrock. The inner islands are those that lie inside a line connecting Point Allerton in Hull and Lovells and Deer Islands to Winthrop. The outer islands are geologically and topographically an extension of the inner harbor islands but cluster just outside of the 4-mile-wide entrance to the harbor.

Native Americans settled in the Boston area about 2,500 years ago and raised crops and fished along the shores of the islands. The Norseman Biarne is said to have visited the area about A.D. 900. English explorers John and Sebastian Cabot noted the harbor when they sailed past it in the fifteeenth century, and many other traders and explorers followed in the sixteenth century. Bartholemew Gosnold, an Englishman, may have explored the

Seawall of Quincy granite on the northeastern shoreline on Georges Island. Storm waves moved blocks from their vertical metal rod moorings. Blocks on the shore include coarse boulders eroded from Georges Island drumlin and riprap that protects the seawall from erosion.

harbor in 1602, and Sieur De Monts, a Frenchman, may have explored it in 1604. Captain Miles Standish and his fellow Pilgrims and Indian guides visited Boston Harbor on September 15, 1621, and a group of Puritans established a permanent settlement in 1630 at Beacon Hill. The rest, as we say, is history.

Drumlins of the Inner Islands

The inner islands and peninsulas are primarily drumlins and tombolos. Drumlins are ice-sculpted mounds of glacial till. A tombolo is a bar of sand or gravel that ties the islands to other islands, to former now-submerged islands, and to the mainland.

Waves and tides erode the drumlins and redeposit the sediment as spits and tombolos. Sea cliffs and piles of large boulders in and beneath the surf are common on the seaward side of the islands. Waves have winnowed away the fine-grained materials.

Viewed from the air on a clear day, drumlins are the dominant, recognizable landform of the Harbor Islands. Geologists have mapped about two hundred drumlins in the Boston Basin and in the surrounding highlands. The long axes of these sturdy, streamlined hills of glacial till are strung out in a southeasterly direction—the direction in which the last continental ice sheet moved. As the ice moved forward, it molded glacial till into these mounds. You can obtain a magnificent view of the island drumlins from many places in and near the Boston Harbor, including the Prudential Tower.

The very best view was probably had in the eighteenth and nineteenth centuries. In those times the land had been cleared of trees for farmland, lumber, and firewood. The elliptical hills spread over the countryside like schools of fish in a pond. Today, these lofty landforms are largely unrecognizable—except in Boston Harbor and along the seacoast—because forests and urban sprawl blanket the land. Because the drumlins commonly rise over 100 feet above the surrounding area, they played a strategic role in colonial historic events.

Castle Island, a headland park that juts out into Boston Harbor across the channel from Logan International Airport, is a small drumlin that resembles a castle. The star-shaped stone walls of Fort Independence, built in 1801, crown this 20-acre recreational park. Edgar Allan Poe, enlisting in the U.S. Army under the name Perry, served here for five months.

Head west from Castle Island on Day Boulevard and you will pass the famous L Street Beach and Baths, known for its hardy "Brownies" who go for a dip every day of the year. To the northwest a short distance from L Street Beach, you can see Telegraph Hill in Boston, an impressive, 3,000-foot-long, east-trending drumlin that rises 145 feet above sea level.

Drumlins in the Boston area typically consist of a lower till and an upper till. The older, lower till is olive gray, compact, somewhat stratified, and weathered. It contains shell fragments of the marine clam *Mercenaria mercenaria*. The lower till breaks easily along parallel planes, and contains cobbles and boulders that are more commonly striated or grooved by scratching than those in the upper till. The sandy upper till is olive, oxidized, somewhat stratified, and unweathered. It contains more boulders than the lower till and lacks marine shells. A layer of sand separates the two tills. The lower till is almost certainly Illinoian in age because of its great depth of weathering and the presence of abundant marine shells, which would have accumulated during the Sangamon interglacial stage. These shells indicate that the Boston Harbor water was warmer than it is today.

You can see the upper and lower till on the southeast side of Long Island. The drumlins on Moon and Rainsford Islands lack the upper till layer.

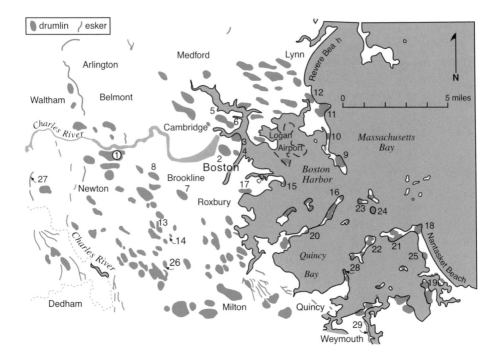

NAME OR LOCATION OF DRUMLINS AND ESKERS

 1 Nonantum, near Soldiers Field Road
 2 Beacon Hill
 3 Copps Hill
 4 Fort Hill
 5 Breeds Hill
 6 Bunker Hill
 7 Parker Hill
 8 Corey Hill
 9 Deer Island
10 Winthrop Head
11 Winthrop Highlands
12 Beachmont
13 Bussey Hill in Arnold Arboretum,
    Green Hill, Mt. Walley, and
    Anderson Park drumlins

14 Peters Hill
15 Castle Island
16 Long Island Head
17 Telegraph Hill, Boston
18 Allerton Hill
19 Sagamore Head
20 Moon Island
21 Telegraph Hill, Hull
22 Peddocks Island
23 Gallops Island
24 Georges Island
25 Strawberry Hill
26 Monterey Hill
27 Auburndale esker
28 Nut Island Hill
29 Great Esker

Distribution of drumlins and eskers in the Boston area and Boston Harbor.
—LaForge, 1932; Skehan, 1979

Twenty to twenty-five feet of weathered lower till is exposed above slumped sediment on Moon Island. On Prince Head on Peddocks Island, you can see the older till, which is leached to beach level, in a 30-foot-high cliff. You can get a good view of drumlins in the harbor from Webb State Park in Weymouth.

Boston Basin

The bedrock in and around Boston Harbor and the surrounding bays is primarily late Precambrian with some Cambrian-age fossiliferous sediments. The late Precambrian sediments accumulated in the Boston Basin, a rift basin in the Avalon terrane. The basin filled with volcanic and sedimentary rocks, including the Roxbury conglomerate—a sequence of conglomerate, shale, sandstone, altered basalt, and tillite that ranges from 1,170 to 5,500 feet thick.

The conglomerate and its associated volcanic rocks form the eastward-plunging core of the Central anticline. The anticline underlies the southern two-thirds of Old Harbor and Columbia Point on the waterfront, where the University of Massachusetts at Boston and the John F. Kennedy Library are situated. The conglomerate also occurs along the southern margin of the Boston Harbor in the Adams Shore and Houghs Neck peninsula of Quincy and along Hingham Harbor. Don't miss a walk through the marvelously rocky and wooded Worlds End Reservation just east of Hingham Harbor. Here, the Roxbury conglomerate rests unconformably on an eroded surface of Dedham granite and is cut by faults that you can easily recognize.

The rest of Boston Harbor and the Harbor Islands eastward to Massachusetts Bay is mainly underlain by the Cambridge slate, a rhythmically layered, fine-grained rock whose light gray silt to fine sand layers alternate with dark layers ranging from clay to fine silt.

You can see well-layered volcanic ash deposits of the altered Brighton volcanic rocks lying below the Roxbury conglomerate at Atlantic Hill at the south end of Nantasket Beach. A basalt dike of Jurassic age cuts through the sea cliff.

Squantum Member at Squantum Head

Squantum Head lies between Dorchester Bay on the north and Quincy Bay and the city of Quincy on the south. A rocky promontory at Squaw Rock Park is the best-exposed and most extensive outcrop of the Squantum member in the Boston area. This world-famous, 400-foot-thick sedimentary sequence within the Roxbury formation contains a wide range of particle sizes and rock types, from felsites—light-colored, fine-grained igneous rocks—and granite to rounded pebbles and cobbles of quartzite.

Well-layered, water-deposited volcanic ashes in the sea cliff on the north-facing slope of Atlantic Hill at the south end of Nantasket Beach. —Taylor Heewon Khym photo

You can also see the Squantum member just 1 mile to the west off Dorchester Street at Squantum Head, at Atlantic Hill at the south end of Nantasket Beach, and at the Arnold Arboretum.

Nearly a century ago, geologists identified glacial deposits of Permian age in South Africa. R. W. Sayles noted that clastic deposits in Squantum resembled the South African ones. For lack of any evidence to the contrary, he interpreted the Squantum deposit as a tillite, consolidated glacial till, of late Paleozoic age. Geologists have tried to find conclusive evidence of the origin of the Squantum member ever since.

By the late 1970s, geologists realized that the Boston Basin sediments were late Precambrian in age, and so the Squantum member must be too. Rhythmically bedded siltstones above the sediments look like varves and bolster the hypothesis of a glacial origin, but geologists have not found any glacially scratched pebbles. The fact that it was deposited while the Gondwanan supercontinent sat near the south pole also supports the glacial theory.

In 1993, geologist Richard Bailey of Northeastern University in Boston interpreted the sediment as a submarine debris flow or a pebbly mudflow

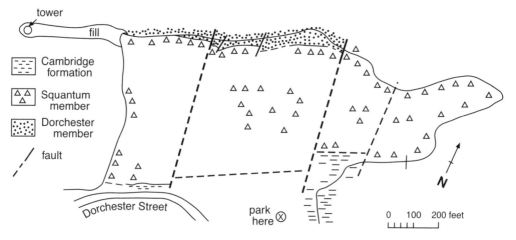

Geology of Squaw Rock Park on Squantum Head in Quincy. —Hepburn and others, 1993

that slid into the sea from a coastal upland. That upland was probably tectonically active, like the Alaskan coast today, and likely contained glaciers. The glaciers plucked fragments and blocks of bedrock, then moved them to lower elevations. Glacial meltwater deposited the coarsest gravels near the seashore while a marine delta of finer-grained gravels, sands, and silts formed offshore.

Nut Island Tunnel

The 4.5-mile-long Nut Island Tunnel connects Deer Island to Nut Island at the end of Houghs Neck in Quincy Bay. Built in the early 1990s, the sewage tunnel burrows through bedrock of Cambridge slate deep beneath the Boston Harbor. Geologists witnessed the tunnel excavation and mapped the rocks. The Cambridge formation in the tunnel commonly ranges from a bright red to maroon to pink and green, cleaved siltstone and sandstone, with beds of conglomerate. The conglomerate consists of pebbles of soft sedimentary rock rather than the hard-rock pebbles, such as granite and felsite, typical of the Roxbury formation on the mainland. You can see similarly bright maroon and gray slate and siltstone beds in shore exposures just north of the Squantum end of the Long Island causeway. Look for them along the shore to the east of the parking area for Squaw Rock Park.

Basalt dikes of Jurassic age intrude the Cambridge slate. Workers encountered numerous dikes during construction of the Nut Island Tunnel. Zones of altered, or bleached, slate are associated with basalt dikes. The abundance of such dikes in the tunnel and on the Brewster Islands suggests that the harbor has been extensively intruded by basalt along northeast-trending fissures as rifting began to form the Atlantic about 190 million years ago in early Jurassic time.

Outer Islands of Bedrock

The outer islands are primarily bedrock. Their long axes trend northeast, the dominant trend of the bedrock bedding, basalt sills and dikes, and major faults. The glaciers probably deposited southeast-trending drumlins on the outer islands as well, but the waves completely eroded them away, leaving the generally northeast-trending bedrock.

The Brewsters, a cluster of islands between Boston Harbor and Massachusetts Bay, have a large number of basalt sills and dikes of Jurassic age. A massive sheet of basalt is exposed on the north side of Middle and Outer Brewster Islands, while the south shore consists of slate similar to that of Calf Island. Below the main sill of basalt are several thinner sills. No bedrock is exposed on Great Brewster Island. On Little Brewster, the bedrock site of Boston Light, both basalt sills and slate are well exposed, with chunks of slate mixed in with the basalt.

Green Island, Little Calf Island, and Calf Island form a line that continues to The Graves. The Graves, a small island east of the Brewster Islands, consists of a massive and coarsely crystalline feldspar- and pyrite-bearing basalt with prominent joints. A fine-grained vertical dike cuts the island's

Boston Light on Little Brewster Island is a 98-foot-high, granite lighthouse constructed in 1716, the oldest lighthouse in North America and a National Historic Landmark. Shoreline is Cambridge slate intruded by basalt. View looking north.

south end. On Calf Island, clearly defined beds of gray slate of the Cambridge formation slant down at a gentle angle to the southeast.

Winthrop Peninsula

The Winthrop Peninsula, a magnificent series of drumlins connected by sand and gravel spits, forms the northern margin of Boston Harbor. Winthrop Peninsula began as an array of ten drumlins. Longshore drift of reworked glacial materials formed the spits and tombolos that tie the drumlins together. Neighboring headlands serve as a source of sediment of sand beaches and gravel bars.

The growth of sand spits and tombolos created the Winthrop Peninsula out of a cluster of drumlin islands. —Kelley and others, 1993

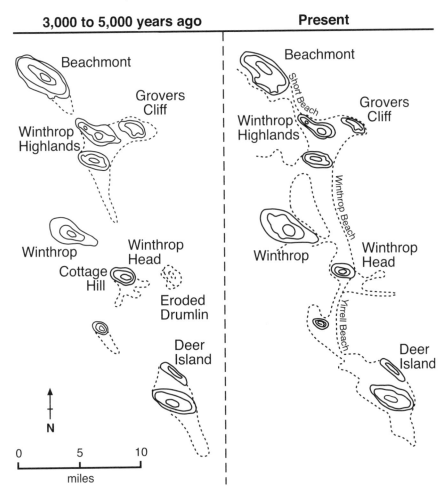

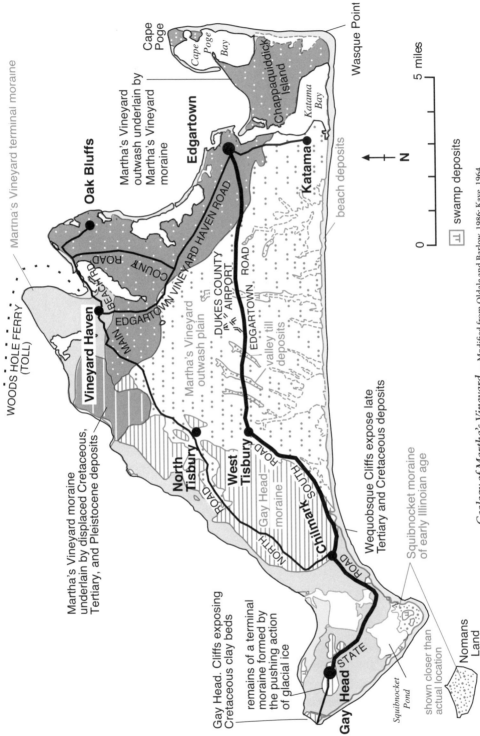

WOODS HOLE FERRY (TOLL)

Martha's Vineyard terminal moraine

Martha's Vineyard moraine underlain by displaced Cretaceous, Tertiary, and Pleistocene deposits

Martha's Vineyard outwash underlain by Martha's Vineyard moraine

Oak Bluffs

Edgartown

Cape Poge

Cape Poge Bay

Chappaquiddick Island

Katama Bay

Wasque Point

Vineyard Haven

BEACH RD

MAIN

COUNTY ROAD

EDGARTOWN VINEYARD HAVEN ROAD

Katama

EDGARTOWN ROAD

DUKES COUNTY AIRPORT

Martha's Vineyard outwash plain

valley till deposits

beach deposits

North Tisbury

West Tisbury

Chilmark

NORTH ROAD

Gay Head moraine

SOUTH ROAD

Gay Head. Cliffs exposing Cretaceous clay beds

remains of a terminal moraine formed by the pushing action of glacial ice

Wequobsque Cliffs expose late Tertiary and Cretaceous deposits

Squibnocket moraine of early Illinoian age

Squibnocket Pond

Gay Head

STATE ROAD

shown closer than actual location

Nomans Land

N

0   5 miles

⊥ swamp deposits

Geology of Martha's Vineyard. —Modified from Oldale and Barlow, 1986; Kaye, 1964

The Islands: Martha's Vineyard and Nantucket

You can get to Martha's Vineyard and Nantucket, known locally as The Islands, by seasonal or year-round ferries from Woods Hole and Hyannis, or by airplane. Both islands have geological features unique in Massachusetts. And where else will you find names like West Chop, Lobsterville, Chilmark, Madaket, Sankaty Head, and Siasconset?

Ancient Coastal Plain

The coastal plain of the eastern United States is a wedge of sediments that thickens seaward. Rivers draining the ancient Appalachian Mountains deposited these sediments, which now form much of the submerged continental shelf. The ancient landward margin of the coastal plain probably reached as far inland as Boston and Providence before the Pleistocene glaciers efficiently bulldozed away or buried much of the Cretaceous and Tertiary coastal sediments. However, Gay Head and Wequobsque Cliffs in western Martha's Vineyard expose rocks laid down on the coastal plain during Cretaceous and Tertiary time.

Surface exposures on Martha's Vineyard and boreholes from both islands reveal three units of coastal plain rocks. The upper unit consists of

Brightly colored Cretaceous and Tertiary strata in Gay Head Cliffs, Martha's Vineyard. —Robert N. Oldale photo, U.S. Geological Survey

Tertiary and Pleistocene sand and gravel. The middle unit, 80- to 65-million-year-old rocks of late Cretaceous age, consists of colorful layers of clay and a thin bed of poor-quality lignite coal. The lowest unit, also of late Cretaceous age, consists mainly of sand, along with colorful clay layers. One of its clay layers contains many fossils of land plants, leaves, pinecones, flowers, and seeds; another clay layer, probably deposited in lagoons, contains fossil seashells of marine mollusks.

We owe our view of these rocks to the huge glaciers that thrust them up in front of the toe of ice. You can see thrust faults at Sankaty Head on Nantucket, in cliffs in Martha's Vineyard moraine, and at Gay Head.

Greensands at Gay Head

Glauconite is an avocado green mineral that closely resembles black biotite mica in its composition and internal crystal structure. It occurs as little green pellets—the size of small to medium sand grains—within sedimentary rocks laid down in seawater. Sandy sediment that contains glauconite is called greensand. Many geologists think glauconite forms in the intestinal tracts of small invertebrate animals.

A layer of greensand in the upper unit of late Tertiary rocks in Gay Head Cliffs contains a magnificent suite of fossils, first described in 1793 and later studied by legendary geologist Sir Charles Lyell. Fossils include fish, shellfish, crabs, reptiles, whales, rhinoceroses, and mastodons. Ocean currents probably carried the bones of land mammals from seashores to this marine deposit.

The greensand also yields polished chert pebbles that contain fossil corals. They are probably stomach stones of seals or walruses. The nearest sources of these coral-bearing pebbles are the Hudson-Mohawk region of upstate New York or Hudson Bay in Canada. Fossil pollen and spores in the greensand suggest that the region had a warm climate until about 5 million years ago, but the climate cooled to temperate conditions during late Miocene time to the beginning of the Pleistocene ice age, 5.3 to 1.6 millions years ago.

Old Glacial and Interglacial Deposits

Martha's Vineyard and Nantucket include places where the glaciers of the last ice age did not scrape the relics of preceeding ice ages off the landscape. Geologists assign the glacial deposits on the islands to three glacial and one interglacial stages: the Wisconsinan ice sheet, Sangamon interglacial stage, the Illinoian ice sheet, and an earlier Pleistocene glaciation. The oldest glacial deposit underlies a fossiliferous interglacial bed containing the bones of an early Pleistocene horse, but no actual age dates are available.

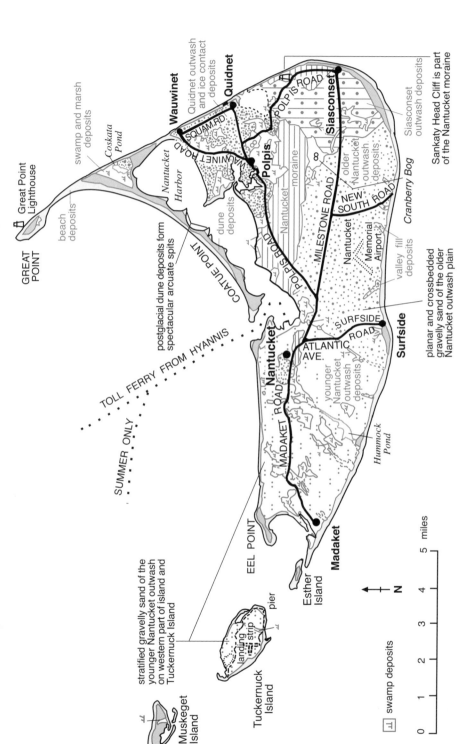

Geology of Nantucket. —Modified from Oldale and Barlow, 1986

GREAT POINT

Great Point Lighthouse

swamp and marsh deposits

Coskata Pond

beach deposits

Nantucket Harbor

COTUE POINT

postglacial dune deposits form spectacular arcuate spits

TOLL FERRY FROM HYANNIS

SUMMER ONLY

Wauwinet

Quidnet outwash and ice contact deposits

Quidnet

Siasconset outwash deposits

Sankaty Head Cliff is part of the Nantucket moraine

SQUAM RD.

POLPIS ROAD

Siasconset

WAUWINET ROAD

Polpis

Nantucket moraine

8

NEW SOUTH ROAD

older Nantucket outwash deposits

Cranberry Bog

POLPIS ROAD

MILESTONE ROAD

Nantucket Memorial Airport

valley fill deposits

dune deposits

planar and crossbedded gravelly sand of the older Nantucket outwash plain

SURFSIDE

ATLANTIC AVE.

SURFSIDE ROAD

Surfside

Nantucket

MADAKET ROAD

younger Nantucket outwash deposits

Hummock Pond

EEL POINT

Madaket

Esther Island

stratified gravelly sand of the younger Nantucket outwash on western part of island and Tuckernuck Island

landing strip

pier

Tuckernuck Island

Muskeget Island

0 1 2 3 4 5 miles

N

⊤⊤ swamp deposits

View looking west at the southern end of Sankaty Head on the eastern shore of Nantucket. Strata are: (a) *Illinoian glacial deposits,* (b) *marine gravel of Sangamon age,* (c) *a bed of sediment stirred by organisms,* (d) *shelly lower part of Sankaty sand,* (e) *a bed with worm tunnels in lower part of Sankaty sand,* (f) *upper part of Sankaty sand, and* (g) *medium to fine sand of Wisconsinan glacial deposits.* —Robert N. Oldale photo, U.S. Geological Survey

Sankaty Head is part of the Nantucket moraine. From top to bottom, the cliff exposes Wisconsinan glacial deposits, Sangamon interglacial deposits, and Illinoian glacial deposits. The Sangamon sediments were deposited in seawater about 140,000 to 120,000 years ago when sea level was higher than during the glacial stages. The Sankaty sand deposits of the interglacial period contain fossils of marine seashells.

Squibnocket moraine, the low hills and sometimes swampy terrain of Squibnocket Point and Nomans Land, is the oldest moraine on Martha's Vineyard, probably of early Illinoian age. The pinkish gray till is compact, thick, and stratified.

Most Recent Ice Age

Many geologists believe that the most recent ice sheet, the Wisconsinan, began overflowing the polar region about 80,000 years ago. The glacier deposited its terminal moraine along this stationary front by 23,000 to 22,000 years ago. By about 20,000 years ago, the advancing ice slowed as its front moved into the more temperate climate of Martha's Vineyard and Nantucket. About 18,000 years ago, the ice was still moving forward, but its front was stationary because it was melting as fast as it advanced.

The arcuate backbone of The Islands is made up of terminal moraines. Martha's Vineyard moraine was deposited where the Cape Cod Bay and Narrangansett Bay–Buzzards Bay ice lobes met. Nantucket moraine was deposited by the Cape Cod Bay lobe. These moraines are segments of a continuous deposit—extending southwest to Block Island and the row of hills on Long Island—that marks the southernmost extent of the Wisconsinan ice sheet. The till of Martha's Vineyard contains rock fragments from bedrock to the northwest in Massachusetts and Rhode Island.

The Narragansett–Buzzards Bay lobe thrust the Gay Head moraine into place in Wisconsinan time. The thrusting interleaved older glacial drift with the younger Wisconsinan drift.

Meltwater streams heavily laden with sediment deposit outwash plains as they flow away from the ice—away from the moraines. The outwash plains of The Islands, one on Martha's Vineyard and two of different ages on Nantucket, were deposited south of the farthest reach of the ice, so blocks of ice could never have been stranded on the plains, thus explaining the absence of kettle ponds.

Siasconset and Quidnet outwash deposits on the eastern side of Nantucket formed as the South Channel lobe melted. The younger Siasconset outwash spilled over the top of the Nantucket moraine.

When The Islands Became Islands

When the ice sheet melted back north of Cape Cod, sea level was about 300 feet lower than it is today, so the shoreline was about 75 miles south of The Islands. Broad areas of the continental shelf were exposed. Peat, bones, and teeth dredged up from the shelf show that plants and animals, including now-extinct elephants, mastodons, and mammoths, lived there as recently as 10,000 years ago. Fossils of animals still present today in North America include deer, bears, wolves, bison, caribou, and musk oxen. As the ice continued to melt, hardwood forests replaced the evergreen forests and tundra that first took root on The Islands, and the tundra animals migrated north with their food supply.

Martha's Vineyard and Nantucket were islands by 2,000 years ago, but the shoreline was probably one-half mile to several miles farther offshore than it is now. In the past 2,000 years, sea level rose about 6 feet. Sea level rose about 1 foot around the Cape and Islands during the past 100 years. It is uncertain whether sea level will continue to rise as tide gauge records suggest, or whether the rate of rise may slow down or even increase. The rise of sea level at The Islands may indicate local subsidence of the land rather than an overall rise in the level of the world's oceans.

Beaches and Marshes

As sea level rose, waves eroded glacial deposits and washed the sediments along the shore, forming beaches and spits. The modern shoreline continues to erode away and build up with the ebb and flow of the sea. At Wasque Point, the exposed southeastern corner of Martha's Vineyard, a wide foreshore of sand usually limits the erosion of the point to an average of 33 feet per year. One year the sandy barrier was absent, and the waves directly attacked and eroded 350 feet of the point.

Marsh deposits consist of fine-grained sand and mud washed into the intertidal zone, where roots of vegetation capture it. Because marsh plants grow near sea level, the sediments bury them, and then new marsh plants grow. The marsh keeps pace with rising sea level and marches landward, engulfing boulders and trees. Coskata Pond, a tidal embayment on Nantucket with salt marshes and tidal creeks, is part of Coskata-Coatue Wildlife Refuge, one of six refuges on the Islands. This refuge covers 1,117 acres of beach and dune.

U.S. 6
Cape Cod
64 miles

When you cross the Cape Cod Canal, you enter a land shaped entirely by ice, water, and wind. Take a moment at the canal viewpoint, at the U.S. Corps of Engineers Visitor Center on U.S. 6 north of the canal, to inspect this waterway. The Boston, Cape Cod & New York Canal Company built the 17.4-mile-long canal to link Cape Cod Bay to Buzzards Bay. The canal is 480 to 700 feet wide and 32 feet deep. It was completed in 1914, and the U.S. Corps of Engineers took over its management in 1928. The canal follows two former rivers: the Scusset River, which flowed north into Cape Cod Bay, and the Monument or Manomet River, which flowed south into Buzzards Bay.

The canal essentially makes Cape Cod an island, but that's nothing new. The Cape was an island at least once before in its recent geologic past. As the last glacier began melting from this area sometime around 17,000 years ago, the valley of the Cape Cod Canal likely overflowed with meltwater spilling from Glacial Lake Cape Cod southward into Buzzards Bay.

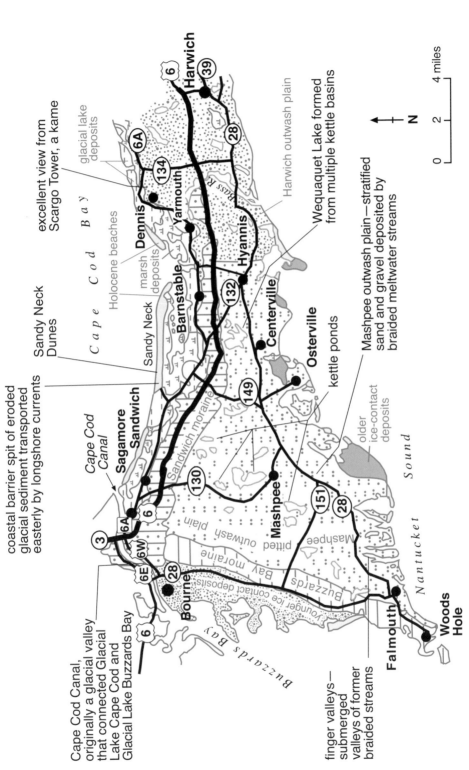

Pleistocene glacial deposits and recent sediments of western Cape Cod. —Modified from Oldale and Barlow, 1986

coastal barrier spit of eroded glacial sediment transported easterly by longshore currents

excellent view from Scargo Tower, a kame

glacial lake deposits

Harwich outwash plain

Wequaquet Lake formed from multiple kettle basins

Mashpee outwash plain—stratified sand and gravel deposited by braided meltwater streams

Sandy Neck Dunes

Holocene beaches

marsh deposits

Sandy Neck

Cape Cod Bay

kettle ponds

older ice-contact deposits

Nantucket Sound

Cape Cod Canal

Cape Cod Canal, originally a glacial valley that connected Glacial Lake Cape Cod and Glacial Lake Buzzards Bay

finger valleys— submerged valleys of former braided streams

Sandwich moraine

pitted outwash plain

Bay moraine

Younger ice contact deposits

Buzzards Bay moraine

Mashpee

Buzzards Bay

Bourne

Sandwich

Sagamore

Barnstable

Dennis

Yarmouth

Bass R.

Hyannis

Centerville

Osterville

Harwich

Falmouth

Woods Hole

N

0 2 4 miles

3 6A 6 6W 6E 28 130 149 151 28 132 134 6A 28 39 6

Seafarers christened the two broad parts of Cape Cod as Upper Cape and Lower Cape—terminology that may confuse a landlubber. The terms *upper* and *lower* take us back to sailing days when vessels moved upwind or downwind relative to the prevailing westerly winds. Visualize Cape Cod as a human arm with the shoulder at Cape Cod Canal. The western Cape is the upper arm and is called the Upper Cape or Inner Cape. The eastern Cape, called the Lower Cape or Outer Cape, is the lower arm with Monomoy Island at the elbow, Nauset Beach the forearm, the narrow stretch near Pilgrim Lake the wrist, and Province Lands the fist. The tip looks like a giant hook. The Cape Cod Bay ice lobe formed much of the Inner Cape. The Outer Cape formed from the complex interactions of two ice lobes—the Cape Cod Bay and South Channel lobes.

Many geologic features on the Cape are more modern than those seen by the Pilgrims who dropped anchor in Provincetown Harbor on November 11, 1620. Cape Cod National Seashore, Nauset Beach, Monomoy Island National Wildlife Refuge, and numerous marshes and kettle lakes dominate this tenuous land of drifting sand and eroding glacial outwash.

Sandwich Moraine

Immediately after crossing the Cape Cod Canal on the Sagamore Bridge, U.S. 6 follows the crest of the Sandwich moraine for nearly 15 miles—an elevated moraine through marshy coastal lowlands is quite convenient for a highway. The moraine is 3 miles broad at its west end and tapers to 0.5 mile in the east. The tops of hills stand as high as 275 feet in elevation. Irregular hills, undrained depressions, and east-west trending, boulder-strewn ridges characterize the moraine. You can recognize its southern margin by the sharp contrast between its high, rough topography and the smooth, flat surface of the Mashpee pitted plain to the south and Harwich plain to the east. The northern boundary is less obvious; it grades into dominantly stratified deposits.

The Cape Cod Bay lobe of the most recent ice sheet deposited the Sandwich moraine. The ice sheet reached as far south as Nantucket, Martha's Vineyard, and Long Island, where it deposited a somewhat linear row of terminal moraines. As the ice sheet retreated, it deposited glacial outwash—debris that the ice had plucked from the earth as it moved across New England. It paused in its retreat, hestitating and even advancing forward slightly, long enough to form another set of moraines, including the Sandwich moraine.

The Sandwich moraine and other coastal end moraines are not simply piles of dumped debris. Geologists used to think moraines formed when glacial ice melted and released its load in one large mound at the edge of the ice. The Sandwich moraine has actually been thrust up in sheets—not unlike a double-deck sandwich—and shoved into place by the ice. Occasionally, the melting ice lobe surged forward and pushed sheets of stratified outwash up in front of it. The series of thrust-faulted shingle blocks dip upstream—the direction opposite to which the ice moved forward. Eventually the ice rode over the thrust sheets and deposited a layer of basal till. When the ice melted back, new outwash was deposited behind the moraine.

Knob-and-kettle topography is well developed along the Sandwich moraine. Watch for conical hills with relatively flat tops, and round to irregular depressions. The knobs are ice-contact deposits, or kames, that formed at the edges of the melting ice blocks. The Sandwich moraine continues through Barnstable but disappears beneath outwash sediments around Bass River in Yarmouth.

Buzzards Bay Moraine

The Buzzards Bay moraine extends southward from just southeast of U.S. 6 and Massachusetts 28 at the Cape Cod Canal to Falmouth and the Elizabeth Islands. Massachusetts 28 follows the crest of this moraine south to Falmouth. The north end of the Buzzards Bay moraine meets the east-trending Sandwich moraine at nearly a right angle.

The coastal moraines in New England are for the most part well-behaved rows of thrust-faulted outwash all the way from eastern Long Island to Cape Cod. Why isn't the Buzzards Bay moraine parallel to the other moraines? It formed along the curving eastern margin of the Narragansett Bay–Buzzards Bay ice lobe. The Sandwich moraine formed along the western margin of the Cape Cod Bay lobe. The junction and position of the moraines reflect the junction and position of the ice lobes.

You can see on the geologic map on page 83 that the Sandwich moraine crosses over the top of the northern end of the Buzzards Bay moraine. This relationship tells us that the Buzzards Bay moraine formed first.

Mashpee Pitted Plain and Harwich Outwash Plain

The elevation of the Mashpee pitted plain decreases from 200 feet at the apex of the Sandwich and Buzzards Bay moraines to sea level at Nantucket Sound. As the Cape Cod Bay ice lobe advanced over the sheets of outwash in the Sandwich moraine, voluminous meltwater at the glacial front drained into a network of braided streams. The water picked up sediment in its path and deposited it in gently south-dipping layers. Meltwater from the Cape Cod Bay lobe also deposited the south-sloping Harwich outwash plain, which extends from Dennis to Harwich and north to Eastham.

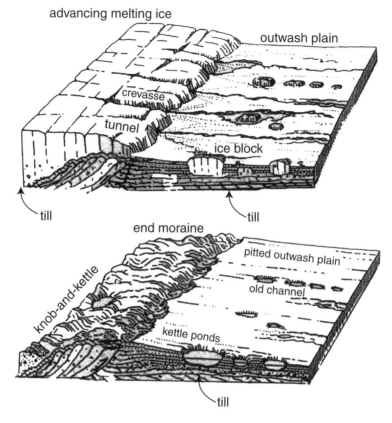

Formation of a pitted outwash plain. Meltwater braided streams bury re-sidual ice blocks that later melt. Overlying outwash sediments collapse into kettle holes and kettle ponds. —Modified from Strahler, 1988; thrust sheets based on Oldale and O'Hara, 1984

Kettles Are Not Just for Cooking

Blocks of ice left behind when the glacier melted back from Martha's Vineyard and Nantucket lay scattered between the terminal moraines on The Islands and the new end moraines—Buzzards Bay moraine and Sand-wich moraine. The braided streams partially or completely buried these ice blocks. When they eventually melted, the overlying layers of outwash sand sagged and collapsed to form kettle depressions. The collapsed de-posits form irregular hilly knobs. Knob-and-kettle topography is well developed between Orleans and Nauset Beach. Most of the numerous, delightful ponds and lakes of Cape Cod are kettles. In forested areas, you can still recognize kettles because the pattern of treetops often mimics the shape of the kettle hole.

The size of a kettle mirrors the size of the ice block or cluster of blocks responsible for the depression. Long Pond, north of U.S. 6 in Brewster and Harwich Townships, is the largest kettle lake on Cape Cod, covering 743 acres. Water depths in kettle ponds typically range from less than 10 feet to more than 80 feet. Cliff Pond, west of U.S. 6 in R. C. Nickerson State Park in Brewster Township, is the deepest of Cape Cod's ponds: the surface is 140 feet above sea level and the bottom is 60 feet below sea level. Slumping sand probably partially filled the bottom of the pond, so the ice block responsible for Cliff Pond may have been more than 200 feet thick.

Cockle Cove in South Chatham is a former kettle. When sea level rose at the end of the last ice age, the inundating ocean connected the kettle to Nantucket Sound. Longshore drift eventually built a barrier of sand that protects the embayment.

Sandy Neck Dunes

The large and unstable dunes on Sandy Neck are exceeded in size in Massachusetts only by those of Province Lands at the tip of Cape Cod. Deforestation and off-road vehicles are largely responsible for the instability of the dunes at Sandy Neck. You can explore the dunes from Sandy Neck Beach.

View looking southeast across Wequaquet Lake, the largest kettle lake on the Mashpee pitted plain, from the shore in Barnstable Township. —Marialice Curran photo

The sea, replenished by the melting glaciers, rose to near its present stand about 7,000 years ago. Ocean waves eroded glacial sediments from the cliffs east of Cape Cod Canal along the south shore of Cape Cod Bay, and longshore currents transported them eastward. When the currents reached Barnstable Bay, they slowed as they encountered deeper water and dropped their load of sand. The sand now forms a coastal barrier spit. As these currents continued to build Sandy Neck eastward, the Barnstable marsh formed in the protected backwaters behind the spit, and Barnstable Harbor inlet became progressively more restricted.

Glacial Lake Cape Cod

As the Cape Cod Bay lobe retreated north of the Sandwich moraine, Glacial Lake Cape Cod formed in an arcuate lowland north of the moraine. The Sandwich moraine impounded the lake's south side, while the South Channel lobe prevented the lake waters from escaping to the east. Most of the lake deposits now lie beneath the waters of southern Cape Cod Bay, but some are present along the north slope of the Sandwich moraine along the shore of the bay. From time to time, you can observe these deposits in newly excavated sand and clay pits near Massachusetts 6A from Cape Cod Canal to Dennis.

Two spillways controlled the level of Glacial Lake Cape Cod: the Monument River near the site of Cape Cod Canal, and the Whites Brook–Parker River outlet that breached the east end of the Sandwich moraine near the present site of Bass River between Yarmouth and Harwich. Glacial Lake Cape Cod drained shortly after the ice dimple between the Cape Cod Bay lobe and the South Channel lobe retreated north of Provincetown but before the Boston area filled with seawater.

Scargo Tower

You can obtain an excellent view of the coast all the way to Provincetown from the 28-foot-high Scargo Tower. The tower sits on a boulder kame hill that rises to 160 feet in elevation east of Dennis and overlooks Scargo Lake, a kettle pond. Take Scargo Hill Road, which intersects Massachusetts 6A 0.35 mile west of the intersection with Massachusetts 134.

Outwash Plains of the Outer Cape

The entire Outer Cape slopes to the west. The land is about 100 to 170 feet or more in elevation near the eastern coastline and slopes down to lower elevations along the western coastline, the east side of Cape Cod Bay. By contrast, the Inner Cape slopes south.

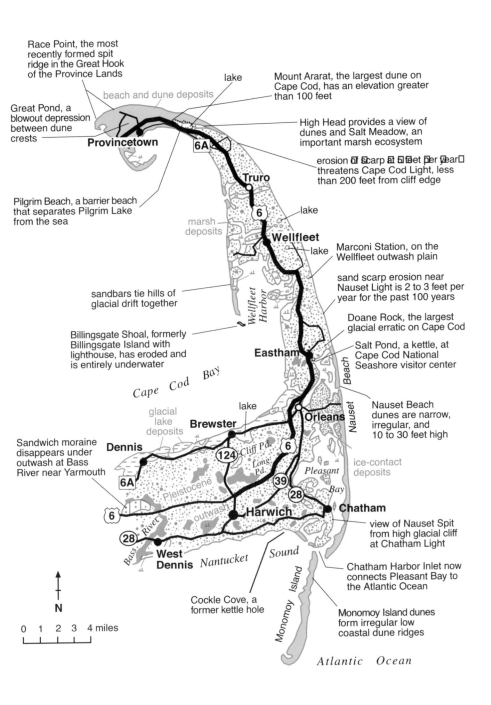

Race Point, the most recently formed spit ridge in the Great Hook of the Province Lands

lake

Mount Ararat, the largest dune on Cape Cod, has an elevation greater than 100 feet

Great Pond, a blowout depression between dune crests

beach and dune deposits

Provincetown

6A

High Head provides a view of dunes and Salt Meadow, an important marsh ecosystem

erosion of scarp at 5 feet per year threatens Cape Cod Light, less than 200 feet from cliff edge

Truro

Pilgrim Beach, a barrier beach that separates Pilgrim Lake from the sea

marsh deposits

6

lake

Wellfleet

lake

Marconi Station, on the Wellfleet outwash plain

sandbars tie hills of glacial drift together

Wellfleet Harbor

sand scarp erosion near Nauset Light is 2 to 3 feet per year for the past 100 years

Doane Rock, the largest glacial erratic on Cape Cod

Billingsgate Shoal, formerly Billingsgate Island with lighthouse, has eroded and is entirely underwater

Eastham

Salt Pond, a kettle, at Cape Cod National Seashore visitor center

Cape Cod Bay

glacial lake deposits

lake

Nauset Beach

Nauset Beach dunes are narrow, irregular, and 10 to 30 feet high

Brewster

Orleans

Sandwich moraine disappears under outwash at Bass River near Yarmouth

Dennis

124

Cliff Pd.

6

ice-contact deposits

6A

Long Pd.

Pleasant Bay

Pleistocene

39

28

6

outwash

28

Harwich

Chatham

view of Nauset Spit from high glacial cliff at Chatham Light

28

Bass River

West Dennis

Nantucket Sound

Monomoy Island

Chatham Harbor Inlet now connects Pleasant Bay to the Atlantic Ocean

N

0 1 2 3 4 miles

Cockle Cove, a former kettle hole

Monomoy Island dunes form irregular low coastal dune ridges

Atlantic Ocean

Glacial and recent deposits of eastern Cape Cod. —Modified from Oldale and Barlow, 1986

The Cape Cod Bay ice lobe retreated earlier than the South Channel lobe. Meltwater from the western edge of the South Channel lobe was able to flow west toward Cape Cod Bay, the depression left by the melted Cape Cod Bay lobe. West-flowing streams of meltwater from the South Channel lobe deposited the three outwash plains on the Outer Cape, which are younger than those of the Inner Cape. Outwash deposits that were closest to the ice lobe are the thickest, so the land elevation slopes down to the west.

The outwash plains on the Outer Cape are the Eastham, Wellfleet, and Truro. The Wellfleet outwash plain is the oldest and, at 150 feet tall, the highest and most extensive meltwater feature deposited from the South Channel lobe. Around North Eastham the outwash plain has a smooth surface, whereas from South Wellfleet to Truro the plain is roughened by kettles. The younger outwash plains are lower and grade into glacial lakebeds.

A Missing Moraine

U.S. 6 follows the crest of the prominent Sandwich moraine on the Inner Cape. The Eastham, Wellfleet, and Truro outwash plains of the Outer Cape bury the eastern end of the Sandwich moraine and the Harwich outwash plain. From Harwich, U.S. 6 heads north, but not on a moraine. Where is the moraine associated with the South Channel outwash plains?

Most experts believe the South Channel lobe did not form a moraine. Meltwater from the glacial lobe deposited gravels and sands into outwash plains, but the lobe may not have remained stationary long enough to form a moraine during its retreat.

Marconi Wireless Site

Marconi Station, an electromagnetic wave transmitter, was built on the edge of a marine scarp of Wellfleet outwash. You can see a contact between the outwash plain and younger glacial deposits in a scarp at Marconi Beach. Guglielmo Marconi, an Italian inventor, made the first wireless transmission from the U.S. to Europe on January 19, 1903. The message was from President Theodore Roosevelt to King Edward VII of England. Four transmission towers, each 200 feet tall, sat atop this Wellfleet cliff. Since 1903, the scarp has steadily retreated. In 1917, the station was closed, and more recently the sea has undermined the foundation of the station's powerhouse, though some of the tower footings are still visible.

Doane Rock and Other Glacial Boulders

Glacial till—unsorted glacial debris—appears in places along the eastern coastline's eroded marine scarps, such as 0.5 mile south of the Pamet River Coast Guard Station and at the Nauset Beach Light. The glacial ice

Concrete remnants of Marconi Station along the eroded scarp of Wellfleet outwash.
—Jackie Perchard photo

sheet transported and then dropped large boulders in the till. If you walk along the shore at Highland Light near North Truro, you can see boulders and clay in the scarp in certain places. Both large and small glacial erratic boulders are exposed at low tide on Nauset Beach.

Near the Salt Pond visitor center is Doane Rock—also called Enos Rock—the largest glacial boulder on Cape Cod. It is 40 feet long, 25 feet wide, and 30 feet high, but it is buried to a depth of 12 feet in outwash and wind-blown sand of the Eastham outwash plain. This greenish gray, layered igneous rock may have come from the Gulf of Maine.

Rising Tides and Ocean Currents

As the great continental glaciers melted, sea level rose rapidly but sporadically at an approximate rate of 50 feet per 1,000 years. The rate of sea level rise slowed gradually about 10,000 years ago, and from about 7,000 to 2,000 years ago sea level rose about 11 feet per 1,000 years. In the last 2,000 years, sea level has risen about 6 feet.

Tide-gauge records indicate a rise in sea level of 1 foot during the past century for Cape Cod, Nantucket, and Martha's Vineyard—we do not yet understand why the rate has increased. With a history like this, a supply of tall rubber boots seems a good idea!

About 3,500 years ago, Outer Cape Cod was almost twice as broad as it is now and much higher on the eastern, seaward side. Waves began to erode the soft sand deposits, carving them into high scarps. Meanwhile, the waves swept the sand along the beach and built spits and islands, linking them together like beads on a string. A nearly continuous sea cliff that ranges from 60 to 170 feet high extends in a smooth curve from east of Orleans to Pilgrim Heights. The lower scarp along Cape Cod Bay on the western side of the peninsula is 40 to 120 feet high.

Solid red line shows the shoreline of Cape Cod as it may have looked 3,500 years ago; dashed line shows the present shore-line. —Strahler, 1988

N

0   10 miles

Bob Oldale, a glacial geologist, standing on Doane Rock, a glacial erratic near Salt Pond visitor center. —Dann Blackwood photo

People who own shoreline property would probably describe the redistribution of glacial sands as disastrous. Outer Cape lighthouses and beachfront property have fallen before the relentless waves. The highlands of the Outer Cape have retreated some 500 feet during the past two centuries. The waves sweep the eroded material north to the Province Lands, especially at Race Point, and south along the eastern shore to Monomoy Island.

Waves that approach the shore at an oblique angle wash up the beach in a slightly diagonal direction. If you look down on the beach from a sea cliff on the Outer Cape, you can see the waves do not break on the beach at right angles to the shore. The onshore swash moves grains of sand along a parabolic path—think of the parabolic curve of the MacDonald's arch—up the beach and back again. Sand moves farther along the beach with

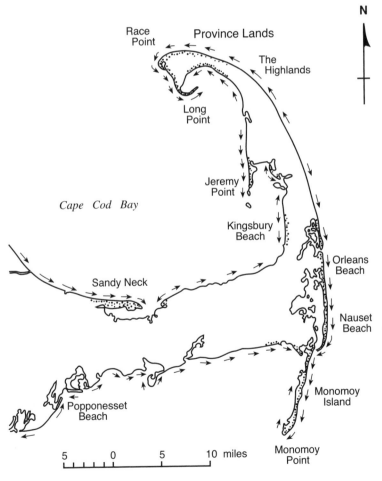

Arrows show the prevailing directions of sand movement. —Strahler, 1988

each curved path. In addition, a river of water—the longshore current—flows parallel to the shore and drags sand with it along the ocean bottom.

Nauset Beach and Monomoy Island—long, narrow barrier islands—formed as sand drifted southward along the oceanward side of the Cape. Sand also moved north along the west shore of Monomoy Island. You can view the barrier islands from the high cliff of glacial debris at Chatham Light.

Spits and Dunes at Province Lands

An enormous amount of sand has accumulated at the Province Lands. Prolonged northward and southward drift of sand on both sides of the Cape produced a succession of five spits. Each of the former spits is now a ridge of sand dunes. The oldest dune ridge extends from southeast of Pilgrim Lake through Provincetown. Four other dune ridges fan out successively from east of Pilgrim Lake to northeast of Provincetown. The youngest sand ridge extends along the northern beach from southeast of Pilgrim Lake to Race Point. Where the shore curves south, the drifting sand continues straight into deeper water, forming a sandbar. The sandbar eventually will rise above water as a spit that curves with the Cape.

Race Point Beach, the youngest and most northerly sand spit ridge on Cape Cod. —Christine Bronchuk photo

Wind and waves have long been partners in building and molding the Province Lands out of fine-grained sands. Wind lifts sand grains from the beach and wafts them inland, where, as the velocity slows, sand grains fall to the dune surface. When grains are carried over the crest of the dune they come to rest, temporarily out of the wind, on the downwind, or lee, slope of the dune. The dunes of the Province Lands are truly spectacular and quite accessible.

Mount Ararat—no place to look for Noah's Ark—is the largest dune in the Province Lands. It rises to an elevation of 100 feet, which makes it the second highest dune on the East Coast, after Jockey Ridge in North Carolina. Mount Ararat is one of several nested, parabolic dunes—those shaped like crescent moons. The open end of the dune faces northwest into the prevailing wind. Great volumes of sand and sparse vegetation along the trailing ends provide ideal conditions for the formation of parabolic dunes.

Great Pond, near the most northwesterly point of Province Lands at Race Point, occupies a blowout depression between dune crests. A blowout forms where wind scours a hollow from areas devoid of vegetation. The pond is fresh water derived from rainfall.

Oblique aerial photo, looking southwest at the parabolic, nested, marching dunes that are encroaching on Pilgrim Lake. —Ned Johnson photo

View from High Head

U.S. 6 passes west of Pilgrim Spring and High Head, the northwest-facing scarp of Pilgrim Heights at the southeast end of Pilgrim Lake. High Head, a former marine sea cliff formed of the northernmost glacial deposits on Cape Cod, is now a half-mile inland from the sea. The Pilgrims drank fresh water from the spring at the base of High Head.

From High Head you can look across Pilgrim Lake to the parabolic dunes and Mount Ararat. The marching dunes, advancing 15 to 20 feet per year, threaten to bury Pilgrim Lake. You can also see the breach that Minot's Gale opened across the spit in 1851. The breach follows Salt Meadow, a marsh that extends from the Atlantic to Pilgrim Lake.

Minot's Gale of 1851

Minot's Gale, a Nor'easter that blew along the coast on April 17, 1851, caused extensive flooding and damage. In addition to sweeping a light-house off the coast near Cohasset, 45 miles to the northwest of Cape Cod, the gale also breached the narrow spit of land separating East Harbor, known today as Pilgrim Lake, from the Atlantic Ocean. Following the breaching, sand built up in Provincetown Harbor as ebb tides carried sand out of East Harbor. To save Provincetown Harbor from further sand buildup, citizens built a dike across East Harbor inlet in 1896, sealing off East Harbor from Provincetown Harbor and creating Pilgrim Lake.

Minot's Gale also breached Nauset Beach for a width of more than 500 feet and produced a channel 11 feet deep. Prior to the storm, Nauset Beach was about 1,250 to 1,650 feet wide at high water.

Interstate 90 (Mass Pike)
Boston—Auburn
47 miles

The Mass Pike—Interstate 90—begins in downtown Boston and heads west across Massachusetts. Between Boston and Auburn, it crosses rocks of the Avalon and Nashoba terranes, liberally covered with a blanket of glacial deposits. The crystalline rocks, however, are well exposed in uplands around Boston and along I-90.

The colonial history of Boston intertwines closely with Pleistocene gla-cial history. The ice sheet modified the landscape, and the seashore shifted in and out as the ice melted and the earth's crust rose. The unusual glacial deposits of Beacon Hill contain abundant and potable spring water—a

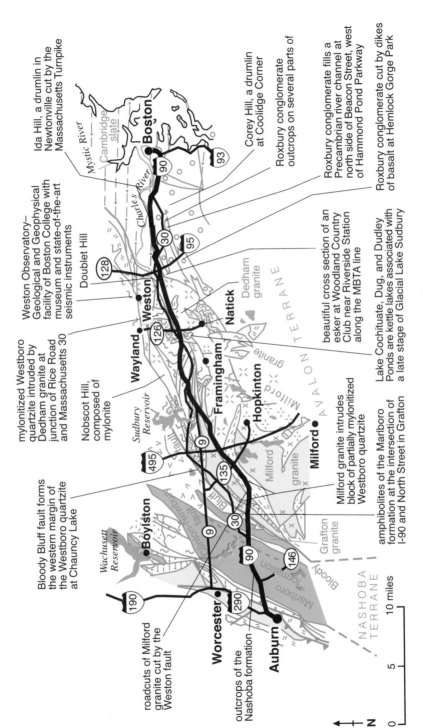

Ida Hill, a drumlin in Newtonville cut by the Massachusetts Turnpike

Corey Hill, a drumlin at Coolidge Corner

Roxbury conglomerate outcrops on several parts of

Roxbury conglomerate fills a Precambrian river channel at north side of Beacon Street, west of Hammond Pond Parkway

Roxbury conglomerate cut by dikes of basalt at Hemlock Gorge Park

Weston Observatory—Geological and Geophysical facility of Boston College with museum and state-of-the-art seismic instruments

Doublet Hill

mylonitized Westboro quartzite intruded by Dedham granite at junction of Rice Road and Massachusetts 30

Nobscot Hill, composed of mylonite

Bloody Bluff fault forms the western margin of the Westboro quartzite at Chauncy Lake

roadcuts of Milford granite cut by the Weston fault

outcrops of the Nashoba formation

beautiful cross section of an esker at Woodland Country Club near Riverside Station along the MBTA line

Lake Cochituate, Dug, and Dudley Ponds are kettle lakes associated with a late stage of Glacial Lake Sudbury

Milford granite intrudes block of partially mylonitized Westboro quartzite

amphibolites of the Marlboro formation at the intersection of I-90 and North Street in Grafton

Boston

Mystic River

Cambridge slate

Charles River

93

90

30

95

128

Weston

Wayland

126

Natick

Dedham granite

Framingham

Hopkinton

Milford granite

AVALON TERRANE

Milford

Grafton granite

9

495

135

30

9

Bluff

Clinton-Newbury fault

Boylston

Wachusett Reservoir

Sudbury Reservoir

Nashua

190

Worcester

290

90

146

Auburn

Marlboro formation

Bloody

NASHOBA TERRANE

N

0   5   10 miles

Geology along the Massachusetts Turnpike between Boston and Auburn. —Modified from Zen and others, 1983

perfect place to build a city. The settlers modified their landscape as well, filling in parts of the harbor with glacial deposits and even some tea!

Glacial deposits blanket the bedrock throughout most of downtown Boston, reaching a thickness as great as 300 feet. These sediments consist of sand, gravel, silt, and clay. They were deposited by four episodes of glacial ice advances and by marine waters that filled the basin during inter-glacial periods.

Original shoreline (in solid black), *modern shoreline, and prominent features of Old Boston.* —From Skehan, 1979; modified from Russell Lenz, cartographer, courtesy of the *Christian Science Monitor*

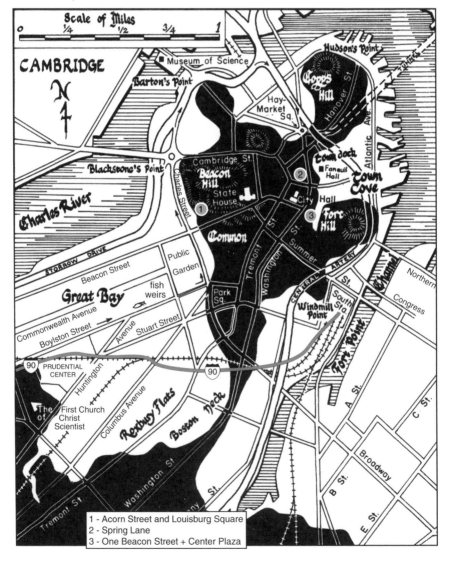

Back Bay

Before colonial times, a narrow neck of land connected the area of present-day Boston to the mainland. Washington Street in the South End follows this neck. Downslope and west of Beacon Hill—the present Charles Street, Public Garden, and Roxbury Flats—lay under marine waters. This large bay was called Great Bay, or Back Bay. Back Bay was filled in 1858 with gravel hauled by train from glacial deposits in Needham. Glacial deposits from the upper slopes of the three hills comprising Tremount were also used for fill.

During excavation for the New England Mutual Life Insurance Company in the Back Bay, workers found molded wooden stakes and sharpened seashells. Archeologists identified them as fish weirs that date to about 4,000 years ago. A newspaper article in 1913 tells of workmen encountering stakes and shells beneath 18 feet of fill and 14 feet of silt during excavations for the Boylston Street subway. The site of the Prudential Center, just east of Massachusetts Avenue, was near the strandline at the landward end of the fish weirs. The weirs were aligned approximately along Huntington Avenue heading toward Beacon Hill.

Diorama of workers filling in the Back Bay and Roxbury Flats in 1858 with gravel brought from Needham by train, on right. —Courtesy of New England Financial. Art by Sarah Annette Rockwell and Henry H. Brooks

Diorama of the Boylston Street fish weir constructed by Native Americans about 4,000 years ago. Prudential Center and Christian Science Mother Church are now situated near the shoreline in the foreground. Tremount (Beacon Hill) in background. —Courtesy of New England Financial. Art by Sarah Annette Rockwell and Henry H. Brooks

The original Boston Peninsula was known as Shawmut, the name the local Indians gave to the area. Boston has been known at various times as Shawmut and Tremount (also spelled Trimount), which refers to three prominent, adjacent hills—Pemberton Hill, Beacon Hill, and Mount Vernon. Of these, colonial excavations left only Beacon Hill, on which the statehouse now sits.

William Blackstone's Spring

When Puritan settlers arrived in Salem in June 1630, they joined a group of one hundred English colonists who had tried to settle there the previous year. They evidently were dismayed to find a rocky landscape with only a thin soil cover and an inadequate water supply for the large group that now numbered 1,100 persons. After only a week, the Puritans pulled up anchor and sailed to Charlestown, where a single spring provided water for the entire community. Several prominent members of the Charlestown community died of an illness attributed to bad water, so they searched for yet another place to settle.

A retired Anglican priest, the Reverend William Blackstone, or Blaxton as his name was sometimes spelled, had settled in solitary splendor on Beacon Hill in 1625. He had an abundant supply of fresh water from a spring on the hill and invited the Charlestown settlers to move south across the Charles River. The population of downtown Boston having exploded to over 1,100, Blackstone set out for greener pastures. He eventually settled near the river that would bear his name and become an important industrial waterway on the Massachusetts–Rhode Island border.

Blackstone's spring was near the west end of Beacon Hill, south of Louisburg Square and near Acorn Street. As time went on, many households in colonial Boston had their own wells, and the town well was, quite fittingly, on Spring Lane. This well was situated at what is now Washington Mall, about 50 feet from the intersection of Washington Street and Court Street.

Beacon Hill

Geologists in the nineteenth and twentieth centuries considered Beacon Hill a drumlin because of its elliptical shape. A drumlin's tightly compacted glacial till has little pore space within it to hold water. How could Beacon Hill emit spring water?

The true identity of Beacon Hill was discovered during construction of the Boston Common Garage beneath the southwest slope of Beacon Hill, at the intersection of Beacon Street and Charles Street. Engineers assumed Beacon Hill was a drumlin, and that the 45-foot-deep excavation would be in deep till and clay and, therefore, dry. As the excavation proceeded, they uncovered the crests of three gravel anticlines. The copious groundwater that flowed from these gravel beds required costly remedial action to drain.

So, what is Beacon Hill? It is primarily a moraine! It is a complexly faulted mass of well-bedded sand, interbedded sand and clay, gravel, and till. A series of north-dipping slabs of glacial material have been thrust over one another like an imbricated package of dominoes.

One possible explanation of how Beacon Hill moraine formed is that the ground was deeply frozen when the glacier overrode it. The frozen ground may have adhered to the bottom of the ice and began moving with it. Slabs of frozen ground sheared away from the underlying unfrozen ground and became part of the ice sheet itself. With continued ice movement, these slabs of frozen sediment moved upward into the ice and were carried along like gigantic dominoes until they piled up at the melting end of the glacier.

Geologists who thought Beacon Hill was a drumlin were not entirely wrong. A small, 0.5-mile-long drumlin of dense till is buried beneath some of the Beacon Hill deposits. It was discovered during excavations at the Old Granary Burial Ground at the corner of Washington and School Streets. Excavations for One Beacon Street and the Center Plaza revealed the topset and foreset beds of a glacial delta deposited against the northwest flank of the drumlin. Similar delta deposits were found in excavations for the development of Charles River Park.

Even though Beacon Hill is not a drumlin, numerous drumlins dot the Boston area. Two prominent hills along the Mass Pike are Cory Hill and Ida Hill. Cory Hill stands between Beacon Street and Commonwealth Avenue at Coolidge Corner. Mass Pike cuts through the northwest end of Ida Hill in Newtonville.

Postglacial Uplift and Marine Deposits

The great ice sheet depressed the earth's crust, so the relative sea level at the end of the ice age was higher than it is now. The rate of rise in the earth's crust from glacial unloading was greater than the rate of rise of sea level from glacial melting. The postglacial sea level fell from 50 feet above present-day level to 80 feet below, even though the sea was rising from the influx of meltwater. When the earth's crust stopped rebounding about 7,500 years ago, sea level finally rose to its present level.

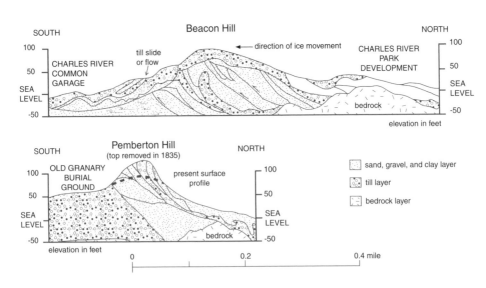

Simplified geological cross sections through Beacon Hill and Pemberton Hill show complex repetition of till, sand, and gravel layers produced by southward thrusting of glacial ice. —Modified after Kaye, 1976

Marine clay deposits, commonly referred to as the Boston blue clay, were deposited widely in the estuaries of the lower Charles River, the Mystic River, and other lowland areas after the ice age glaciers receded. As the continental crust of eastern North America rebounded faster than the sea rose, the Boston blue clays were uplifted and exposed to erosion. Some were redeposited unconformably on the floor of Boston Harbor and Massachusetts Bay. As the sea began encroaching on land again, the forests of the Back Bay that had flourished during falling sea level were buried by estuarine black muds full of marine seashells, and by salt marshes that thrived along the edge of Back Bay.

Gray clays, organic-rich estuarine sediments, and salt-marsh peat, mostly less than 45 feet thick, were also deposited unconformably on older deposits of the Boston Harbor estuary. A radiocarbon date of 11,650 years ago on organic sediments from under the John Hancock Tower approximates the time of glacial warming—the final disappearance of glacial ice from Boston.

Charles River and Glacial Lake Charles

The Charles River originates at Echo Lake in Hopkinton, heads south through Milford, and then winds a circuitous northeastern route to Boston Harbor. The Mass Pike follows the Charles River lowland west out of Boston to I-95 in Newton.

As the continental ice sheet receded north about 14,000 years ago, Glacial Lake Charles formed in the Charles River lowland in the western part of the Boston Basin and beyond. The bedrock hills of the Newton Highlands formed the eastern shore of the lake and divided the encroaching marine waters in the basin from the fresh glacial meltwater. Glacial Lake Charles, one of the largest lakes in eastern Massachusetts, spread from near Hopkinton to Medway, Medfield, Wellesley, Dedham, Newton, and Waltham. The Mass Pike passes over glacial lake deposits between Newton and Southborough. Delta sands and gravels of Glacial Lake Charles are present in Wellesley and Newton. Before the lake formed, a river beneath the ice sheet deposited the prominent esker south of the Mass Pike just east of the I-95 junction at Woodland Country Club near Riverside Station.

Glacial Lake Sudbury, a large lake in the Sudbury River valley, extended from near Mass Pike north to near Concord. Lake Cochituate, Dug Pond, and Dudley Pond in Natick and Wayland are kettle lakes associated with a late stage of Glacial Lake Sudbury.

Boston's Bedrock

The bedrock beneath the glacial cover in Boston is sedimentary rock that filled a rift in the Avalon microcontinent in late Precambrian time.

The Precambrian sedimentary rocks of the Boston Basin are younger than 595 million years and older than 540 million years. The bedrock underlying Beacon Hill and the adjacent waterfront consists of Cambridge slate, which has yielded microfossils of late Precambrian age. We know this rock primarily from tunnels and borings. Another basin rock, the Roxbury conglomerate, outcrops more extensively than the Cambridge slate.

Though the Roxbury conglomerate underlies much of downtown Boston, it and the underlying Brighton volcanic rocks are better exposed in Roxbury, Brighton, and the suburbs. The conglomerate outcrops in several places on the Boston College campus. Some buildings on campus are made with it, complemented by more notable architectural stones such as the Weymouth, Quincy, and Chelmsford granites.

The Roxbury conglomerate fills a Precambrian river channel of finely stratified siltstone in Newton. The channel is tilted to the north in a fold and is well displayed on the north side of Beacon Street west of the Hammond Pond Parkway. You can also see Roxbury conglomerate, cut by basalt dikes, at Hemlock Gorge National Historic Park on the Charles River. The park is just downstream from Echo Bridge near the junction of Massachusetts 9 and I-95. As the earth's crust rose after the ice sheets melted, the Charles River cut through the thin glacial deposits and exposed bedrock at Newton Upper Falls at Hemlock Gorge, and also at Newton Lower Falls Park on Park Road west of I-95.

Hard Rocks of the Uplands

The intersection of Mass Pike with I-95/Massachusetts 128 marks the western boundary between the Boston Basin rocks and the somewhat older, harder rocks of the uplands that surround the basin. Near Newton Corner, note the Newton Highlands visible to the south. Heartbreak Hill of the Boston Marathon marks the upland boundary along Massachusetts 30.

The 610-million-year-old Dedham granite is the most abundant rock formation in the 5-mile stretch between Weston and Natick. Along the Mass Pike and other roadways, and in the hills north of the Weston tollbooths, you can see numerous outcrops of maroon to pink Dedham granite and black basalt typical of the Avalon terrane. As the Avalon volcanic island chain formed along the edge of the Gondwanan supercontinent, masses of magma billowed up into the chain and formed large batholiths such as the Dedham granite.

A fairly well-rounded glacial erratic of coarse gray granite containing light blue quartz is perched on top of an outcrop of gabbro at Doublet Hill in Weston. Doublet Hill is about a half-mile north of the Mass Pike tollbooths at the Massachusetts 128/I-95 interchange.

Doublet Hill glacial erratic, a granite boulder resting on gabbro bedrock in Weston.
—Dan Graham photo

Between Natick and Framingham, the main rocks are schist and mylonitized and unmylonitized Westboro quartzite and gabbro. You can see superbly exposed outcrops of highly sheared quartzite and schist injected with veins of granite along Massachusetts 30, which parallels the Mass Pike, and near the intersection of Massachusetts 30 with Rice Road in Wayland.

You will cross the 610-million-year-old Milford granite and associated rocks, also called the Rhode Island batholith, along the Mass Pike between Framingham and Grafton. Several phases of Milford granite and granodiorite intruded former continental shelf sediments of Gondwana—Westboro quartzite and Blackstone group metasedimentary rocks. The Milford granite is a sparkling, pale pink to white rock composed of sugary aggregates of quartz and feldspar sprinkled with black lumps of biotite.

Even when driving the speed limit along the Mass Pike or pausing at the Framingham tollbooth, you can see clearly that the Milford granite intrudes blocks of black amphibolite. Near the Framingham tollbooth, boudinaged lenses of granite are enclosed in amphibolite schist. The older amphibolite is metamorphosed gabbro and/or basalt, remnants of former dikes and lavas. Relatively unaltered gabbro bodies of mid-Paleozoic age, and basalt dikes, probably of Jurassic age, intrude all of the older rock formations.

Magma of the Dedham granite intruded the mylonitized Westboro quartzite, giving rise to migmatized mylonite. At junction of Massachusetts 30 and Rice Road in Wayland.

The Hope Valley alaskite gneiss, a pale orange to cream phase of the Milford granite, is well exposed along and near I-90 between the intersection of I-90 with I-495 and the North Street overpass in Upton near the Westborough-Upton line.

A highly deformed phase of the Milford granite contains elongated aggregates of recrystallized quartz and plagioclase with large, irregular pink crystals of alkali feldspar. It is well exposed in roadcuts on I-90 from the North Street overpass in Upton to the North Street overpass in Grafton. This phase of the Milford granite contains a large, 6-mile-by-0.4-mile block of partially mylonitized Westboro quartzite. It is exposed along the Old Westboro Road overpass.

Nobscot Hill

The east-west-trending, 500-yard-wide Nobscot mylonite zone appears to have formed independently of the older Burlington mylonite, the sheared margin of the Gondwanan supercontinent. The Nobscot mylonite sheared the 610-million-year-old Milford granite, indicating it is younger than the granite. You can examine splendid exposures of mylonites and other rocks near Lookout Tower on Nobscot Hill in Framingham.

Outcrop of the east slope of Nobscot Hill, Framingham, exposes dark mylonite adjacent to gabbro of the Burlington mylonite zone.

A Slender Nashoba Terrane

I-90 crosses the Bloody Bluff fault, which follows the boundary between the Avalon and the Nashoba terranes, near the North Street overpass in Grafton. The Clinton-Newbury fault zone, the western margin of the Nashoba terrane, is at the U.S. 20 underpass of the Mass Pike. In Auburn, the Clinton-Newbury fault zone is close to the Providence & Worcester Railroad, southwest of Pondville Pond, but is covered by stratified Pleistocene sands and gravels. Strike-slip movement on the Clinton-Newbury fault dramatically narrowed the Nashoba terrane at its southwestern end.

The main rocks at the southwestern tip of the Nashoba terrane are the Marlboro formation and the Nashoba formation. Marlboro strata are mainly dark green to black amphibolite schists and dark gneisses interlayered with colorful quartzites. You can see amphibolites of the Marlboro formation at the intersection of I-90 and North Street in Grafton. The Nashoba formation outcrops along the Mass Pike east of the East Millbury interchange and on both sides of the Mass Pike between Dorothy Pond and the Blackstone River.

The Marlboro formation encloses the dark gray Grafton granite gneiss, a body of rock 10 miles long and 0.75 mile wide. Geologists do not know whether the gneiss was a body of magma that intruded the Marlboro formation or if it is the metamorphosed remains of an ancient volcano. It is probably about the same age as the Fish Brook gneiss, about 500 million years. The Bloody Bluff fault truncates its southern end.

Interstate 93
Methuen—Medford
25 miles

I-93 runs from Boston straight to the White Mountains of New Hampshire. It crosses three Gondwanan terranes, skirts the sediments of a glacial lake, and passes by one of the biggest and best-studied basalt dikes in New England. We'll begin at Methuen on the New Hampshire border and head south.

Glacial Lake Shawsheen-Merrimack

I-93 crosses the floor of Glacial Lake Shawsheen-Merrimack, which occupied the valleys of the Shawsheen and Merrimack Rivers. Its sediments cover large areas on both sides of I-93 and along the east-west Hampshire Road that straddles the New Hampshire state line. Its sediments underlie Peat Meadow west of I-93 and Mystic Pond near Methuen. Varved sediments in west-central Methuen township contain dark and light layers that record a surprisingly brief period—thirty years. Some geologists attribute these sediments to a body of water called Glacial Lake Methuen, but it was probably just an arm of the larger Glacial Lake Shawsheen-Merrimack, which was about 20 miles long and 5 miles wide. The arm was about 10 miles long and 3 miles wide.

Nicely formed kames exist around Haggetts Pond, southwest of the intersection of I-93 with I-495. Pitted outwash plains stretch south of Haggetts Pond to the Essex and Middlesex County boundary and beyond.

Varved sediments in the Shawsheen River valley supplied raw clay for a nineteenth-century brickwork industry. The business closed because the rock-flour clays were of poor quality for ceramics. The peat that underlies the swamp deposits at Peat Meadow was once used for fuel and is now used as a soil conditioner.

Merrimack Terrane

I-93 crosses through the Rockingham belt of the Merrimack terrane. Most of the rock is the Berwick formation—calcareous sandstones and

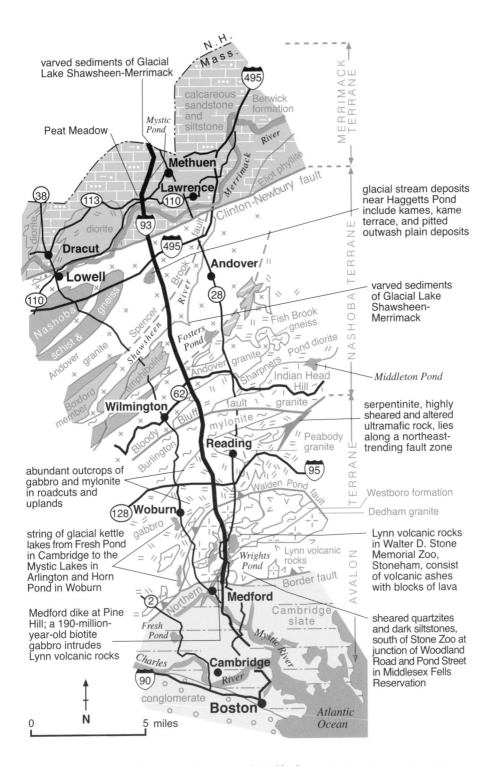

varved sediments of Glacial
Lake Shawsheen-Merrimack

Peat Meadow

Mystic Pond

Methuen

Lawrence

calcareous
sandstone
and
siltstone

Berwick
formation

River

38

113

93

diorite

diorite

Dracut

Lowell

110

495

Andover

28

MERRIMACK
TERRANE

glacial stream deposits
near Haggetts Pond
include kames, kame
terrace, and pitted
outwash plain deposits

varved sediments
of Glacial Lake
Shawsheen-
Merrimack

Merrimack River

Eliot phyllite

Clinton-Newbury fault

fault

Brook River

Spencer
Shawsheen

Shawsheen

amphibolite

Nashoba
schist &
Andover
granite

Boxford
member

Fosters
Pond

Andover granite

Sharpners

Fish Brook
gneiss

Pond diorite

Indian Head
Hill

granite

Middleton Pond

NASHOBA TERRANE

Wilmington

62

fault

mylonite

Bluff

Bloody

Burlington

Reading

Peabody
granite

95

serpentinite, highly
sheared and altered
ultramafic rock, lies
along a northeast-
trending fault zone

abundant outcrops of
gabbro and mylonite
in roadcuts and
uplands

128 **Woburn**

gabbro

Walden Pond fault

TERRANE

Westboro formation

Dedham granite

string of glacial kettle
lakes from Fresh Pond
in Cambridge to the
Mystic Lakes in
Arlington and Horn
Pond in Woburn

Lynn volcanic
rocks

*Wrights
Pond*

Border fault

AVALON

Lynn volcanic rocks
in Walter D. Stone
Memorial Zoo,
Stoneham, consist
of volcanic ashes
with blocks of lava

Medford dike at Pine
Hill; a 190-million-
year-old biotite
gabbro intrudes
Lynn volcanic rocks

2

Northern

*Fresh
Pond*

Medford

Cambridge
slate

Mystic River

sheared quartzites
and dark siltstones,
south of Stone Zoo at
junction of Woodland
Road and Pond Street
in Middlesex Fells
Reservation

Charles

90

conglomerate

Cambridge

River

Boston

*Atlantic
Ocean*

0   5 miles

N

Geology along I-93 between Methuen and Medford. —Modified from Zen and others, 1983

siltstones of Silurian age. The Silurian Dracut diorite intrudes the Berwick formation near Dracut. Elongated masses of Ayer granite also intrude the Rockingham belt. A several-mile-long pluton of Ayer granite lies within a mile of I-93 in New Hampshire, but the granite is not present along our route in Massachusetts. The southern boundary of the Merrimack terrane is the Clinton-Newbury fault zone, which I-93 crosses near the Chandler Road overpass in Andover. A cap of diorite west of Massachusetts 38 in Lowell and Dracut may be older than the Ayer granite, although the two rocks might be related.

Nashoba Terrane

I-93 crosses into the Nashoba terrane and onto a large pluton of Andover granite south of the Clinton-Newbury fault zone. Although the Andover granite may have formed over a long period of time, the late pegmatitic phase is dated at 412 million years. Molten magma that became the Andover granite nearly engulfed great blocks of amphibolite of the Boxford member of the Nashoba formation and the Fish Brook gneiss, probably during the long interval from late Ordovician time to early Devonian time. The Fish Brook gneiss, the oldest rock in the Nashoba terrane at 500 million years of age, is a mottled to swirled gray metamorphic rock that probably originated as a volcanic rock. The Sharpners Pond diorite intruded all of these formations during Silurian time.

Burlington Mylonite of the Avalon Terrane

A substantial part of the Burlington mylonite zone, an important tectonic feature of the Avalon terrane, formed in Precambrian time during intense shearing along the margin of Gondwana. A younger mylonite in the Burlington zone is probably Silurian to Devonian in age and formed when the Avalon and Nashoba plates ground together. The Burlington mylonite is a magnificently sheared zone at least 50 miles long. The Dedham granite of the Avalon volcanoes intruded and consumed part of it about 610 million years ago.

A mass of gabbro intruded the mylonite about 375 million years ago, during Devonian time. The gabbro is broadly exposed north and south of the junction of I-93 with I-95. A careful examination of these rocks will reveal stray blocks of mylonite liberally scattered through a sea of gabbro.

The southern half of the Burlington mylonite zone lies south of the Walden fault in Woburn and north of the Northern Border fault in Medford. This block consists mainly of three rock types: a granite equivalent in age to the Westwood granite, the Lynn volcanic rocks, and the pink Dedham granite that intruded sheared quartzites of the Westboro formation. Both

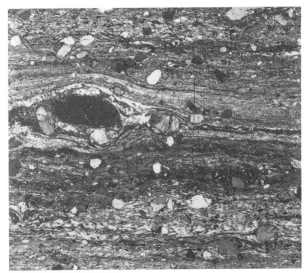

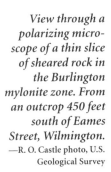

View through a polarizing microscope of a thin slice of sheared rock in the Burlington mylonite zone. From an outcrop 450 feet south of Eames Street, Wilmington.
—R. O. Castle photo, U.S. Geological Survey

the granites and volcanic rocks are cut by late Paleozoic basalt dikes that in turn are cut by the Medford dike of Jurassic age.

Medford Dike

The breakup of the supercontinent Pangaea began with rifting during middle Triassic time that continued into early Jurassic time. About 1,000 to 2,000 basalt dikes intruded rocks of the northern Avalon terrane near Boston during active rifting. The famous Medford dike was injected as gabbro magma, a coarse-grained equivalent of basalt, about 190 million years ago. It is 375 feet across and trends northeast. An outcrop of the dike looks like a brown wall for at least 3.5 miles in the Middlesex Fells Reservation and along I-93 at Wrights Pond in Medford.

All those dikes, along with the contemporaneous rift opening of the Connecticut Valley, are evidence that the continental crust was stretching as the Atlantic Ocean began to form. The fissures that the dikes now fill were the plumbing for the basalt lava flows that erupted across enormous areas along the east coast during early Jurassic time.

Buried Bedrock Valley

A string of kettle ponds marks the location of the buried preglacial bedrock valley of the Merrimack River. Horn Pond in Woburn lines up with the Mystic Lakes in Arlington and Fresh Pond in Cambridge. The meltwater from the receding glacier probably followed this buried topography, dumping glacial material and burying chunks of ice in a linear path. Large holes, or kettles, formed as the ice blocks melted. Glacial material may have blocked the Merrimack's southern course, so it turned northeast near Chelmsford.

Cambridge Slate

The Cambridge slate underlies the northern half of the Boston Basin, which is south of the Northern Border fault in Medford. The elusive Cambridge slate outcrops south of Medford in Somerville and northeast of Medford in Melrose. You can see excellent exposures of rhythmically bedded Cambridge slate in the Winter Hill section of Somerville, three blocks west of Mystic Valley Avenue and two blocks northwest of St. Polycarp's Church, whose steeple is visible from I-93. Take exit 4 from I-93 south or exit on Massachusetts 28 from I-93 north. The former quarry, now a housing project, exposes rhythmically bedded slate intruded by basalt dikes. The heat of the intruding basalt baked the adjacent rock into a hard, fine-grained metamorphic rock.

Interstate 95
Canton—Attleboro
22 miles

Interstate 95 crosses the Avalon terrane, two Pennsylvanian-age coal basins, and myriad glacial deposits from the receding Narragansett Bay–Buzzards Bay glacial lobe. Along the route you'll see a classic location for the study of early Cambrian fossils, learn about sedimentary bedding structures, and experience the modern ambiance of eighteenth-, nineteenth-, and early-twentieth-century coal mining towns.

Red Rocks of the Norfolk Basin

Watch for the roadcut of maroon to pink and gray sedimentary beds, north-dipping to nearly vertical, along the northbound lane of I-95, just 1.7 miles south of the junction with Massachusetts 128. A short distance east, the Conrail railroad cuts through these beds near warehouses of the Shawmut Industrial Park.

These beds are in the north limb of the trough of the Norfolk Basin and contain a great variety of sedimentary features. Four sets of exposed beds, or cycles, become more fine-grained near the top. A magnificent sequence of pebble conglomerate beds passes upward into sandstone, siltstone, and finally into maroon slate at the top of the cycle. The fining-up cycles, as

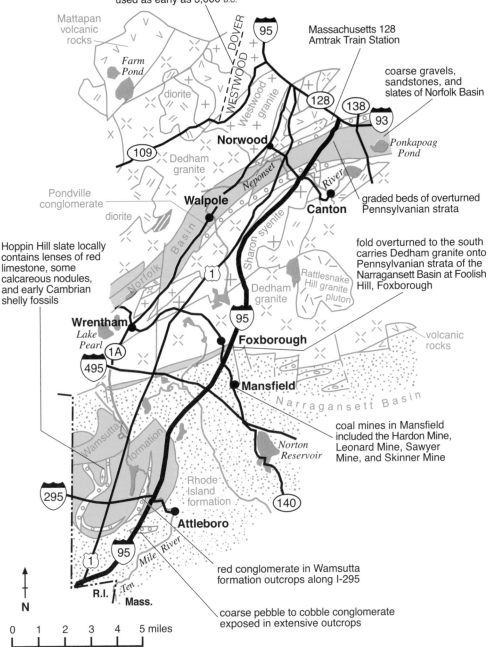

Hale Reservation; ancient quarries yielded fine-grained rock that Native Americans fashioned into arrowheads and other objects; archeologists estimated the quarries were used as early as 3,000 B.C.

Massachusetts 128 Amtrak Train Station

coarse gravels, sandstones, and slates of Norfolk Basin

Mattapan volcanic rocks

Farm Pond

diorite

Westwood granite

DOVER

WESTWOOD

95

128

138

93

Ponkapoag Pond

109

Dedham granite

Norwood

Neponset

Canton

River

graded beds of overturned Pennsylvanian strata

Pondville conglomerate

diorite

Walpole

Norfolk Basin

Sharon syenite

fold overturned to the south carries Dedham granite onto Pennsylvanian strata of the Narragansett Basin at Foolish Hill, Foxborough

Hoppin Hill slate locally contains lenses of red limestone, some calcareous nodules, and early Cambrian shelly fossils

1

Rattlesnake Hill granite pluton

Dedham granite

Wrentham

Lake Pearl

1A

495

95

Foxborough

volcanic rocks

Mansfield

Narragansett Basin

Wamsutta formation

Norton Reservoir

coal mines in Mansfield included the Hardon Mine, Leonard Mine, Sawyer Mine, and Skinner Mine

295

Rhode Island formation

140

Attleboro

1

95

Mile River

red conglomerate in Wamsutta formation outcrops along I-295

Ten

R.I.

Mass.

coarse pebble to cobble conglomerate exposed in extensive outcrops

N

0 1 2 3 4 5 miles

Geology along I-95 between Canton and Attleboro. —Modified from Zen and others, 1983

Pennsylvanian redbeds of the northern Norfolk Basin at the Shawmut Industrial Park adjacent to the Conrail railroad tracks in Canton. Thick sandstone on the right grades up into siltstone. A river eroded a channel in the siltstone and then deposited conglomerate, which grades into the sandstone on the left. —Heewon Taylor Khym photo

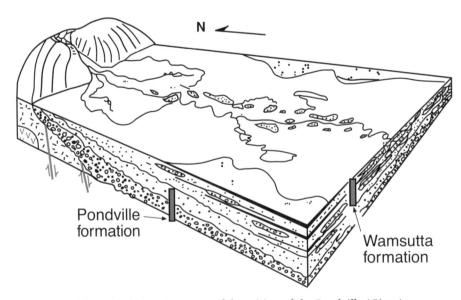

The probable environment of deposition of the Pondville (Giant) conglomerate of the Pennsylvanian Norfolk Basin south of the Blue Hills. —Modified from Hepburn and others, 1993

well as other sedimentary structures that formed when these beds were horizontal, tell us that the top of these steeply dipping beds is to the south.

The north-trending Stony Brook fault, associated here with a basalt dike of Mesozoic age, cuts through the railroad outcrop. The block on the west side of the Stony Brook fault has moved down as much as 2,100 feet.

A short distance farther south on I-95, near the Neponset Street overpass, the Pennsylvanian sedimentary rocks in the middle of the Norfolk Basin consist of purple slates and massive conglomerate and sandstone. Look south from this vantage point across the Norfolk Basin lowland to hills of older plutonic igneous rocks in the Sharon Upland.

Sharon Upland

The crystalline rocks of the Sharon Upland between Canton and Foxborough are mainly Precambrian granite, syenite, and gabbro of the Avalon terrane. In late Precambrian time, the Avalon terrane was a chain of volcanic islands on the margin of Gondwana at the south pole. Igneous activity resulted in the extrusion of volcanic rocks and intrusion of plutonic rocks between 610 and 589 million years ago.

Except for the Sharon syenite, the rocks of the Sharon Upland are similar to those elsewhere in the Avalon terrane. The mass of syenite is 13 miles long and as much as 1.5 miles wide. It formed along the border of a gabbro body that intruded granite. Syenite is a pale rock composed mostly of feldspar with very little or no quartz.

Watch in the median at the Mechanic Street overpass at interchange 8 in Sharon for a contact where gabbro intrudes older granite. Veins of pale gray granite fill fractures that surround blocks of gabbro the size of bowling balls. This sort of thing—subterranean fireworks to a hardrock geologist—is common in the Avalon terrane but perhaps is as well displayed here as anywhere.

Geologists have two competing theories for how this contact formed. In the first theory, two magmas with different chemistries intrude simultaneously, but crystallize at different temperatures and thus different times. The granite magma crystallizes later and fills in the fractures of the cooling gabbro magma. In the second theory, the hot magma of the gabbro melts a surrounding, already-crystallized granite, which then squirts explosively into the intruding gabbro.

A mass of magma intruded these rocks in Devonian time and crystallized into the Rattlesnake Hill granite. You can see unusually fresh exposures of this granite in the abandoned Welch quarry situated southwest of Rattlesnake Hill and about 1 mile south-southeast of Massapoag Lake.

Glacial Lake Neponset

Neponset Reservoir is a small lake centered in the area once inundated by Glacial Lake Neponset, which flooded three shallow basins to the south of the receding Narragansett Bay–Buzzards Bay ice lobe. The melting ice front lay near the Neponset River, between South Norwood and East Walpole.

The sediments associated with Glacial Lake Neponset record an earlier stand of the glacial front than that which impounded Glacial Lake Charles to the north. The sediments include sands and gravels near the junction of U.S. 1 and I-95 in Walpole and Sharon. Deposits of the western arm of Glacial Lake Neponset exist along and west of U.S. 1 near CMGI Field and the Foxborough Raceway.

In the 1950s, I saw sand pits along U.S. 1 between Massachusetts 128 and Neponset Street in Norwood that exposed layers of pebbly sand. The small faults that broke them suggested that they were deposited on a floor of ice that later melted and broke apart. Perhaps they are part of a delta laid down where the meltwater streams emptied into the northern margin of Glacial Lake Neponset. The Automile, Norwood Memorial Airport, and the Massachusetts 128 Amtrak Railroad Station now stand on the remains of this delta plain.

The road south of Foxborough crosses sand and gravel delta plains associated with meandering meltwater streams flowing south from Glacial Lake Neponset. They extend south of I-495.

Mansfield Coal Mines

I-95 passes just west of Mansfield, where a number of small mines and prospects access coal in the Narragansett Basin. Peat deposited in the basin was baked into coal about 275 million years ago during early Permian time. Thick beds of coal, ranging from anthracite to meta-anthracite, were mined in the Rhode Island formation for fuel, to raise steam, heat households, smelt copper, and for other foundry activities in the eighteenth, nineteenth, and early twentieth centuries.

Rhode Island Formation at Attleboro

An outcrop in Attleboro is a must-see for any budding geologist interested in sedimentary structures. A glacially polished outcrop of the Rhode Island formation shows now-vertical graded beds, clastic dikes, crossbedding, and channel fillings. Clastic dikes form when pressure forces unconsolidated sand and silt upward into cracks in overlying rocks or when granular sediments fill in cracks in underlying rocks.

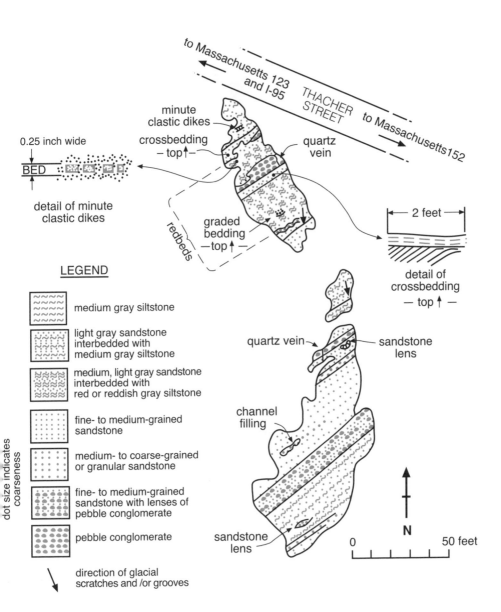

0.25 inch wide

BED

detail of minute
clastic dikes

minute
clastic dikes

crossbedding
– top↑–

quartz
vein

graded
bedding
–top↑–

redbeds

2 feet

detail of
crossbedding
– top ↑ –

LEGEND

medium gray siltstone

light gray sandstone
interbedded with
medium gray siltstone

medium, light gray sandstone
interbedded with
red or reddish gray siltstone

fine- to medium-grained
sandstone

medium- to coarse-grained
or granular sandstone

fine- to medium-grained
sandstone with lenses of
pebble conglomerate

pebble conglomerate

direction of glacial
scratches and /or grooves

dot size indicates coarseness

to Massachusetts 123
and I-95

THACHER STREET

to Massachusetts 152

quartz vein

sandstone
lens

channel
filling

sandstone
lens

N

0   50 feet

Outcrop map of vertical beds and sedimentary features of the Pennsylvanian Rhode
Island formation on Thacher Street along the west side of Conrail Railroad tracks in
Attleboro. —Modified from Lyons and Chase, 1976

Looking southeast at glacially polished outcrop of vertical beds of the Pennsylvanian strata. On Thacher Street along the west side of Conrail Railroad tracks in Attleboro.

Hoppin Hill

Early Cambrian conglomerate, quartzite, and green and red slates with fossiliferous limestone beds exist at Hoppin Hill Reservoir, south of the junction of U.S. 1 and Massachusetts 1A, in North Attleboro. Hoppin Hill is the classic locality for these early Cambrian strata, which were deposited in a nearshore marine environment along the edge of the Avalon terrane or on the fringe of an Avalon volcano.

The early Cambrian rocks contain fossils of small shelly animals that are somewhat older than the earliest Cambrian trilobites. The rocks are about 545 million years old. The sudden appearance in the rock record of life-forms with hard parts is called the Cambrian explosion. It occurred just after the formation of the Avalon terrane. No land plants lived then, but shallow waters were full of primitive algae and bacteria. These fossiliferous rocks are similar in age, fossils, and type to the Weymouth formation exposed southeast of Boston.

The Cambrian strata at Hoppin Hill rest on eroded granite of Precambrian age. Though several of us geologists saw the nonconformable contact of the quartzite with the granite when it was periodically exposed by digging, construction of condominiums in the late 1980s permanently covered the contact. A 25-square-mile area of Pondville conglomerate and Wamsutta redbeds entirely surrounds the outcrops at Hoppin Hill.

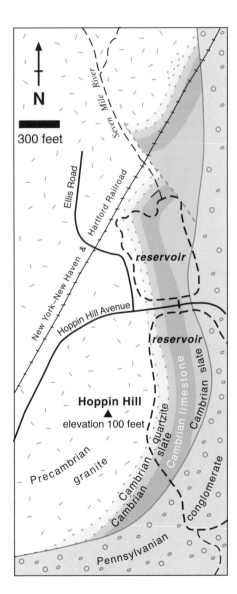

Rocks in the Hoppin Hill area.
—Modified from Anstey Jr., 1979

Where Two Ice Lobes Met

Patterns of glacial scratches on the bedrock between Attleboro and Pawtucket, Rhode Island, suggest that the western margin of the Narragansett Bay–Buzzards Bay lobe sat near here. Glacial scratches on bedrock 10 miles to the east of Attleboro trend southeast, which means that the Narragansett Bay–Buzzards Bay lobe produced them. Glacial scratches that trend southwest on bedrock in Attleboro and Pawtucket were probably left by a small lobe to the west. An alignment of sediments laid down along the edge of the ice, as well as two areas of glacial sediments of different composition, supports this conclusion.

Interstate 95
Salisbury—Peabody
28 miles

Seaside towns with long seafaring histories, miles of beaches, wildlife preserves, and the largest salt marsh in the state lie tucked away in the northeastern corner of Massachusetts. A trip through the region will also bring you in contact with more than ten types of rocks. In a mere 28 miles between Salisbury and Peabody, I-95 crosses segments of three Gondwanan terranes: the northeasternmost part of the Merrimack terrane, the northeastern tip of the Nashoba terrane, and the northern Avalon terrane.

Merrimack Terrane

I-95 crosses the Newburyport batholith of the Merrimack terrane between the New Hampshire border and the Clinton-Newbury fault zone. The outer mile or two of the northern and western part of the batholith consists of medium-grained gray granite with feldspar crystals the size and shape of a horse's tooth—the French call it a *dent de cheval* granite. The inner part of the batholith is granodiorite and tonalite. Tonalite is somewhat darker and more mafic in composition than granite. These darker crystalline rocks are exposed north and south of the junction of I-95 with I-495. All the rocks in the Newburyport batholith are about 420 million years old, a late Silurian age.

The Kittery quartzite is the oldest formation in the Rockingham belt of the Merrimack terrane. It is probably late Ordovician or Silurian in age. Deposited as a limy beach sand, this lime silicate quartzite contains the minerals actinolite and biotite. Named for splendid outcrops in southernmost Maine, the Kittery quartzite also surfaces in nearby New Hampshire

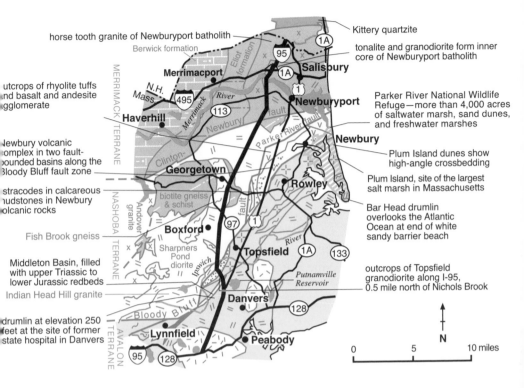

horse tooth granite of Newburyport batholith

Berwick formation

Kittery quartzite

tonalite and granodiorite form inner core of Newburyport batholith

MERRIMACK TERRANE

Merrimacport

Salisbury

utcrops of rhyolite tuffs ind basalt and andesite igglomerate

Mass.
N.H.

River

Newburyport

Parker River National Wildlife Refuge—more than 4,000 acres of saltwater marsh, sand dunes, and freshwater marshes

Haverhill

Merrimack River

Newbury fault

Parker River fault

Newbury

Jewbury volcanic complex in two fault-bounded basins along the Bloody Bluff fault zone

Clinton-

Georgetown

Plum Island dunes show high-angle crossbedding

Plum Island, site of the largest salt marsh in Massachusetts

stracodes in calcareous nudstones in Newbury olcanic rocks

NASHOBA TERRANE

Andover granite

biotite gneiss & schist

Rowley

Bar Head drumlin overlooks the Atlantic Ocean at end of white sandy barrier beach

Fish Brook gneiss

Boxford

River

Topsfield

Middleton Basin, filled with upper Triassic to lower Jurassic redbeds

Sharpners Pond diorite

Ipswich

Putnamville Reservoir

outcrops of Topsfield granodiorite along I-95, 0.5 mile north of Nichols Brook

Indian Head Hill granite

Danvers

drumlin at elevation 250 feet at the site of former state hospital in Danvers

Bloody Bluff

AVALON TERRANE

Lynnfield

Peabody

N

0   5   10 miles

Bedrock geology along I-95 between Salisbury and Peabody.
—Modified from Zen and others, 1983

at Hampton Beach near Great Boars Head. At the Seabrook exit from I-95, go east on New Hampshire 107 to Hampton Falls and then southeast on New Hampshire 101 to New Hampshire 1A at Hampton Beach. Outcrops of the quartzite are exposed in the triangle of one-way roads at that junction.

Clinton-Newbury Fault Zone

I-95 crosses the Clinton-Newbury fault zone, which separates the Merrimack and the Nashoba terranes, in Newbury. This fault zone, including a splay called the Scotland Road fault, strikes out to sea beneath Plum Island in easternmost Newbury, about 600 feet south of the township line. The Scotland Road fault in Newbury and Newburyport is probably the northeasternmost break of the Clinton-Newbury fault zone. It follows the north side of Scotland Road east of I-95, and U.S. 1 crosses it at the Newburyport-Newbury township line.

Nashoba Terrane

This northeastern part of the Nashoba terrane consists mainly of three rock units that range in age from Cambrian to Devonian: Fish Brook gneiss, Andover granite, and Sharpners Pond diorite.

The Sharpners Pond diorite is the largest calc-alkaline pluton in the Nashoba terrane, and a lenticular mass of this diorite fills much of the terrane along our route. Sharpners Pond diorite, a very dark igneous rock composed mainly of black biotite mica and equally black hornblende, crystallized about 430 million years ago in Silurian time. The pluton is 13 miles long and 2 miles wide. Splays of the Clinton-Newbury fault zone define its north and south margins.

You can see Sharpners Pond diorite on the north side of Scotland Road just west of the exit from I-95 on the Newbury–West Newbury line. The diorite is brecciated, and granite surrounds fragments and pillows of the dark rock. The hot mafic magma of the Sharpners Pond diorite intruded previously crystallized granite. Temperatures of 1,100 degrees Celsius melted the granite, which in turn intruded the crystallizing diorite.

I-95 crosses the Fish Brook gneiss, the oldest rock in the Nashoba terrane, in Georgetown. A metamorphosed volcanic rock, the Fish Brook gneiss crystallized 500 million years ago. It forms a cliff along Massachusetts 133 west of I-95 and east of Georgetown Center. A splendid outcrop is present on the southbound lane of I-95 at the exit ramp from Massachusetts 133.

The Clinton-Newbury fault zone and the Bloody Bluff fault converge at Plum Island. The Nashoba terrane appears to pinch out at this point, and the Merrimack terrane offshore to the north may be in contact with the Avalon terrane to the south.

Avalon Terrane

I-95 crosses the Bloody Bluff fault, which separates the Avalon and Nashoba terranes, just south of the Ipswich River in Middleton. Fault blocks of 600-million-year-old Topsfield granodiorite separate Newbury sedimentary and volcanic rocks of Silurian and Devonian age from Triassic sedimentary rocks in Danvers. The pink to orange, coarse-grained granodiorite surfaces in a belt one-half mile wide and 2 miles long. Crystals of quartz, plagioclase, hornblende, and biotite make up about 10 to 20 percent of the rock. Oval crystals of quartz are up to 1 centimeter long. I-95 passes outcrops of the granodiorite about one-half mile north of Nichols Brook, west of Putnamville Reservoir.

I-95 crosses about 4 miles of sheared black gabbros south of the junction of I-95 and U.S. 1 in northwestern Danvers Township.

Newbury Volcanic Rocks

The 418-million-year-old Newbury volcanic complex is present in two fault-bounded basins along the Bloody Bluff fault zone. Grayish green to dark green andesites and yellowish brown rhyolites interbed with dark gray to olive gray sedimentary rocks. Calcareous mudstones in the Newbury volcanic complex contain ostracodes, shelly bivalved crustaceans from 2 to 18 millimeters long, that identify the rocks as late Silurian and early Devonian in age. Because the Newbury complex is in a fault block, we know very little about its deposition or origin. Similar rocks with similar fossils in coastal Maine belong to the Avalon terrane, so it seems reasonable to suppose that the Newbury rocks do as well.

Fine-grained siltstones and an ostracode-bearing limestone bed of the Newbury complex surface in outcrops along both sides of I-95 in Topsfield near where the Rowley Bridge Road veers close to the interstate. You can see the northeast-trending Newbury volcanic rocks and sediments near U.S. 1 south of the Parker River to Doles Corner in Rowley. Look for fossiliferous red mudstones between Doles Corner and the Rowley-Ipswich town line.

Middleton Basin

The Middleton Basin, a Triassic fault block, pulled apart at the same time the Atlantic Ocean began to open. The basin, 3.6 miles long and 0.3 mile wide, is filled with redbeds, sediments of Triassic age as indicated by diagnostic plant fossils. I-95 crosses the basin in Danvers Township. The redbeds are in fault contact with Precambrian gabbro. The contact is exposed in the Essex Bituminous Quarry. Unfortunately, it is difficult to get permission to enter this quarry, which is north of Russell Street in Peabody.

Marine Sediments

A blanket of marine sediments extends inland from Cape Ann west to I-95. The irregular western margin of these deposits records the western shore of the Atlantic Ocean in Pleistocene time. Seawater encroached on the coast north of Boston while the Cape Cod Bay ice lobe was melting more rapidly than the onshore lobe to the west. Delta sediments deposited at the melting front of the ice lobe interfinger with fossiliferous deposits laid down in seawater.

The Northern Shore

The shoreline between the New Hampshire border and Cape Ann is a splendid place to examine sand deposits laid down between high and low tide. The tidal range is large, and the coastal waters are clear. Wind plays an

Aerial view looking north at Plum Island, with the Merrimack River and Joppa Flats at its northern end. Longshore currents refracting around the south end of Plum Island transport and deposit sand that partially buries the fringe of South Plum drumlin in the foreground. —Photo courtesy of Peabody-Essex Museum

important role in shaping the coast, especially during Nor'easters, storms that bring 15-foot and larger swells.

Plum Island, a barrier island more than 8 miles long and as much as 1 mile wide, protects the largest saltmarsh in Massachusetts. Its southern end is attached to a group of drumlins. The island took shape between 7,000 and 6,000 years ago.

Dunes on Plum Island have high-angle crossbeds formed by onshore winds that carry sand above the storm high-tide line. The dunes migrate landward, shifting the shoreline to the west. Longshore currents generated by northeast winds carry sand southward along the coast. The island has grown southward, connecting to Bar Head, a drumlin at the southern tip of the island. Over the past 175 years, human development has added to the actions of wind and waves in shaping Plum Island.

Merrimack River Estuary

Newburyport, a significant port during and after the Revolutionary War, sits on the south shore of the Merrimack River. Clipper ships sailed into this port in the nineteenth century.

Just east of Newburyport and south of the estuary of the Merrimack River is Joppa Flats, 300 acres of land exposed at low tide. The main channel of the Merrimack River is along the northern margin of the estuary. As salt water pushes up the channel during the rising tide, it deflects the fresh water southward over the flat. The salinity of the flat rarely rises above 10 to 15 parts per thousand, allowing freshwater sedges to survive.

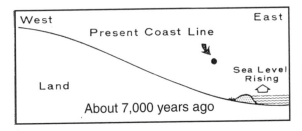

About 7,000 years ago sea level was substantially lower than now, and the barrier beach lay well off the modern shore.

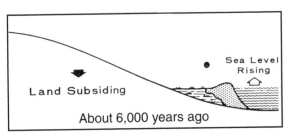

About 6,000 years ago the land subsided, sea level rose, and the barrier beach approached the modern coast.

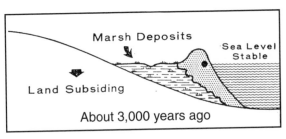

About 3,000 years ago, the land was still subsiding but sea level was stable. The trailing edge of Plum Island was a short distance off the modern shore.

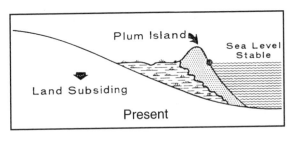

Today, the land is still subsiding slowly.

The evolution of the Plum Island barrier beach from about 7,000 years ago to the present. —Modified from McIntyre and Morgan, 1962

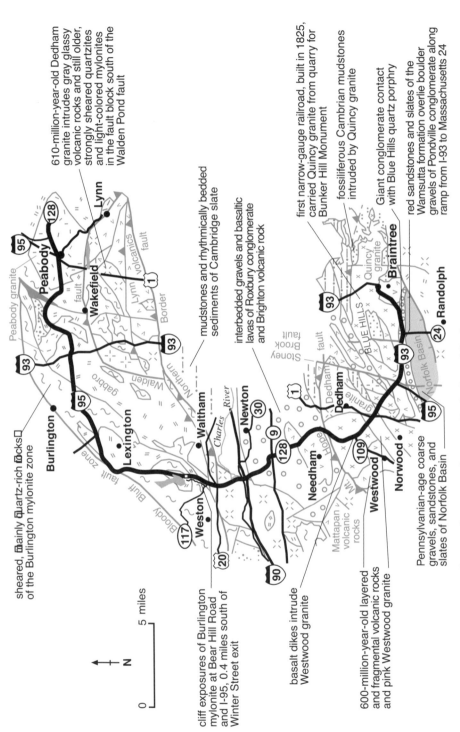

sheared, mainly quartz-rich rocks
of the Burlington mylonite zone

610-million-year-old Dedham
granite intrudes gray glassy
volcanic rocks and still older,
strongly sheared quartzites
and light-colored mylonites
in the fault block south of the
Walden Pond fault

mudstones and rhythmically bedded
sediments of Cambridge slate

interbedded gravels and basaltic
lavas of Roxbury conglomerate
and Brighton volcanic rock

first narrow-gauge railroad, built in 1825,
carried Quincy granite from quarry for
Bunker Hill Monument

fossiliferous Cambrian mudstones
intruded by Quincy granite

Giant conglomerate contact
with Blue Hills quartz porphry

red sandstones and slates of the
Wamsutta formation overlie boulder
gravels of Pondville conglomerate along
ramp from I-93 to Massachusetts 24

cliff exposures of Burlington
mylonite at Bear Hill Road
and I-95, 0.4 miles south of
Winter Street exit

basalt dikes intrude
Westwood granite

600-million-year-old layered
and fragmental volcanic rocks
and pink Westwood granite

Pennsylvanian-age coarse
gravels, sandstones, and
slates of Norfolk Basin

N

0   5 miles

Geology along I-95 and Massachusetts 128 between Peabody and Braintree. —Modified from Zen and others, 1983; Rast and others, 1993

Interstate 95/Massachusetts 128
Peabody—Braintree
46 miles

The hills of eastern Massachusetts are famous as the rockbound home of the computer industry. You can see plentiful bedrock exposures in roadcuts and public parks throughout this densely populated region. The Avalonian rocks along I-95 tell fascinating stories of their wanderings during more than 600 million years from somewhere near the south pole to Laurentia, the early North American continent.

Bostonians modestly refer to their city as the Hub of the Universe because roads radiate from Boston Harbor like the spokes of a wheel. I-95 and I-495 form large semicircular wheels around and away from Boston Harbor. People who drive I-95 between Peabody on the North Shore and Braintree on the South Shore see a 5-mile-wide zone of sheared rocks—mylonites—as well as great masses of pale granite and dark gabbro that intruded as molten magmas.

The band of crystalline upland that I-95 follows between Peabody and Braintree is the northern and western rim of the Boston lowland, which held seawater for a time during the last ice age. Traced south, I-95 skirts the eastern margins of Glacial Lakes Concord, Sudbury, Charles, and Neponset, successively.

Peabody Granite

Peabody sits atop the aptly named Peabody granite pluton that crystallized about 370 million years ago in Devonian time. I-95 crosses a smaller, boot-shaped body of Peabody granite in Wakefield. The alkaline magma of the Peabody granite formed during a rifting episode as the Avalon terrane migrated from Gondwana to Laurentia. This greenish gray granite is one of several large alkaline plutons of early to middle Paleozoic age north of Boston.

The Peabody granite intrudes older dark gray to black plutonic rocks—gabbros. Gabbros are typical of volcanic island chains, such as Avalon, but have long been a puzzle here because gabbros of different ages look alike. Although some of the gabbros here are Precambrian age, others intruded 375 million years ago in Devonian time. You can see roadside cliffs of dark gabbro on either side of I-95 between Peabody and Wakefield.

Burlington Mylonite

Between Peabody and U.S. 20 in Weston, the highway passes through alternating patches of dark gabbro and light mylonites. The finely layered

Veins of granite cut through gabbro. Crystallized granite veins melted when hot gabbro contacted previously crystallized granite. The newly melted granite reacted explosively with the gabbro, fragmenting it and shooting veins of granite back into the opened nooks and crannies. Outcrop at Weston Observatory.

mylonites have a strong foliation. For many years, geologists thought these light and dark rocks were welded volcanic ash deposits. We now recognize them as sheared rocks formed in very broad fault zones. Temperature, pressure, and deformational forces control the mineral development in mylonites. Here, the mineral layers consist of brittley deformed minerals such as feldspar that broke into pieces and ductilely deformed minerals such as quartz that recrystallized and smeared out like stiff taffy.

The early phase of the Burlington mylonite zone formed during Precambrian shearing in a broad fault zone along the margin of the Gondwanan supercontinent. The shearing action ground gray to white quartzites and dark schists of the Westboro quartzite into mylonite. Parts of the Westboro quartzite, which was deposited as quartz sands along the shores of Gondwana well before 610 million years ago, escaped the intense shearing of the mylonite zone. You can see Westboro quartzite in Wayland along Massachusetts 126 between U.S. 20 and Massachusetts 30.

You can view the Burlington mylonite in a cliff along I-95, 0.4 miles south of the Winter Street exit near Bear Hill Road in Waltham. Tiny glassy veins intrude the mylonite here. The veins formed by melting of rock during intense earthquake activity.

Avalon's Granite Core

South of U.S. 20 in Weston and Waltham, I-95 crosses granites that formed the core of the Avalon volcanic chain. The 610-million-year-old Dedham granite, the 599-million-year-old Westwood granite, and the 600-million-year-old Mattapan volcanic rocks form the highlands of Needham, Westwood, Norwood, and Dedham. Roadcuts immediately south of U.S. 20 expose the pink to gray, coarse-grained Dedham granite. Basalts of Jurassic age intrude both the Dedham and Westwood granites.

You can see Westwood granite along I-95 in Westwood near Massachusetts 109. The Westwood granite intrudes and often contains large chunks of the Dedham granite. These two granites are difficult to distinguish. The Dedham granite is older and typically more highly deformed, and the alteration of feldspars give rise to very fine-grained green minerals such as epidote and chlorite.

Mattapan volcanic rocks crop out near I-95 between Highland Avenue and Great Plain Avenue (Massachusetts 135). You can see splendid exposures of these hard, glassy, variously colored volcanic rocks east of I-95 on Second Avenue and along the approach to and grounds of the Sheraton Inn south of Highland Avenue in Needham.

Geologist Newton Chute examines the sharp contact between the Westwood granite and a basalt dike of Jurassic age at interchange 55A in Westwood. —Newton Chute photo, U.S. Geological Survey

Explosive breccia formed when molten Westwood granite intruded gabbro and Dedham granite. The dark, mainly angular fragments are gabbro, and the light fragments are Dedham granite. —Newton Chute photo, U.S. Geological Survey

Boston Basin

South of Massachusetts 30, on the Newton-Weston township line, the Charles River meanders along the western margin of the Boston Basin, where sedimentary rocks deposited between 590 and 540 million years ago have been complexly faulted along with the older Dedham granite. These late Precambrian sedimentary rocks, derived by erosion from the Mattapan volcanoes and other older rocks, were deposited on top of the Dedham and Westwood granites.

The Roxbury conglomerate is the dominant sedimentary basin rock along I-95 south of I-90. It contains basaltic volcanic rocks at its base covered by coarse sedimentary rocks eroded from glacier-covered highlands. The Avalon terrane was near the south pole when these rocks were deposited, so we believe that ice was probably involved in the erosion process.

Glacial Lake Charles

When the Narragansett Bay–Buzzards Bay lobe of the last ice sheet melted to a position near the Mass Pike in Newton and Weston, it impounded Glacial Lake Charles to the south. That lake extended south nearly to where Massachusetts 128 and U.S. 1 are joined. Stream sediments were deposited in outwash fans south of Glacial Lake Charles near the present site of the Massachusetts 128 Amtrak Railroad Station and just west of the Blue Hills highlands. Large parts of these thick deposits have been mined for sand and gravel.

Blue Hills Quartz Porphyry

The Blue Hills volcano erupted 440 million years ago at the end of Ordovician time. I-95 follows the southern margin of the Blue Hills quartz porphyry, the extrusive volcanic equivalent to the Quincy granite, which is the volcano's crystallized magma chamber. The two rocks have the same chemical composition, but the porphyry cooled on the earth's surface whereas the granite cooled at depth. For more than 9,000 years, Native Americans have chipped blades for arrowheads and tools from the very fine-grained volcanic rocks in the Blue Hills.

You can see a contact between the quartz porphyry and the overlying sedimentary rocks in the northwest cloverleaf at the junction of I-93 with Massachusetts 28, just east of Massachusetts 24 in Randolph. The rocks just above the contact include the Giant conglomerate of Pennsylvanian time, about 315 to 300 million years ago. It is the oldest sedimentary unit of the Norfolk Basin.

The conglomerate contains boulders eroded from the Blue Hills quartz porphyry. Because weathered quartz porphyry boulders of the conglomer-

Close-up of Blue Hills quartz porphyry. —Heewon Taylor Khym photo

Looking east at the contact and fault between the Giant conglomerate and the deeply weathered Blue Hills quartz porphyry. —Heewon Taylor Khym photo

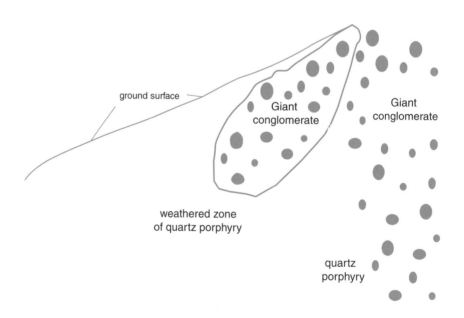

Interpretive section for photo.

ate rest upon the weathered surface of the Blue Hills quartz porphyry, the contact may be confusing upon first inspection. The rounded residuals of the porphyry in the weathered surface are shades of maroon and green due to iron pigments. The old erosion surface was approximately horizontal until crustal movements jammed the Boston Basin against the Blue Hills and the Norfolk Basin during late Permian time.

Quincy Granite and the First American Railroad

East of Norwood and Dedham, the Blue Hills rise sharply to the north of I-93. The Blue Hills Reservation is on the distinctively greenish Quincy granite, which is about 440 million years old. It owes its unusual color to the minerals riebeckite, a dark blue amphibole, and aegerine, a greenish pyroxene. Both minerals are rich in sodium and do not commonly occur in granite. The Quincy granite has long been used as an architectural stone. Quarry owners financed the first commercial narrow-gauge railway in the United States during construction of the Bunker Hill monument in Charlestown in 1826. The railway hauled granite from the quarry to the harbor.

The Quincy granite of the Blue Hills intrudes lower and middle Cambrian sedimentary rocks. The middle Cambrian formation contains the oversize *Paradoxides* trilobites of Braintree. These fossils are typical of middle Cambrian rocks of Gondwanan terranes—fragments of the great Gondwanan supercontinent of the Southern Hemisphere. It is amazing to find the trilobites in Massachusetts and to realize that they and the rocks associated with them probably started their journey many thousands of miles to the south.

Interstate 195
Wareham—Seekonk
44 miles

Deeply indented harbors, tidal marshlands, and gentle headlands characterize this scenic coastal route across southeastern Massachusetts. The interstate crosses nine coastal rivers with such names as Sippican, Mattapoisett, Acushnet, and Paskamanset. They discharge into Buzzards Bay, Mount Hope Bay, and Narragansett Bay. Glacial meltwater streams established these southward-trending river valleys during the recession of the last continental ice sheet.

Precambrian bedrock underlies the glacial material from Wareham to Fall River. Many of the cobbles and other rock fragments in the Buzzards

Bay moraine on Cape Cod and in the Martha's Vineyard moraine came from bedrock exposed along or near I-195. West of Fall River, I-195 crosses onto the Narragansett Basin—a large basin of coal-bearing strata of Pennsylvanian age.

Trail of Moraines

The Narragansett Bay–Buzzards Bay lobe deposited the arcuate Buzzards Bay moraine that forms the hills along the eastern shore of Buzzards Bay. This lobe melted back earlier than did the Cape Cod Bay lobe to the east, which deposited the Sandwich and Ellisville moraines. The western end of the Sandwich moraine was deposited over the top of the Buzzards Bay

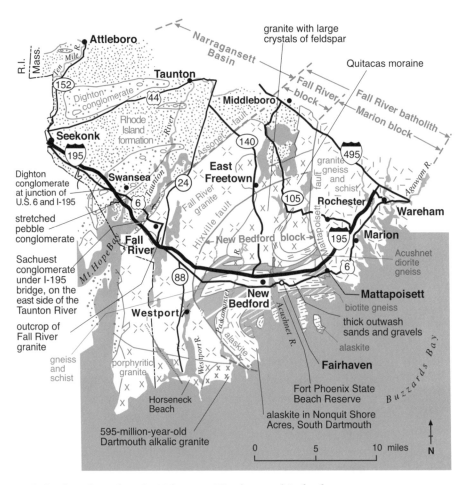

Bedrock geology along I-195 between Wareham and Seekonk. —Modified from Hermes and Zartman, 1992; Zen and others, 1983

moraine. The shape of the intersection of the Sandwich and Buzzards Bay moraines gives you an idea of the shape of the intersecting lobes.

A less prominent series of morainal hills to the northwest parallels the Buzzards Bay moraine. These morainal deposits represent places where the ice stood still for awhile during its retreat—long enough to deposit a mound of debris. The Narragansett Bay–Buzzards Bay ice lobe deposited the Quittacas, Hog Rock, Snipatuit, and Middleborough moraines as it melted northward. The Quittacas moraine, west of Wareham and just south of Great Quittacas Pond in Rochester township, is an east-west trending line of irregular hills about 100 feet high.

Wareham Pitted Plain

Wareham rests on the southern edge of the Wareham pitted outwash plain. The plain is a large fan of outwash that formed between the Narragansett Bay–Buzzards Bay and Cape Cod Bay ice lobes. The elevation of the plain decreases from nearly 200 feet in the north to sea level in the south. Outwash sediments grade from coarse gravel with rounded boulders up to 3 feet in diameter near the Ellisville moraine down to fine sand near Buttermilk Bay and the Agawam River. The pits are kettles—holes where blocks of ice were embedded within the outwash.

Fall River Spillway

As the Narragansett Bay–Buzzards Bay lobe melted back to the north, several glacial lakes formed in low-lying areas of the recently deglaciated land surface. Most lakes were small and short-lived, but two large lakes developed in succession in the Taunton River, Hockomock Swamp, and Jones Rivers areas as the ice front retreated. The lakes—known collectively as Glacial Lake Taunton—covered a combined area of about 500 square miles. For many years, a spillway at Fall River on the Taunton River controlled the overflow of Glacial Lake Taunton into Narragansett Bay. As the Cape Cod Bay lobe to the east receded northward, a new, lower channel was exposed—the Jones River spillway—that enabled the lake to drain eastward into Cape Cod Bay.

Marshes of Buzzards Bay

Douglas Johnson, a geologist at Columbia University, studied the salt marshes along Buzzards Bay and summarized his findings in *The New England–Acadian Shoreline,* published in 1925. This classic treatise describes features that reflect changes in sea level during glacial and postglacial times.

The New England tidal marsh, a common type described by Johnson, has two layers of peat that usually rest on glacial till, sand, or bedrock. The

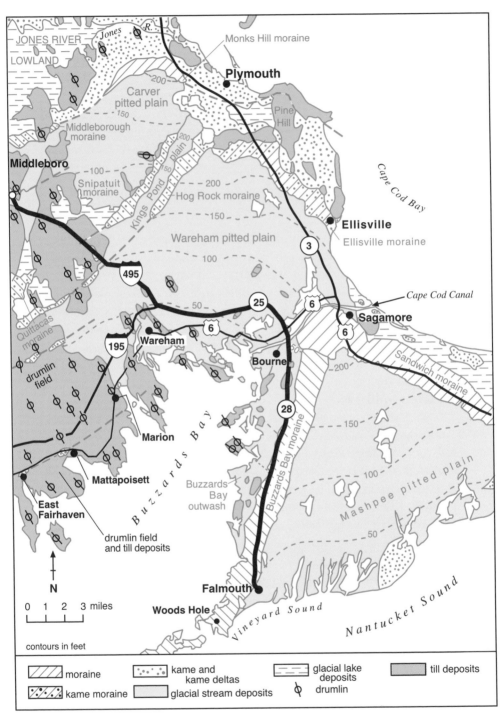

Glacial geology of the Buzzards Bay area. —Modified from Larson, 1982

top layer consists of grayish to brownish peat with variable amounts of silt deposited in salt water, and the lower layer is deep brown to black peat deposited in brackish to fresh water. This sequence indicates a progressive encroachment of marine water—either because of a rise in sea level or because of land subsidence.

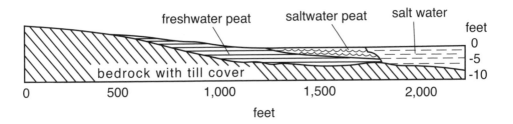

Cross section of salt marsh on southwest side of Mattapoisett Neck, Buzzards Bay.
—Modified after Johnson, 1925

Fall River Batholith

Between Wareham and Fall River, I-195 crosses the Fall River batholith, a complex of 600-million-year-old igneous rocks of the Avalon terrane. Geologists divide the batholith into three blocks: the Marion, the New Bedford, and the Fall River. The batholith intrudes even older metamorphic rocks. These scattered metamorphic rocks form curved strips called roof pendants because they ornament the roof of certain parts of the batholith. Outcrops of bedrock in this region are extremely sparse.

Marion Block. A thick layer of glacial deposits blankets the bedrock from Wareham to Mattapoisett. Information on this bedrock, the Marion block of the Fall River batholith, is so fragmentary that geologists simply lump the rock together as "Precambrian granite, gneiss, and schist." The Marion block may contain plutonic and volcanic rocks of Paleozoic age, but we cannot yet say for sure.

New Bedford Block. At Mattapoisett, I-195 crosses the Mattapoisett Harbor fault, which divides the Marion block from the New Bedford block. The New Bedford block is a complex, deformed assemblage of Precambrian rocks that continues to enlighten geologists—once they puzzle over it long enough—on the tectonic history of the Avalon terrane.

View from Angelica Avenue in Mattapoisett, looking southeast across the marshland and Pine Island Pond to Strawberry Point. Frost heaving booted the scattered boulders up into the peat from the underlying glacial till.

The New Bedford block has been highly deformed in a ductile manner, not unlike the way you can fold warm taffy. Collisions of microcontinental blocks stretched these rock bodies into east-west trending bands. You can see taffylike swirls on the geologic map.

This east-west pattern suggests that a microcontinent moving up from the southeast collided with and deformed these rocks. The collision of the Meguma microcontinent with the Avalon microcontinent likely created the deformation; the Meguma microcontinental plate probably sank beneath the Avalonian plate in a subduction zone.

The New Bedford block consists of gneisses and schists of Precambrian age intruded by several kinds of igneous plutons. It also contains lenses and blocks of altered dark rocks that may be related to the Acushnet diorite gneiss. A large part of the New Bedford block north of I-195 consists of a deformed and extensively mylonitized granite with large microcline feldspar crystals. Extensive outcrops surface along Massachusetts 140 north of I-195.

The Acushnet diorite gneiss forms a fishhook-shaped mass on the map. It consists of medium-grained hornblende diorite that is metamorphosed in part to amphibolite and hornblende gneiss.

Alaskite, a light gray to pinkish gray to tan granitic rock that contains almost no dark minerals except magnetite, surfaces along the coast near New Bedford. You can see it at Nonquit Shore Acres in South Dartmouth.

A strip of gneiss and schist that parallels I-195 near New Bedford consists of dark gray to black hornblende and biotite schist, black-and-white striped gneiss, and black to greenish black amphibolite. Outcrops are well exposed along the eastbound lane and median between the I-195 junction with Massachusetts 140 and the Hixville fault.

Fall River Block. The Hixsville fault, with both ductile and brittle movement, separates the New Bedford and Fall River blocks. I-195 crosses the fault west of Reed Road (interchange 11) and east of Highland Avenue. The Fall River block extends westward from the fault to Mount Hope Bay.

The most abundant rock formation in this boomerang-shaped wedge is the Fall River granite, a light gray, medium-grained granite with only a small content of dark, mafic minerals such as biotite and hornblende. Many such light-colored granites in the southern part of the Avalon terrane have age dates that cluster around 600 million years. The Fall River granite is probably about 600 million years old, too. It has been deformed brittlely—shattered along distinct planes, or faults.

You can see the 595-million-year-old Dartmouth granite of the Fall River block between Horseneck Beach and Potomska Point, 7 miles south of New Bedford. It is an alkalic granite dominated by quartz and potassium- and sodium-rich feldspar. It also contains minor amounts of such minerals as riebeckite, sphene, zircon, and fluorite.

Such alkaline granites are geologically significant because they typically form by deep rifting of the earth's crust rather than by compression during continental collisions. A rift may have formed in the Avalon terrane while it was still part of the Gondwanan supercontinent. The Dartmouth granite is the only Precambrian alkaline granite discovered in the Avalon terrane of southeastern New England.

Pegmatite at Fort Phoenix

Whatever tectonic gyrations distributed the blocks of the Fall River batholith, they probably took place about 275 million years ago in Permian time. At Fort Phoenix State Reservation, along the coast south of Fairhaven and New Bedford, a coarse-grained pegmatitic granite that is probably of Permian age intrudes a dark biotite gneiss of the New Bedford block. This splendid granite resembles, and thus may be an eastern extension of, the Narragansett Pier batholith of Permian age, which crystallized about 275 million years ago. The Narragansett Pier granite surfaces along the south-

Buff pink pegmatitic granite dikes in Fort Phoenix State Reservation that intrude north-dipping, banded biotite gneiss. View to the north along the shore of Buzzards Bay.

western margin of Narragansett Bay and underlies the southern coast of Rhode Island.

Narrangansett Coal Basin

Between the eastern shore of the Taunton River and Seekonk, I-195 crosses stratified coal-bearing rocks of the Narragansett Basin formed about 310 to 290 million years ago during Pennsylvanian time. The strata are highly contorted from at least two episodes of Alleghanian deformation in Permian time that baked the basin's thick peat deposits into coal.

The coal-bearing deposits of the Narragansett Basin originated as freshwater swamps and marshlands into which flowed rivers and streams—an area similar to today's Mississippi delta. The basin was probably actively subsiding as the surrounding mountains rose. Rivers and streams deposited channels of gravel, sand, and silt, as well as levee silts and shales.

The Narragansett Basin rocks along I-195 consist of three major formations: the distinctive smoky quartz–bearing Sachuest conglomerate, which rests on an erosional surface of the Fall River granite; the coal-bearing Rhode Island formation, which makes up the bulk of the Pennsylvanian strata; and the Dighton conglomerate, which forms the top of the sequence.

The Sachuest conglomerate is exposed under the interstate bridge on the east side of the Taunton River in the city of Fall River. The Sachuest conglomerate was one of the first sediments deposited in the coal basin. It is exposed on the southeastern margin of the basin but is deeply buried in the interior. The Assonet fault, which underlies Mount Hope Bay and the Taunton River, has removed parts of the basal conglomerate here along the eastern margin of the basin.

Between the Taunton River and Seekonk, I-195 crosses mainly through the Rhode Island formation. Peat and stream sediments deposited in the ancient Narragansett Basin are now conglomerate, sandstone, siltstone, slate, carbonaceous slate, and anthracite and meta-anthracite coal.

I-195 crosses a 4-mile-wide zone of Dighton conglomerate—the top of the Narragansett Basin sediments—near Swansea. The northeast-south-west trending band is the axis of a synclinal fold produced during the Alleghanian mountain building event.

The gray conglomerate consists mainly of rounded quartzite, ranging in size from pebbles to cobbles and even boulders. It also contains some granite and slate fragments as well as lenses of sand. In the southeastern part of

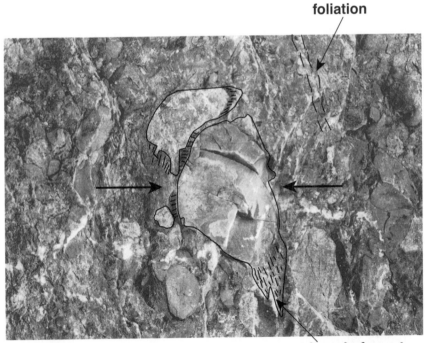

Stretched quartzite pebbles of the Dighton conglomerate in outcrop on U.S. 6 east of the junction with I-195 in Swansea. Pressure solution produced beards of fibrous quartz at ends of the pebbles. Arrows indicate direction of compressional forces.

the basin, deep burial with elevated temperatures and pressures deformed the Dighton cobbles by pressure solution. Northwest-to-southeast compressional forces squeezed these cobbles, causing the quartz to dissolve in water and migrate to the northeast and southwest ends of the cobbles, where it precipitated as fibrous crystals. The middle parts of the cobbles are now slender, and the northeast and southwest ends display beards of quartz.

You can see stretched pebbles in Swansea along U.S. 6 east of the junction with I-195. Purgatory Chasm in Middletown, Rhode Island, is another classic locality for viewing pressure solution features. You can see more typical Dighton conglomerate—with little or no pressure solution features—along the exit and entrance ramps from U.S. 6 to I-195 west of Swansea.

Interstate 495
Salisbury—Chelmsford
36 miles

I-495 between Chelmsford and Salisbury generally parallels the Merrimack River, which flows northeast through Massachusetts to the sea at Newburyport. *Merrimac* is a Native American word meaning "swift waters." Between Chelmsford and Lawrence, the river follows the Clinton-Newbury fault zone, the border between the Merrimack and the Nashoba terranes.

I-495 begins in the Merrimack terrane at Salisbury, crosses the fault just south of the Merrimack River at the Lawrence–North Andover township boundary, and enters the Nashoba terrane. Glacial deposits blanket the region. There are widely scattered outcrops of bedrock in the region, but you cannot see many from I-495.

Salisbury Beach

Salisbury Beach, a great recreational beach, is about as far north in Massachusetts as you can get. The beach has a big tidal range, and the clear coastal waters make it possible to study the intertidal environment. The sands and gravels of the beach began to build up after the recession of the last continental ice sheet. The formation of Salisbury Beach, a barrier beach, is similar to that of Plum Island to the south. The continual deposition of sand along the barrier beach protects the marsh behind the beach from erosion.

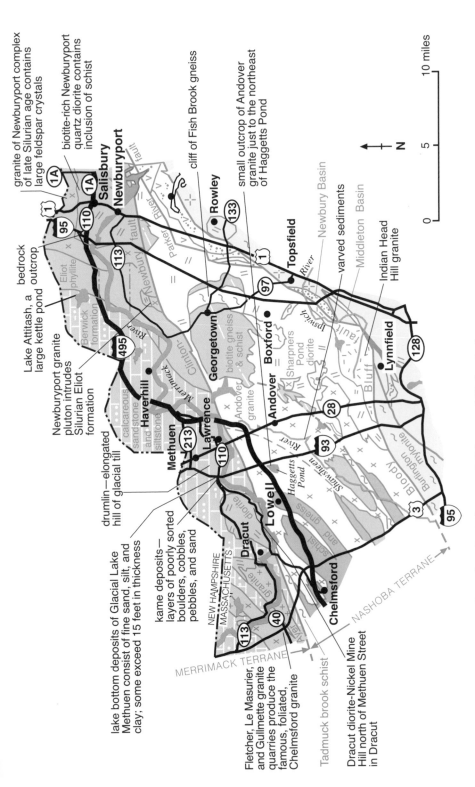

granite of Newburyport complex of late Silurian age contains large feldspar crystals

biotite-rich Newburyport quartz diorite contains inclusion of schist

cliff of Fish Brook gneiss

small outcrop of Andover granite just to the northeast of Haggetts Pond

Salisbury

Newburyport

Rowley

Topsfield

Newbury Basin

varved sediments

Middleton Basin

Indian Head Hill granite

Lynnfield

bedrock outcrop

Lake Attitash, a large kettle pond

Newburyport granite pluton intrudes Silurian Eliot formation

Eliot phyllite

Berwick formation

Parker River

Newbury fault

fault

Ipswich fault

River

Clinton-Newbury fault

biotite gneiss & schist

Georgetown

Boxford

Sharpners Pond diorite

Bluff

Burlington mylonite

Bloody

Haverhill

Methuen

Lawrence

Martimack River

calcareous sandstone and siltstone

Andover granite

Andover

Haggetts Pond

Shawsheen River

drumlin—elongated hill of glacial till

lake bottom deposits of Glacial Lake Methuen consist of fine sand, silt, and clay; some exceed 15 feet in thickness

kame deposits— layers of poorly sorted boulders, cobbles, pebbles, and sand

diorite

Dracut

Lowell

Chelmsford

schist and gneiss

granite

NEW HAMPSHIRE

MASSACHUSETTS

NASHOBA TERRANE

MERRIMACK TERRANE

Tadmuck brook schist

Fletcher, Le Masurier, and Gullmette granite quarries produce the famous, foliated, Chelmsford granite

Dracut diorite-Nickel Mine Hill north of Methuen Street in Dracut

N

0   5   10 miles

Geology along I-495 between Salisbury and Chelmsford. —Modified from Zen and others, 1983

Glacial Marine Sediments

A blanket of seafloor sediments extends inland from the coast as far west as I-95. Marine waters encroached on the coastal region as the Cape Cod Bay ice lobe retreated. The waters deposited marine sediments between approximately 17,000 and 14,000 years before present. The sediments are present along I-495 for about 2 miles west of its junction with I-95. You might see the seashell-bearing silts or sands if you happen upon an excavated construction site. Ice-contact delta sediments are deposited over and interfinger with the marine sediments. In this region, the Cape Cod Bay lobe retreated more rapidly to the north relative to the onshore lobe to the west. When no longer constrained by the Cape Cod Bay lobe, the onshore ice sheet began flowing to the southeast.

Merrimack Terrane

The Merrimack terrane, lying north and west of the Clinton-Newbury fault zone, extends westward from the Atlantic Ocean near Salisbury some 70 miles to Athol in central Massachusetts. I-495 passes over the Rockingham belt of the Merrimack terrane between Salisbury and Lawrence. The predominant rocks are the Newburyport batholith, which intruded the terrane in late Silurian time, and the Berwick and Eliot metamorphosed sedimentary formations, also of Silurian age.

View looking northwest at east-dipping beds of the Berwick formation. At ramp to Massachusetts 28 from the southbound lane of Massachusetts 213 in Methuen near the New Hampshire border. —Carol Skehan photo

The strata and the elongate plutonic bodies in the Rockingham belt, such as the Chelmsford granite, trend to the northeast. Many of these rock units have been chopped off at the Clinton-Newbury fault zone where they abut the east-northeast-trending formations of the Nashoba terrane.

Salisbury sits on the Newburyport batholith. A late-phase pluton of the batholith was dated at 418 million years, near the Silurian-Devonian time boundary. The Newburyport batholith has several igneous rock types: a gray, medium-grained granite with large crystals of microcline; a granite porphyry; a gray, medium-grained granodiorite; a biotite-rich quartz diorite; and a tonalite. Tonalite is somewhat darker and more mafic in composition than granite. The quartz diorite contains inclusions of schist. You can see inclusions aligned like a school of fish in outcrops on Seabrook Road on either side of the Massachusetts–New Hampshire border and within the circle of roads on Rings Island on the north shore of the Merrimack River just east of U.S. 1.

I-495 crosses the Eliot formation, which is composed of phyllite, a low-grade metamorphic rock, for about 1 mile east of Merrimacport. I-495 between Merrimacport and Lawrence crosses the Berwick formation, a muscovite schist composed of metamorphosed thin- to thick-bedded limy sandstone and siltstone. The Eliot and Berwick formations are older than the 418-million-year-old Newburyport rocks that intrude them.

View to the northeast from University Avenue Bridge over the Merrimack River in Lowell. The broad pavement of Berwick formation, called Pawtucket Falls, is a folded sequence of calcareous sandstone, siltstone, and shale overturned to the southeast. The dam in the right distance is part of the Lowell National Historic Park.

Clinton-Newbury Fault Zone

The east-northeast trending, northwest-dipping Clinton-Newbury fault zone extends from Newbury on the coast to well inland at Clinton and then into southern Connecticut. I-495 runs parallel to the fault zone, crossing it in North Andover and Lawrence, a few blocks south of the Merrimack River. The latest movements on this fault zone probably took place toward the end of the Paleozoic era during the collision of continental blocks that marked the assembly of the Pangaean supercontinent. The Clinton-Newbury fault zone abruptly truncates the south end of the formations and plutons of the Rockingham belt.

Nashoba Terrane

I-495 passes into the Nashoba terrane after crossing the Clinton-Newbury fault in North Andover. The latest phase of the Andover granite, which intruded the Nashoba terrane about 412 million years ago in early Devonian time, underlies much of our route between Andover and Lowell.

Because the pink to buff Andover granite contains a lot of muscovite and biotite mica, it probably formed from the melting of deeply buried clay-rich sedimentary rocks. Clay and mica are chemically similar. The subduction of the Avalon terrane beneath the Nashoba terrane melted the sedimentary rocks plunging into the subduction trench, creating the magma that became the Andover granite.

The Nashoba formation, which consists of interlayered gneisses and sulfidic schists, underlies I-495 near Lowell. These schists, which contain sillimanite, an alumina-rich mineral, were probably once sulfidic black shales. Weathering and erosion from a granite-rich terrane produced sediments that were then deposited in a low-oxygen, marine environment. The shales were subsequently buried to great depths at high pressure, probably when sheets of rocks were thrust over them.

The Tadmuck Brook schist of early Silurian age has been traced only as far northeast as Lowell. It is superbly displayed farther to the southwest between Chelmsford and Shrewsbury.

Dracut Pluton and Nickel Mine Hill

The 4-mile-broad Dracut pluton near Dracut is Silurian in age and intrudes the Berwick formation of the Merrimack terrane. The Clinton-Newbury fault zone truncated and deformed the pluton, and its shape suggests that the north side of the fault moved northeast relative to the south side.

The dark rock with the euphonic name, Dracut diorite, is actually norite, a gabbro that contains the mineral hypersthene. The name *Dracut diorite* is not a total misnomer, though. The pluton varies in composition, and in places is a fine-grained hornblende diorite. Dracut was once the center of the Pawtucket Indian tribe, whose chief, Passaconaway, was friendly to the white settlers.

The Nickel Mine Hill, north of Methuen Street in easternmost Dracut and a half-mile north of the Merrimack River, was among the earliest mining ventures in the United States. It may have been worked as early as the graphite mines at Sturbridge in south-central Massachusetts, which were mined beginning in about 1640.

Adit to the Nickel Mine in the Dracut pluton. The first shaft was 8 feet wide and 43 feet deep; when reopened in 1876, miners deepened the shaft to 61 feet.
—Carol Skehan photo

North face of the highly jointed and faulted Dracut pluton in the quarry at Nickel Mine Hill, north of Methuen Street in easternmost Dracut. —Carol Skehan photo

During the Revolutionary War, Joseph Varnum extracted iron ore for cannonballs from the Dracut area. Nickel was found there in 1875, and in 1881, the Dracut Nickel Mine established stamping and refining mills. The nickel was extracted from the minerals pyrrhotite, pentlandite, and chalcopyrite. Today, crushed stone products are quarried for the production of asphalt and concrete.

Chelmsford Granite

The 430-million-year-old Chelmsford granite is one of the best-known architectural stones in the northeast. Masses of granite are elongated in a northeasterly direction, probably formed during the collision between the Nashoba and the Merrimack microcontinents in Silurian time. The granite is typically striped with flowing bands of alternating dark clusters of biotite mica and light-colored, silvery muscovite mica, quartz, and feldspar. Flow bands formed from remnant layers or lenses of sedimentary rock within magma. The lenses are thoroughly transformed into igneous rock

View looking west at the LeMasurier Brothers Quarry, circa 1959. Nearly horizontal joints, or natural fractures, mark levels utilized by the quarry workers in extracting blocks of Chelmsford granite. Note the north-dipping flow bands below the derrick and to right of the author. —John LeMasurier Sr., photo

but preserve the direction of flow within the magma chamber. The tectonic deformation of the Chelmsford granite probably happened about 370 million years ago during the Acadian mountain building event.

The Chelmsford granite splits readily along desired directions because it was stressed during mountain building activity. Just as you may tear a piece of paper neatly by first creasing it, the Chelmsford granite is tectonically predisposed to split along smooth planes with a minimum of drilling.

The Chelmsford granite is exceptionally well suited for use as curbstone because the light gray minerals reflect light from automobile headlights. Quarry workers can extract this quartz-rich granite by applying a hot flame to it. The flame heats and expands the quartz and essentially cuts the stone.

A polished slab of light gray Chelmsford granite from the Fletcher Quarry in Westford township is on display at Weston Observatory. The slab captures a spectacular event in the granite's explosive history. I believe the granite had largely crystallized when it was intruded by a nearly contemporaneous pegmatite dike. Hot, silica-rich gaseous fluids associated with

Slab of Chelmsford granite with contemporaneous pegmatite along the left side. Large feldspar crystals were ripped apart by gaseous fluids that explosively escaped along the pegmatite conduit. Flow bands in upper right corner of granite dip down to right.

the crystallizing pegmatite explosively escaped to the surface via the pegmatite conduit. The pressurized gas and fluids ripped the coarse, 3- to 4-inch-long potash feldspars apart at right angles to their length. Quartz then filled the 1-millimeter-thick cracks.

The magma chamber acted like a pressure cooker that was suddenly cracked, and the enormous explosion of gas and fluids ripped apart everything in its path.

Glacial Lake Shawsheen-Merrimack

Glacial Lake Shawsheen-Merrimack occupied the valley of the Merrimack River to about Merrimacport and the valley of the north-flowing Shawsheen River, which enters the Merrimack River at North Andover. Since northeastern Massachusetts has few well-defined valley walls, meltwater lakes were not as continuous, widespread, nor as long-lived as other meltwater lakes in eastern Massachusetts. Varved sediments, representing thirty years of lake bottom deposition, are present in North Andover, and lake bottom deposits are present in northwestern Methuen near the New Hampshire line. Some people attribute these deposits to a body of water called Glacial Lake Methuen. This lake was probably an arm of the much larger lake of regional proportions, Glacial Lake Shawsheen-Merrimack.

Varved lake bottom sediments in the Shawsheen River valley supplied clay for a nineteenth-century brickwork industry. The rock-flour clay was of poor quality for ceramics, though, so the business closed.

Merrimack River

The Merrimack River, the fourth largest river in New England, flows south from New Hampshire into the Chelmsford area. At the town boundary between Lowell and Chelmsford, it turns abruptly northeast and flows nearly parallel to and along the Clinton-Newbury fault as far as Lawrence. Here, the river winds along a northeasterly course across the Merrimack terrane to the sea at Newburyport.

The preglacial Merrimack River may have continued southeast at Lowell and entered the sea at Boston. Geologist Irving Crosby prepared a map of buried bedrock valleys during his search for groundwater aquifers in the glacial sands and gravels of eastern Massachusetts. Preglacial rivers formed these valleys, but in some places glacial meltwater streams buried them with sands and gravels. Crosby inferred from bore samples and seismic data that the ancient Merrimack Valley originally followed a course south to Woburn, where it was at least 60 feet deep, and on to Boston Harbor, where the valley was about 200 feet deep. The Mystic Lakes, Spy Pond, and Fresh Pond lie along the course of this buried valley.

We don't know exactly why the Merrimack River abandoned its south-easterly course to Boston for its present route to the sea at Newburyport. Perhaps glacial material dammed the valley. Once diverted in a northeasterly direction, the river followed the fractured rocks of the Clinton-Newbury fault zone until it found the softer rocks of the Rockingham belt near Lawrence.

Glacial outwash deposits within the buried bedrock valley provide enormous economic resources for the region. Wells that supply Lowell, Winchester, and Woburn with abundant groundwater are situated in such deposits. These sediments, formed of hard, durable fragments of crystalline igneous and metamorphic rocks, are mined for concrete aggregate and other building purposes.

Interstate 495
Chelmsford—Westborough
32 miles

The Nashoba terrane, a lens sandwiched between the Clinton-Newbury fault zone to the west and the Bloody Bluff fault zone to the east, extends from Newbury on the Atlantic coast north of Boston to Chester, Connecticut, near Long Island Sound. I-495 crosses the Nashoba terrane between Chelmsford and Hudson, tracing a course parallel to and about 1 mile southeast of the Clinton-Newbury fault zone. I-495 heads south at Hudson, crosses the Bloody Bluff fault zone and Burlington mylonite, and passes into the Avalon terrane near Westborough. Glacial deposits cover much of the surface, but many roadcuts expose bedrock along the route.

Clinton-Newbury Fault Zone
The Clinton-Newbury fault zone separates the Nashoba and Merrimack terranes. The Nashoba microcontinent slid under the Merrimack microcontinent, presumably along an oceanic subduction zone. This relationship makes the Clinton-Newbury fault zone one of the more interesting tectonic features in New England because it provides a view of the deep interior of an oceanic trench.

The fault has since experienced several episodes of intense shearing, folding, and metamorphism since it first developed in Silurian time. The rocks of the deep subduction zone eventually migrated back to the earth's surface by about Mississippian time. Much later, after late Paleozoic time, the

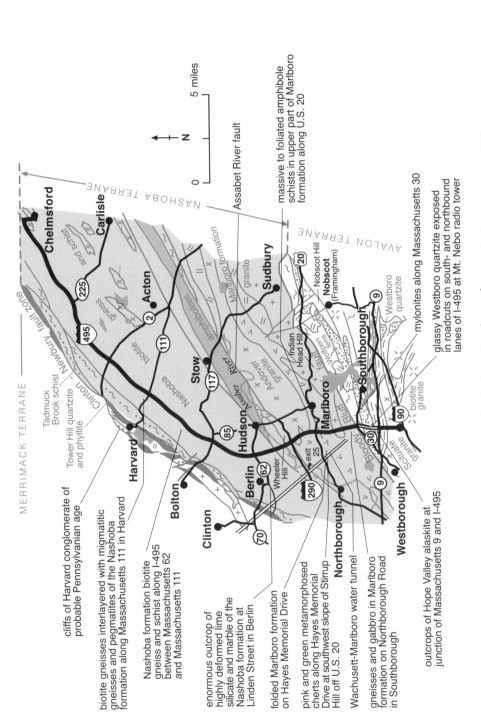

MERRIMACK TERRANE

NASHOBA TERRANE

AVALON TERRANE

Chelmsford

Carlisle

Acton

Stow

Sudbury

Harvard

Hudson

Marlboro

Southborough

Berlin

Bolton

Clinton

Northborough

Westborough

225

495

2

111

111

117

85

62

290

70

20

9

9

90

30

Nobscot Hill

Nobscot
(Framingham)

exit 25

Wheeler
Hill

Indian
Head Hill

Bloody
Bluff

gneiss

biotite

gneiss

Nashoba

amphibolites

Marlboro formation

Assabet River

granite

Andover
granite

Wolpe
Wolpe Wolpe

Westboro
quartzite

Scituate
granite

biotite
granite

and schists

Newbury fault zone

Clinton — Newbury fault zone

Tadmuck
Brook schist

Tower Hill quartzite
and phyllite

Assabet River fault

massive to foliated amphibole
schists in upper part of Marlboro
formation along U.S. 20

N

0   5 miles

cliffs of Harvard conglomerate of
probable Pennsylvanian age

biotite gneisses interlayered with migmatitic
gneisses and pegmatites of the Nashoba
formation along Massachusetts 111 in Harvard

Nashoba formation biotite
gneiss and schist along I-495
between Massachusetts 62
and Massachusetts 111

enormous outcrop of
highly deformed lime
silicate and marble of the
Nashoba formation at
Linden Street in Berlin

folded Marlboro formation
on Hayes Memorial Drive

pink and green metamorphosed
cherts along Hayes Memorial
Drive at southwest slope of Stirrup
Hill off U.S. 20

Wachusett-Marlboro water tunnel

gneisses and gabbro in Marlboro
formation on Northborough Road
in Southborough

outcrops of Hope Valley alaskite at
junction of Massachusetts 9 and I-495

mylonites along Massachusetts 30

glassy Westboro quartzite exposed
in roadcuts on south- and northbound
lanes of I-495 at Mt. Nebo radio tower

Geology along I-495 between Chelmsford and Westborough. —Modified from Zen and others, 1983

View northeast of a relatively treeless drumlin, circa 1925. This is probably Birch Hill, now covered with trees, 1 mile north of Gleasondale along Massachusetts 62 between Hudson and Maynard. —W. C. Alden photo, U.S. Geological Survey

rocks were cool enough to break rather than flow ductilely, and brittle faults formed.

Rocks of the western Nashoba terrane along I-495 were pulled down to depths of at least 10 miles in the subduction zone, then returned to the surface. Their minerals tell the story of this journey. The original volcanic and sedimentary rocks of the Nashoba microcontinent recrystallized as they sank into regions of high temperature and pressure, transforming the original rocks into metamorphic ones as new minerals formed.

Tadmuck Brook Schist

The shearing between the Nashoba and Merrimack terranes produced the broad zone of the Tadmuck Brook schist. The schist is distinctive, a dark gray rock that looks like a silky slate because it is full of minute flakes of mica that confer the sheen. But this rock is more complicated than just that. It contains crystals of gray sillimanite grown on crystals of pink andalusite as much as 18 inches long. Andalusite and sillimanite have the same chemical composition but they crystallize under different conditions of temperature and pressure. It seems that the andalusite crystals grew while the rocks sank quickly deep beneath the Merrimack terrane into regions of high pressure. The rocks then absorbed enough heat to raise their temperature to a level at which sillimanite could form.

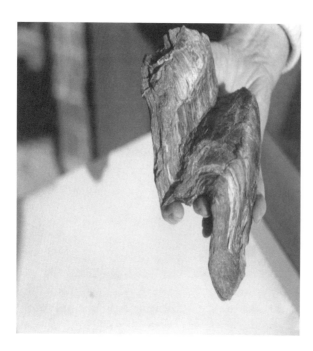

Intensely folded and sheared Tadmuck Brook schist.

Nashoba Formation

The Nashoba formation, to the east of the Tadmuck Brook schist, consists of layers of metamorphosed volcanic rocks in the east and sedimentary rocks to the west. The volcanic rocks are rich in biotite and hornblende. The sedimentary rocks probably began their careers as sediments eroded from a volcanic chain during Cambrian or Ordovician time, quite possibly the same chain that produced the Marlboro volcanic rocks. The layers of the Nashoba formation—dark and pale mica schists, quartzite, marble, and very dark amphibole schists—all dip steeply down to the west. The abundance of mica and sillimanite indicates that the original sediment was rich in aluminum, and was perhaps a clay.

I-495 follows the trend of those tilted and strongly metamorphosed layers between Chelmsford and Hudson. You can see biotite gneisses and schists along I-495 between Massachusetts 111 and Massachusetts 62. At an outcrop along Massachusetts 111 in Harvard, look for large flakes of muscovite in the schists. You can also see sillimanite and biotite along the edges of rock heated by intruding granite and pegmatites. Granitic magma, apparently from the Andover batholith, intimately intruded them to make a mixed metamorphic and igneous rock—a migmatite. The dark parts are the metamorphic component, and the pale parts, the granite.

Dark migmatitic gneiss of the Nashoba formation intruded by pink to buff Andover granite at junction of I-495 with Massachusetts 117 in Stow.

Assabet River Fault

Near Hudson, I-495 turns southeast and passes over the Assabet River fault just north of the junction with I-290. The Assabet River follows the northeast-trending fault, which cuts across and omits a part of the eastern Nashoba formation to the south. The fault forms the boundary between the Nashoba and Marlboro formations between Hudson and Shrewsbury. You can see a splay of the fault in the exit ramp to Massachusetts 62 from I-495 north. Mylonites with large feldspar crystals occur within some splays of the fault. Large 0.5- to 1-inch diameter, rounded to S- or Z-shaped feldspar crystals grew in the mylonite during deformation.

At the junction of I-290 and I-495, a superb and enormous set of outcrops contain fault breccia created as movement along the fault fractured the brittle rocks into angular fragments. Light-colored calcite mineralization in the fractures throws the dark breccia into high relief. *Breccia* literally means "broken" in Italian.

Marlboro Volcanic Rocks

I-495 crosses Marlboro volcanic rocks between the junction with I-290 and the junction with the Wachusett Aqueduct in Southborough near Massachusetts 30. These volcanic rocks consist mainly of dark green to black amphibole schists. You can see folded Marlboro formation along Hayes Memorial Drive south of U.S. 20. Amphibole schists from the upper part of the formation surface along I-495 at the junction with U.S. 20. Gneisses and gabbro of the Marlboro formation surface along Northborough Road in Southborough Township.

The volcanic rocks include beautiful, thinly layered, pink and green quartzite. The pink quartzite consists mostly of minute grains of quartz and pink, manganese-containing garnet; the green quartzite consists of quartz and green epidote. These layers were probably clayey cherts on the ocean floor near active volcanoes that periodically spewed basalt lava over the top of the cherts.

The 412-million-year-old Andover granite, the 400-million-year-old Straw Hollow diorite, and the 349-million-year-old Indian Head Hill granite intrude the Marlboro formation. The most prevalent, the Andover granite, is a pale pink, generally foliated rock. It contains white mica and garnet, very coarse feldspar pegmatites, and gray or pink fine-grained granitic dikes. Granites of this sort probably form through melting of deeply buried sedimentary rocks.

Bloody Bluff Fault Zone

I-495 crosses the Bloody Bluff fault, which separates the Nashoba and Avalon terranes, just south of the Wachusett Aqueduct and Conrail tracks in Southborough. The Bloody Bluff fault zone formed toward the end of

Layered rocks of the Marlboro formation. The light-colored layers are distinctive, colorful quartzites. Specimen is from the Cosgrove Tunnel, formerly called the Wachusett-Marlboro tunnel, in Northborough.

Paleozoic time. It follows the subduction zone between the Nashoba terrane and the Avalon terrane but is much younger than, and is not directly related to, the subduction.

The Wolfpen lens, south and east of Marlboro, appears to be a piece of Avalon terrane that is trapped between splays of the Bloody Bluff fault zone. It is difficult to determine the history of such a fault-bounded block. Outcrops of the Wolfpen lens are not exposed along I-495, but you can see it near Massachusetts 30 in Southborough and near U.S. 20 in Marlboro.

Burlington and Nobscot Mylonite Zones

The western margin of the Avalon terrane along I-495 consists mainly of mylonitized, light-colored rocks. The Burlington mylonite zone of intensely sheared, layered, and ground-up rocks is commonly over 3 miles wide. It formed at depth in the shear zone along the margin of the Gondwanan supercontinent. Lenses of quartzite, schist, basalt, and gabbro are preserved within the mylonite, giving clues to the kinds of rocks originally ground up in the subduction zone.

The Nobscot mylonite zone, up to 500 yards wide, lies within the broad Burlington mylonite zone and appears to have formed independently of the Burlington mylonite. The Nobscot mylonite not only cuts the foliation

Close view of sheared Milford granite showing two sets of crossing fractures. On Massachusetts 30 east of Southborough.

of the Burlington mylonite, but it also cuts the 610-million-year-old Milford granite. You can examine splendid exposures of mylonites and other rocks near Lookout Tower on Nobscot Hill in Framingham.

Along Massachusetts 30 just east of I-495, you can see magnificent mylonites with thin layers, streaky or banded structures, lens-shaped inclusions, and brown glassy, chertlike rocks that formed during earthquake movement. Near the junction of Massachusetts 30 and Johnson Road, 0.1 mile east of I-495, strongly mylonitized gabbro contains stretched and fractured feldspars up to 1 inch across.

Avalon Rocks

I-495 crosses bands of Precambrian Westboro quartzite and Milford granite as it approaches Westborough. At the Mount Nebo radio tower on the Northborough-Southborough township line on I-495, glassy Westboro quartzite is exposed in roadcuts along both the north- and southbound lanes.

Hope Valley alaskite, a gneissic phase of the 610-million-year-old Milford granite, contains few dark minerals but lots of quartz, including linear crystal rods of smoky quartz and magnetite. It is visible at the junction of I-495 with Massachusetts 9.

Interstate 495
Westborough—Norton
37 miles

The bulk of this trip crosses bedrock that formed in Precambrian time when the Avalon microcontinent sat alongside the Gondwanan supercontinent. An early collision sheared the continental shelf of Gondwana, and when the shearing was barely finished, enormous masses of molten granite from the Avalon volcanic chain moved into the continental crust and swallowed the shelf edge. These 610-million-year-old granites, the Milford and Dedham batholiths, encompass a large part of the Avalon terrane in Massachusetts.

Milford Granite

Rocks between Massachusetts 9 in Westborough Township and exit 18 in Bellingham Township are primarily the Milford batholith—strongly deformed granite of several phases. It is folded into an arch and sheared into a platy gneiss.

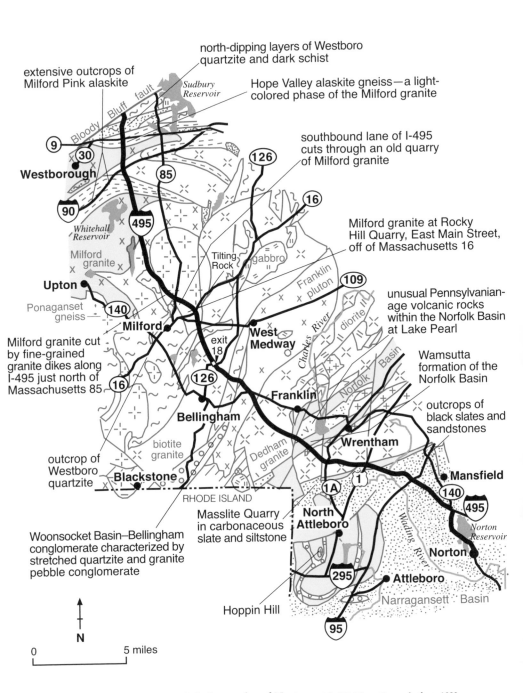

extensive outcrops of
Milford Pink alaskite

north-dipping layers of Westboro
quartzite and dark schist

Hope Valley alaskite gneiss—a light-
colored phase of the Milford granite

southbound lane of I-495
cuts through an old quarry
of Milford granite

Milford granite at Rocky
Hill Quarry, East Main Street,
off of Massachusetts 16

unusual Pennsylvanian-
age volcanic rocks
within the Norfolk Basin
at Lake Pearl

Wamsutta
formation of the
Norfolk Basin

outcrops of
black slates and
sandstones

Milford granite cut
by fine-grained
granite dikes along
I-495 just north of
Massachusetts 85

outcrop of
Westboro
quartzite

Woonsocket Basin–Bellingham
conglomerate characterized by
stretched quartzite and granite
pebble conglomerate

Masslite Quarry
in carbonaceous
slate and siltstone

Hoppin Hill

Sudbury
Reservoir

Bloody Bluff fault

9
30
Westborough
85
90
Whitehall
Reservoir
Milford
granite
Upton
Ponaganset
gneiss
140
Milford
16
Bellingham
biotite
granite
Blackstone
RHODE ISLAND

495
126
16
Tilting
Rock
gabbro
exit
18
West
Medway
126
Franklin
Dedham
granite
Charles River
Franklin
pluton
109
diorite
Norfolk Basin
Wrentham
1A
1
North
Attleboro
295
Attleboro
95

Mansfield
140
495
Norton
Reservoir
Norton
Narragansett Basin
Wading River

0   5 miles

N

Geology along I-495 between Westborough and Norton. —Modified from Zen and others, 1983

A glacially polished outcrop of Milford granite is the foundation for the Irish Round Tower in St. Mary's Cemetery, Milford. This tower, modeled after but smaller than the St. Kevin's Glendalough Tower in Ireland, was constructed in 1894 of Milford granite. —Dolores A. Skehan photo

You can see alaskite gneiss, a light-colored, quartz-rich phase of the Milford granite, in Westborough township at the junction of Massachusetts 9 and I-495, and also along I-90 west of I-495. Farther south, an outcrop of Milford granite cut by fine-grained granitic dikes surfaces along the southbound lane of I-495 just north of Massachusetts 85. The southbound lane continues through an old quarry near the junction with Massachusetts 85. The stone, known by the trade name Milford Pink, was quarried for architectural uses. Milford granite extracted from the quarry farther south at Rocky Hill was used for the Boston Public Library and for Pennsylvania Station in New York.

Roof Pendants

The 610-million-year-old Milford granite swallowed abundant blocks of Westboro strata and left chunks on the surface. These rocks, deposited as beach sands and muds on the continental shelf of Gondwana more than 650 million years ago, are the oldest constituents of the Avalon terrane in Massachusetts. The Westboro formation consists of interbedded quartzite and black schist. The massive to bedded, mainly glassy quartzite is white to gray on fresh surfaces and buff on weathered surfaces. From a distance, the quartzite appears similar to buff and pink varieties of Milford granite. A large outcrop of interbedded quartzite and black schist is well exposed in the southbound lane of I-495, 0.5 mile north of Massachusetts 9 at Mount Nebo on the Westborough-Southborough township line.

Franklin Pluton

The Franklin pluton separates the sheared Milford granite to the west from the less-deformed granites of the Dedham and Fall River batholith to the east. The 25-mile-long pluton extends from Millis to the Rhode Island border and is about 417 million years old, a date on the boundary of Silurian and Devonian time. The Franklin pluton, an alkaline granite similar to the Quincy granite south of Boston, appears to have formed during rifting of the Avalon terrane.

Dedham Batholith

Southeast of the Franklin pluton, I-495 traverses a narrow extension of the Dedham batholith, essentially the same age as the Milford, 610 million years old. Although chemically similar, the Dedham granite is less deformed

Tilting Rock along Massachusetts 16 in Holliston is a glacial erratic of granite characterized by worm-shaped quartz grains. Tilting Rock perches on the edge of a bedrock outcrop of a different type of granite—the Milford. This foliated phase of the Milford granite contains segregations of biotite and other dark minerals.

than the Milford granite. The 1,000-square-mile Narragansett coal basin covers much of the enormous Dedham batholith.

Norfolk Basin

The Norfolk Basin extends southwest to Franklin and Wrentham. This 2.5-mile-wide basin of Pennsylvanian age is filled with Wamsutta redbeds, which consist of red layers of conglomerate, interfingering with gray, black, and maroon layers of slate, and red crossbedded sandstone. In the southwestern part of the Norfolk Basin, basalt and rhyolite volcanic flows are deposited on an erosional surface of Westwood granite, a rock similar to the Dedham granite. An arch in the basin rocks exposes the basal volcanic rocks and the Precambrian Westwood granite. You can see these rocks near Lake Pearl.

Fossiliferous Cambrian Rocks

I-495 passes close to Hoppin Hill in North Attleboro, where early Cambrian rocks were deposited directly on deeply eroded Precambrian granite (possibly Dedham or Westwood granite)—perhaps the only place in eastern Massachusetts where people have seen this erosional contact. Though that contact, situated in a housing development at Hoppin Hill, is no longer accessible to the public nor still visible, you can see the granite and sedimentary rocks on either side of the contact on the western shore of the Hoppin Hill Reservoir.

The Cambrian quartzite and conglomerate grade upward into fossiliferous beds of green, maroon, and gray slates, also of Cambrian age. An unconformity separates these rocks from the overlying, and much younger, sedimentary rocks of Pennsylvanian age. The unconformity—a gap in the depositional rock record in which a large chunk of time is not represented, probably because of erosion—is angular. That is, the beds of the younger overlying rocks do not lie at the same angle as those of the underlying rocks. The younger rocks include the Wamsutta redbeds and volcanic rocks, all overlain by the Pondville conglomerate of the Narragansett Basin. A fault block permits us to see these Cambrian rocks at Hoppin Hill; elsewhere, Pennsylvanian strata of the Narragansett Basin bury them.

Narragansett Coal Basin

Between Wrentham and Norton, I-495 crosses the Narragansett coal basin, which contains Pennsylvanian-age rocks, about 315 million years old and the youngest in the Avalon terrane except for rocks of the Middleton Basin. The Narragansett Basin occupies a lowland area of almost 1,000 square miles, much of it swampland. The strata consist of a complete range of terrestrial sediments from conglomerate, sandstone, and siltstone to shale and coal. Our route passes through the one-time coal-mining towns of Mansfield and Norton, not far from the middle of the basin.

Mansfield Mines

Coal mining in the Narragansett Basin apparently began in 1736 with the Leonard's Mine in Mansfield, then a part of Norton. Five coal mines operated in Mansfield at various times between 1835 and 1923. The Hardon Mine produced 1,200 to 1,500 tons of coal in 1838, mostly used for steam engines and household heating, but closed for lack of funds. The lobby of the Massachusetts State House was heated for an entire week in 1839 with coal from the Hardon Mine while the General Court considered—and ultimately rejected—a bill to fund a project to reactivate the Hardon Mine. The Sawyer Mine, worked from a shaft sunk to 162 feet, produced 5,000 tons between 1848 and 1854. Exploration in the Sawyer Mine located thirteen coal beds, but only two of these were worked.

Masslite Quarry

The Masslite Quarry is on the northwestern margin of the Narragansett Basin in Plainville. Bird and Son Company quarried 440,000 tons of slate for roofing granules between 1934 and 1945. Some associated coal was used for heating. Between 1961 and 1982, Masslite Company quarried as much as 400 tons per day of mixed coal and carbonaceous shale. The rock was combined with fuel oil and ignited to fuse into a product used in making light aggregate for construction of tall buildings.

Interstate 495
Norton—Cape Cod Canal
40 miles

Norton, home to Wheaton College, nestles pleasantly in the northwestern part of the Narragansett Basin, a large coal basin of Pennsylvanian age. I-495 heads southeast across the basin. Near Middleboro the basin sediments thin and peter out, and the late Precambrian Fall River batholith emerges as the underlying bedrock. A drainage divide near Middleboro separates streams that flow into the Narragansett Basin, which is primarily drained by the Taunton River, from those that flow into Buzzards Bay.

Only a few outcrops of bedrock exist in the Taunton River and Buzzards Bay drainage basins. A broad and variably thick cover of glacial deposits

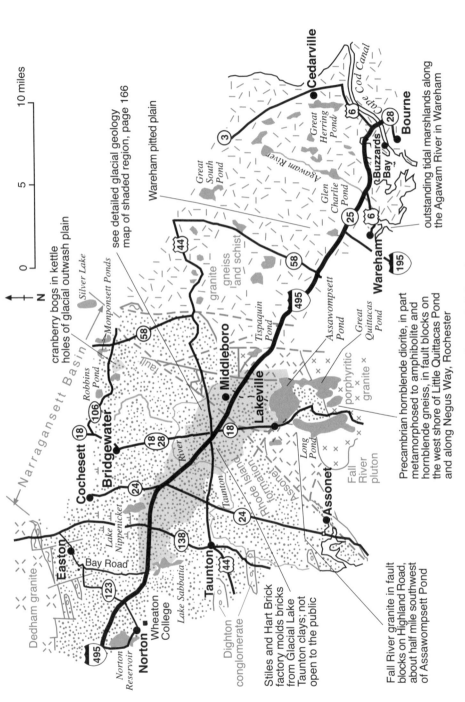

N

0   5   10 miles

cranberry bogs in kettle
holes of glacial outwash plain

see detailed glacial geology
map of shaded region, page 166

Wareham pitted plain

outstanding tidal marshlands along
the Agawam River in Wareham

Precambrian hornblende diorite, in part
metamorphosed to amphibolite and
hornblende gneiss, in fault blocks on
the west shore of Little Quittacas Pond
and along Negus Way, Rochester

Fall River granite in fault
blocks on Highland Road,
about half mile southwest
of Assawompsett Pond

Stiles and Hart Brick
factory molds bricks
from Glacial Lake
Taunton clays; not
open to the public

Narragansett Basin

Dedham granite

Easton

Bay Road

123

Wheaton
College

Norton

495

Norton
Reservoir

Cochesett

18

106

Bridgewater

24

18
28

Lake
Nippenicket

138

Lake Sabbatia

Dighton
conglomerate

Taunton

44

24

Taunton River

Rhode Island
formation

Assonet River

Assonet

24

Middleboro

18

Lakeville

18

Tispaquin Pond

Long
Pond

Fall River
pluton

porphyritic
granite

Assawompsett
Pond

Great
Quittacas
Pond

granite

gneiss
and schist

58

495

Wareham

195

25

6

28

6

Buzzards
Bay

Bourne

Agawam River

Glen
Charlie
Pond

Great
Herring
Pond

Cedarville

3

Great
South
Pond

Cape Cod Canal

44

fault

Robbins Pond

58

Monponsett Ponds

Silver Lake

106

Bedrock geology along I-495 between Norton and Cape Cod Canal. —Modified from Zen and others, 1983

drapes the bedrock. The continental ice sheet carried rocks and sediment that meltwater rivers later distributed into outwash plains and lake deltas. The slaty rocks of the Narragansett coal basin were soft, easy fodder for the bulldozing glaciers, but the much older rocks of the Fall River batholith resisted erosion and form a hilly landscape where they come to the surface between Middleboro and Bourne.

Taunton River Basin

The northern half of our route sweeps across the Taunton River basin, which extends from Foxborough to Middleborough. The basin is the second largest drainage area in Massachusetts, with an area of 1,053 square miles. With only a 20-foot difference in elevation along its 40-mile length, the Taunton River has one of the flattest gradients in the state. This gradient explains the extensive wetlands within the basin, including the 6,000-acre Hockomock Swamp, one of the largest wetlands in New England. The basin supports a great diversity of plants and animals, including the upland sandpiper, a bird that is in decline in its eastern range.

The Taunton River flows south into Mount Hope Bay. Salt water intrudes as far upstream as U.S. 44, about 12 miles upstream of the bay. Tidal effects are noticeable 18 miles upstream to about the Winnetuxet River in Halifax township.

Till

The glacial till in these parts comes in shades of pale gray, yellowish gray, and yellowish brown. It is a mixture of sand, gravel, and boulders, with minor amounts of clay and silt. Some of the boulders are as large as 20 feet across. The thickest till—up to 90 feet—is in streamlined drumlins. Most of the material in the till was picked up along the base of the ice sheet from sandstones of the Rhode Island formation in the Narragansett Basin, and from the granite, schist, and gneiss of the crystalline rocks of southeastern Massachusetts.

Glacial Lake Taunton

If you were to go back in time to southeastern Massachusetts about 15,000 to 14,000 years ago, you would need to take a canoe to cross the numerous glacial lakes. As the Buzzards Bay lobe of the ice sheet melted back to the north, a series of glacial lakes flooded parts of the region. The largest, Glacial Lake Taunton, evolved in two stages—an early, higher stage to the south and a later, lower northern stage.

Rivers of glacial meltwater swept sediments into the lakes and dropped them in flat deposits on the lakebed and in deltas. The upper parts of the

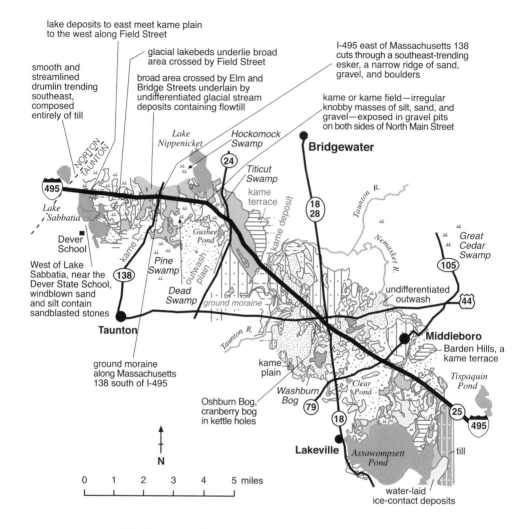

lake deposits to east meet kame plain
to the west along Field Street

glacial lakebeds underlie broad
area crossed by Field Street

smooth and
streamlined
drumlin trending
southeast,
composed
entirely of till

broad area crossed by Elm and
Bridge Streets underlain by
undifferentiated glacial stream
deposits containing flowtill

I-495 east of Massachusetts 138
cuts through a southeast-trending
esker, a narrow ridge of sand,
gravel, and boulders

kame or kame field—irregular
knobby masses of silt, sand, and
gravel—exposed in gravel pits
on both sides of North Main Street

NORTON
TAUNTON

Lake
Nippenicket

Hockomock
Swamp

24

Bridgewater

495

Titicut
Swamp

Taunton R.

Lake
Sabbatia

kame
terrace

18
28

kame deposit

Dever
School

Gushee
Pond

Nemasket R.

Great
Cedar
Swamp

West of Lake
Sabbatia, near the
Dever State School,
windblown sand
and silt contain
sandblasted stones

138

Pine
Swamp

kame

outwash plain

105

Dead
Swamp

ground moraine

undifferentiated
outwash

44

Taunton

Taunton R.

Middleboro

ground moraine
along Massachusetts
138 south of I-495

kame
plain

Barden Hills, a
kame terrace

Washburn
Bog

79

Clear
Pond

Tispaquin
Pond

Oshburn Bog,
cranberry bog
in kettle holes

18

25

495

N

Lakeville

Assawompsett
Pond

till

0   1   2   3   4   5 miles

water-laid
ice-contact deposits

Glacial geology along I-495 between Norton and Middleboro.
—Modified from Hartshorn, 1967; Koteff, 1964

deltas typically consist of tilted foreset beds deposited on nearly horizontal layers of sand, gravel, or lake bottom sediments.

The early stage of Glacial Lake Taunton formed when the receding ice front was about 2.5 miles northeast of Taunton and just northeast of Pine Swamp. During this stage, a kame delta formed just north of Raynham Center where the lake lapped against ice. East-dipping foreset beds indicate that part of the lake lay between Pine Swamp and Massachusetts 24. Massachusetts 24 passes through the middle of the southern stage. The spillway was near Fall River and drained into Mount Hope Bay.

Slanting foreset beds in the kame delta east of Pine Swamp, west of Massachusetts 24 and north of Taunton. The section is about 20 feet high. —J. H. Hartshorn photo, U.S. Geological Survey

Glacial lake varves 1.4 miles south of Taunton Center. The pale layers record summers of the last ice age, the dark layers, winters. —J. H. Hartshorn photo, U.S. Geological Survey

I-495 crosses the floor of the northern, late-stage Glacial Lake Taunton. The receding ice lobe had deposited kame deltas near Brockton and Bridgewater. During this stage, the glacial lake extended from west of Hockomock Swamp and Lake Sabbatia east to Lake Nippenicket, Monponsett Pond, and Silver Lake. The Jones River formed the glacial lake's spillway. The Stiles and Hart Brick factory just west of Middleboro molds bricks from the lake bottom clay deposited during the late stage of Glacial Lake Taunton. It is the only remaining brick factory in operation in Massachusetts.

Hockomock Swamp

The postglacial landscape must have looked like parts of the Canadian Shield, probably complete with a luxuriant population of black flies and mosquitoes. If your experience with swamps and marshlands has been mainly during the buggy season, you may have overlooked some of the myriad wonders of this interesting environment.

The Hockomock Swamp of Taunton, Easton, West Bridgewater, and Bridgewater Townships is one of the largest swamps in the lowlands of the Narragansett Basin. A souvenir of the ice age, Hockomock Swamp exists in the lowland of the northern, late stage of Glacial Lake Taunton. Peat bogs develop in the swamp as decaying vegetation accumulates on the foundation of glacial deposits.

Assonet Fault

The Assonet fault, the northern continuation of the Beaverhead fault of Rhode Island, runs along the southeastern margin of the Narragansett Basin. The jagged boundary of the basin is due to brittle faulting in the

Section of bedrock hill covered with ground moraine, kame terraces, and glacial lake bottom sediments near Dean Street, on the Norton-Easton township line, and Hockomock Swamp. —Modified from Hartshorn, 1967

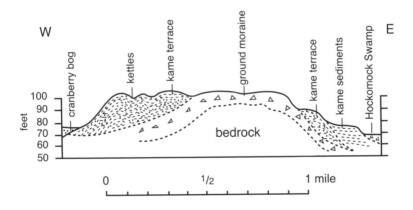

Pennsylvanian-age formations and underlying granite, probably during rifting, about 200 million years ago in Mesozoic time. I-495 crosses the fault just west of Middleboro. To the east of the fault, the buried bedrock is part of the Fall River batholith, which you can see about 0.5 miles southwest of Assawompsett Pond, and on the west shore of Little Quittacas Pond, which is south of Assawompsett Pond.

Buzzards Bay Basin

I-495 between the center of Middleborough Township and the Cape Cod Canal crosses the Buzzards Bay drainage basin, 380 square miles that supply water to seven coastal rivers and a number of small streams that flow into Buzzards Bay. The Weweantic River is the longest at 15.6 miles and the Wareham River the shortest at 2.1 miles. The lower valleys, now inundated with seawater, form a sawtoothed coastline with many harbors and coves important for both commercial fisheries and recreation. The elevation of the basin averages about 100 feet, and the sandy soils are well drained. Kettle ponds and lakes dot the landscape, and freshwater and saltwater marshes abound.

One Great Sand Plain

Simply put, southeastern Massachusetts is one great sand plain. I-495 crosses a large area where torrents of glacial meltwater dumped loads of sediment as the last ice age ended. The Wareham pitted outwash plain, extending in a triangular area from south of Plymouth to West Wareham and Bourne, formed between the Buzzards Bay moraine to the south and the receding ice sheet. The Myles Standish State Forest, an expanse of pine barrens, encompasses much of the Wareham outwash plain.

Kame hills and kettle ponds punctuate the sand and gravel plains. The kames are deposits of outwash sediment that were laid down along the front of the melting glacier, then slumped into small hills as the ice that supported them finally disappeared. Most of the kettles mark places where large pieces of ice were buried in the outwash, then melted. They left depressions that now commonly hold water. Assawompsett Pond, a gigantic kettle south of I-495, is the largest natural body of fresh water in Massachusetts.

Rain percolates too rapidly through the sandy soils to be fully available to plants. Only species adapted to low-nutrient, acidic, and drought conditions can survive on these sand plains. Nevertheless, a great variety of plant and animal species finds the pine barrens congenial. Occasional fires help maintain the natural environment.

Kettle holes in outwash plains make ideal locations for cranberry bogs. The Oshburn Bog, west of Middleborough where I-495 crosses U.S. 44, is a collection of cranberry bogs in kettle holes.

Sandblasted pebbles—ventifacts—with polished faces that have an orange-peel texture.

Wind Deposits

The wind picked up sand and silt from the outwash plains and blew it across southeastern Massachusetts to a depth of several feet. The deposits, mostly fine to medium sand, are commonly oxidized to shades of pale brown, yellow, and reddish brown. They also contain sandblasted stones.

Severe frost action at the end of the ice age stirred these windblown sediments so much that in places they now resemble the glacial deposits below them. The cold wind sandblasted numerous boulders, cobbles, and pebbles that the freeze-thaw activity heaved to the surface, scouring their surfaces into concave facets that meet in sharp edges and corners. Many sandblasted stones—ventifacts—have a polished surface with an orange-peel texture. They were once considered man-made artifacts, but eventually people recognized their true origin. You can see windblown sand and silt deposits up to 6 feet thick that contain sandblasted ventifacts in sands pits west of Lake Sabbatia near the Dever State School.

Bourne

Bourne stands on a thick blanket of glacial sands and gravels that covers Precambrian granitic gneisses. The Bourne Bridge, where Massachusetts 28 crosses the Cape Cod Canal, is on the western edge of the spectacular right-angle junction of the Sandwich and Buzzards Bay moraines. Aptucket Trading Post Museum, on the south side of the Bourne Bridge, features a trading post that dates to 1627, possibly the first on the continent. The museum's many historic relics include a runestone. This 200-pound-slab of rock, the Bourne Stone, is believed to have been found near the site of the Cape Cod Canal. Historians have attributed the inscriptions on the runestone to various people: Phoenicians, Norsemen, and the Native American Sachem Wamsutta.

Massachusetts 2
Cambridge—Leominster
40 miles

Massachusetts 2 between Cambridge and Leominster passes through three great blocks of the earth's crust—the Avalon, Nashoba, and Merrimack terranes. Each terrane experienced a different geological history prior to its final amalgamation with the Laurentian continent in Paleozoic time. Massachusetts 2 takes you across a multitude of ancient rocks as well as by Thoreau's Walden Pond and historic sites commemorating American Indians, colonial times, and the Revolutionary War.

Boston Basin

Our trip begins in the Boston Basin, a Precambrian rift basin within the Avalon volcanic island chain. The Boston Basin filled with volcanic and sedimentary rocks in late Precambrian time. Cambridge slate, a very fine-grained, dark green, layered metamorphic rock, underlies the northern part of the basin at Cambridge. Thick glacial deposits cover most of the bedrock of the northwestern Boston Basin lowland. The Cambridge slate is not exposed near Massachusetts 2, but you can see it in cliff exposures near I-93 in Somerville.

Fresh Pond

Fresh Pond, a kettle pond near Fresh Pond Circle and Shopping Mall along Massachusetts 2, is tucked away in the northwestern corner of Cam-

East-west cross section of the bedrock geology along Massachusetts 2 between Cambridge and Leominster. —Modified from Zen and others, 1983

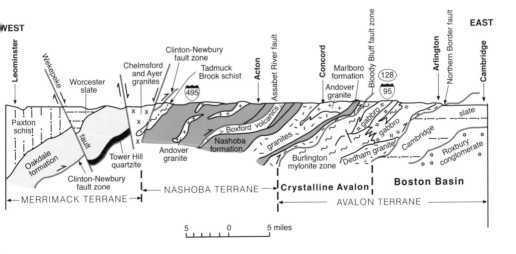

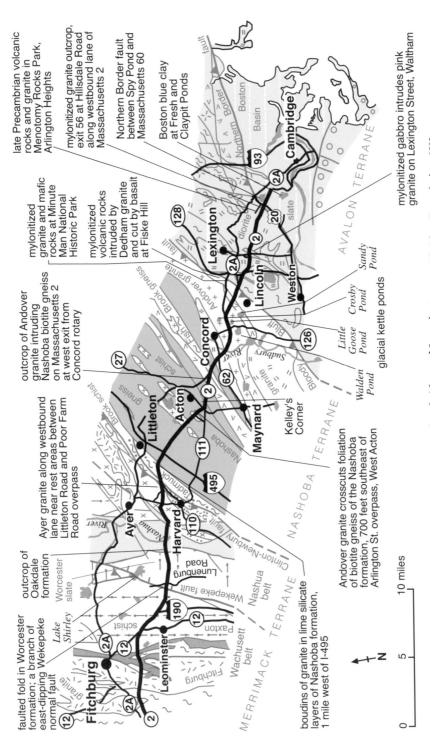

late Precambrian volcanic rocks and granite in Menotomy Rocks Park, Arlington Heights

mylonitized granite outcrop, exit 56 at Hillsdale Road along westbound lane of Massachusetts 2

Northern Border fault between Spy Pond and Massachusetts 60

Boston blue clay at Fresh and Claypit Ponds

mylonitized gabbro intrudes pink granite on Lexington Street, Waltham

mylonitized granite and mafic rocks at Minute Man National Historic Park

mylonitized volcanic rocks intruded by Dedham granite and cut by basalt at Fiske Hill

outcrop of Andover granite intruding Nashoba biotite gneiss on Massachusetts 2 at west exit from Concord rotary

Ayer granite along westbound lane near rest areas between Littleton Road and Poor Farm Road overpass

outcrop of Oakdale formation

faulted fold in Worcester formation; a branch of east-dipping Wekepeke normal fault

Andover granite crosscuts foliation of biotite gneiss of the Nashoba formation, 700 feet southeast of Arlington St. overpass, West Acton

boudins of granite in lime silicate layers of Nashoba formation, 1 mile west of I-495

Geology along Massachusetts 2 between Cambridge and Leominster. —Modified from Zen and others, 1983

bridge in the northwestern Boston Basin lowland. Fresh Pond is at the southern end of a series of kettle lakes extending north through Spy Pond, the Mystic Lakes, and ponds in the high ground in Woburn. Borings and seismic profiles indicate the presence of a preglacial bedrock river channel in the Fresh Pond–Mystic Lakes valley. Although far from certain, geologists believe this may have been the valley of the preglacial Merrimack River. The present-day Merrimack River turns northeast near North Chelmsford. The southern margin of Fresh Pond is a 2,000- to 3,000-foot-wide moraine that the ice sheet pushed into a curved mound. Harvard University Smithsonian Astronomical Observatory sits on the crest of the moraine.

Seismic profiles indicate that glacial lake bottom clay occupies the buried river channel. A glacial lake may have lapped against the highlands northwest of Fresh Pond. Till directly overlies these clays and is in turn overlain by outwash sands and gravels.

Drumlins and the approximate location of a large buried bedrock valley in the greater Boston region. —Modified from Crosby 1939; Chute, 1959

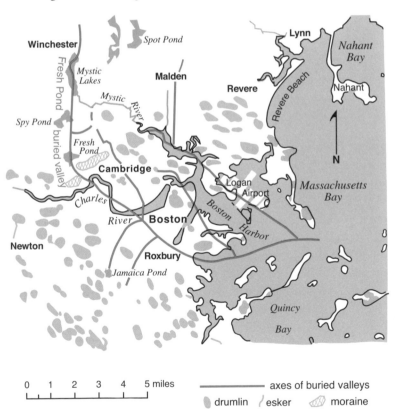

0 1 2 3 4 5 miles

——————— axes of buried valleys

drumlin / esker / moraine

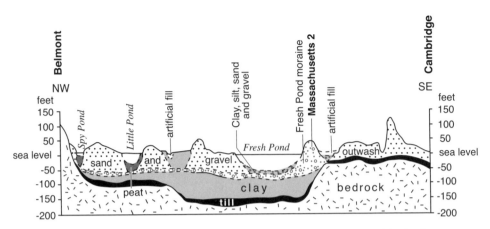

Cross sections from Belmont to Cambridge showing the buried bedrock valley beneath and northwest of Fresh Pond. —Modified from Chute, 1959

Seawater deposited a distinctive glacial and postglacial marine sediment, the Boston blue clay, on top of the glacial outwash sands and gravel. Boston blue clay from Claypit Pond, on Concord Road northwest of Fresh Pond, and from other sites was used in brick making. Burial sites of the Beothuk Indians of Newfoundland have been excavated from the blue clay at Fresh Pond.

Northern Border Fault at the Highlands

On a clear day you can see a magnificent view of Cambridge, downtown Boston, and Boston Harbor from the highlands of Arlington and Belmont. These highlands are Precambrian volcanic and plutonic rocks of the Avalon terrane that were thrust over the Cambridge slate of the Boston Basin along the Northern Border fault. The break in slope where the highlands fall away and meet the Cambridge lowland marks the trace of the west-dipping Northern Border thrust fault. Spy Pond is on the east side of the fault and Massachusetts 60 on the west side. You can trace the thrust fault along the base of the rocky slope for several miles both east and west.

The rocks between the Northern Border fault in Belmont and Arlington and the Bloody Bluff fault zone in Waltham and Lexington consist dominantly of gabbro. These dark plutonic rocks intruded the Avalon terrane in Silurian time. Just west of the Northern Border fault, gabbro-diorite of Devonian age intruded the 610-million-year-old Dedham granite and still-older, light-colored, mylonitized quartzite and other Precambrian rocks.

Menotomy Rocks Park

Menotomy Rocks Park in Arlington Heights features some unusually interesting rocks. To get there, take the Massachusetts 60 (Pleasant Street) exit just west of Spy Pond and go west on the north side of Massachusetts 2 to Hillsdale Road. Turn right on Hillsdale, left on Gray, and then left on Churchill to the park entrance near Hills Pond. A jumble of rocks from a late Precambrian volcanic crater of the Avalon terrane includes fine-grained pink granite crosscut by a brown-weathering dike of felsite—an extremely fine-grained, light-colored rock. Basalt dikes crosscut lumpy, fragmental felsic lavas.

Burlington Mylonite Zone

West of the Northern Border fault, Massachusetts 2 crosses a 1.5-mile-wide zone of highly sheared, alternating dark and light rocks called mylonites. The Burlington mylonite zone formed during major faulting along the edge of the Gondwanan supercontinent. The mylonite contains intensely sheared but recognizable lenses of quartzite and schist from the Gondwanan continental shelf, as well as basalt and gabbro. These residual blocks give us an idea of what the rocks were before they were ground up. The 610-million-year-old Dedham granite of the Avalon volcanic island chain intrudes the Burlington mylonite zone in many places west and north of Boston, so we know the mylonite is slightly older than 610 million years.

Mylonitized dark quartzite in the Burlington mylonite zone at the corner of Hillsdale and Spring Streets, just west of junction of Massachusetts 2 and Massachusetts 60 in Belmont.

Bloody Bluff Fault Zone

Do not confuse the Bloody Bluff fault zone with the Burlington mylonite zone. The mylonite formed in Precambrian time and deep within the earth. The Bloody Bluff fault formed much later, toward the end of Paleozoic time, at a shallow depth, so the deformation was brittle. Rocks to the west of the fault belong to the Nashoba terrane.

Bloody Bluff in Minute Man National Historic Park in Lexington consists of broken-up and leached, reddish orange Indian Head Hill granite of Mississippian age. This granite ranges in color from orange to pink and acquires a rusty stain as it weathers. The 349-million-year-old granite, which is just northwest of the fault zone, is part of the Nashoba terrane. Though faults truncate the granite to the northeast of Bloody Bluff, it is broadly exposed to the southwest. You can see the granite just west of the Bloody Bluff fault at Drumlin Farm Wildlife Sanctuary on Massachusetts 117 in Lincoln. Bloody Bluff is not named so much for the reddish orange granite as for the gory spot where the British attempted to regroup during one of the first battles of the Revolutionary War.

East of the fault zone at Fiske Hill, also in Minute Man National Historic Park, you can see mylonitized volcanic rocks. Dedham granite intrudes the volcanic rock and is in turn intruded by basalt dikes.

Glacial Lakes Concord and Sudbury

Near Concord, Glacial Lake Sudbury occupied the lowland generally south of Massachusetts 2, and Glacial Lake Concord occupied the area mainly north of Massachusetts 2. Rivers flowing into these lakes produced large delta deposits of sand and gravel. When the melting glacier front stood

West-dipping foreset beds covered by topset beds in delta sediments of Glacial Lake Concord. Exposed in gravel pit just south of Massachusetts 2 in Acton. —George Ehrenfried photo

just north of Massachusetts 2 in Concord, it deposited the great kame delta in Glacial Lake Sudbury. These deposits now form the substrate of Walden Woods on the north side of Walden Pond. As the ice receded farther north, the kame deposit became a small dam that began to impound water of a new lake—Glacial Lake Concord. Another kame delta built out into this lake and now forms Authors' Ridge in Concord.

Walden Pond

Henry David Thoreau recounts without comment a Native American legend that I consider to be quite plausible: a mountain once stood where Walden Pond is now. The story goes that one night a group of Native Americans were camped at the base of the mountain. It collapsed, killing all of them except a woman, Walden, who escaped.

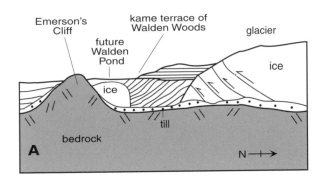

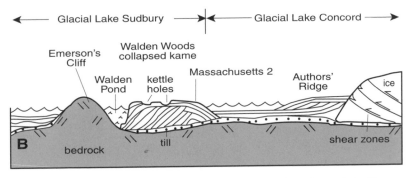

Formation of Walden Pond. (A) Water deposits sands and gravels in Glacial Lake Sudbury, burying large block of ice. (B) Ice block melts. Glacial Lake Concord forms behind the Walden Woods kame. Water deposits kame sands and gravels of Authors' Ridge.

There was a mountain there once—a mountain of ice. It was probably filled with and covered with enough sand, gravel, and rock fragments to support vegetation! The huge, towering block of ice rested against a bedrock hill called Emerson's Cliff. Upon melting, the ice block left a deep depression filled with water, a kettle. Walden Pond is 158 feet deep.

Other kettles that formed near here include White's Pond and Sandy Pond. In Walden Woods, dozens of small, dry depressions that are too shallow to reach the water table record the former presence of small blocks of ice that were buried in the outwash plain.

Nashoba Terrane

The Nashoba terrane lies immediately west of the Bloody Bluff fault zone. Steeply dipping strata consist mainly of metamorphosed volcanic rocks in the east—the Marlboro formation and Fish Brook gneiss—and metamorphosed sedimentary rocks in the west—Nashoba formation and Tadmuck Brook schist. Granites of Ordovician, Silurian, Devonian, and Mississippian age intrude the strata.

Walden Pond, a kettle. View looking northwest toward the site of Henry David Thoreau's cabin from the eastern shore on Massachusetts 126, near the Metropolitan District Commission State Park Headquarters. —Cecelia Santa photo

"Listening Stone" is a carved 5-foot-long glacial erratic of granite excavated from the glacial till of DeCordova Museum and Sculpture Park.

The DeCordova Museum and Sculpture Park is on a bedrock hill overlooking Sandy Pond, a kettle, south of Massachusetts 2 in Lincoln. The hill is Marlboro volcanic rock. Boulders and rock sculptures fill the museum's sculpture park.

Andover Granite and Nashoba Gneiss

The Andover granite of very early Devonian age intrudes the sedimentary rocks of the Nashoba terrane. It is associated with pink, coarse-grained pegmatites with large crystals of microcline feldspar. Look for an outcrop of Andover granite intruding Nashoba gneiss on the west exit from the Concord rotary. Look for another outcrop 0.85 mile northwest of the Massachusetts 2 intersection with Massachusetts 27, about 700 feet southeast of the Arlington Street overpass in West Acton. Andover granite and pegmatite crosscut the Nashoba gneiss.

Tadmuck Brook Schist

You can see an outcrop of Tadmuck Brook schist along Massachusetts 2 in Littleton, 0.15 mile west of Whitcomb Avenue overpass. Tadmuck Brook schist is a highly sheared, mylonitized mica schist. It is rusty when weathered; when fresh, dark gray. Shiny crystals of the mineral marcasite form on the surface of weathering blocks.

The Tadmuck Brook schist looks deceptively like a slate—a low-grade metamorphic rock—but contains crystals of sillimanite, a mineral that

Outcrop along Massachusetts 2, 1 mile west of I-495. Gray granite and veins of quartz intrude light-colored lime silicate rocks of the Nashoba gneiss. A crosscutting, folded dike of Andover granite mimics a fold in the Nashoba strata. Football-shaped lenses of granite— boudins—were once a dike that intruded parallel to the foliation.

forms under high pressures and temperatures. Overgrowths of sillimanite as long as 1.5 feet have replaced knotlike andalusite crystals as the temperature and/or pressure increased during metamorphism. The Tadmuck Brook schist is present only along the western margin of the Nashoba terrane and is closely associated with the Clinton-Newbury fault zone.

Pin Hill

Pin Hill, south of Massachusetts 2 at Harvard, is a bedrock hill of Harvard conglomerate that contains dark gray beds of mica schist. Though the rock is of Pennsylvanian age, no coal is present at Pin Hill. The rocks rest on an erosion surface of the Oakdale formation, a part of the Nashua belt in the Merrimack terrane.

Drumlin Farms

A mantle of glacial and postglacial deposits covers the bedrock formations between Arlington and Leominster. Drumlins and clusters of drumlins are the most conspicuous glacial feature—a visitor might think we farm them here. These drumlins, aptly named whalebacks, are elongate in a southerly direction, the same direction the ice sheet traveled. The continental ice sheet rose up and over great mounds of unsorted till, streamlining them into drumlins. The Metropolitan State Hospital on Trapelo

Tadmuck Brook schist along Massachusetts 2 in Littleton, 0.15 mile west of Whitcomb Avenue overpass.

Road just south of Massachusetts 2 in Waltham sits on a drumlin. You can see drumlins at a distance at Drumlin Farm Wildlife Sanctuary in Lincoln and at the Harvard Street–Mechanic Street exit of Massachusetts 2 on the Lancaster-Leominster township line.

Merrimack Terrane

Massachusetts 2 crosses the Clinton-Newbury fault zone near Harvard. The fault marks the subduction boundary between the Nashoba terrane to the southeast and the Merrimack terrane to the northwest. The Merrimack terrane consists of several distinct belts of rocks. We'll cross the Rockingham, Nashua, and Wachusett Mountain belts between Littleton and Leominster.

The Rockingham belt along Massachusetts 2 consists of the Berwick formation of Silurian age, which is intruded by the distinctive Ayer granite, also of Silurian age. This belt of abundant granite intrusions, chopped up by younger brittle faults, occupies a wedge whose southern point is just north of Clinton, about 10 miles south of Massachusetts 2.

A brittle fault separates the Rockingham belt from the Nashua belt of sedimentary rocks, which includes the Oakdale and Worcester formations. These low-grade metamorphic rocks are essentially devoid of igneous intrusions. They contain well-developed sedimentary structures that enabled geologists to recognize them as submarine fans—masses of loose rock fragments deposited at the mouths of canyons on the ocean floor. Turbidity

Low-angle normal faults in folded Oakdale formation in outcrop west of Lunenburg Road. The faults are part of the major east-dipping Wekepeke normal fault. Black lines show folds and faults.

currents, dense flows of sediment suspended in water, probably flowed down the continental slope and deposited some of the sediment.

The Wekepeke normal fault zone forms the western margin of the Nashua belt. The Oakdale formation was downfaulted along the east-dipping Wekepeke normal fault. An intrusion of granite in Pennsylvanian time may have lifted up the rocks of the Wachusett Mountain belt on the west relative to the Oakdale sediments on the east.

Between the Wekepeke fault zone and Leominster, Massachusetts 2 crosses metamorphic rocks of the Wachusett Mountain belt. The sedimentary rocks of high metamorphic grade extend westward to South Street in Fitchburg and to Merriam Avenue in Leominster at Pierce Pond. These moderately high-grade metamorphic rocks, the Paxton schists, are similar in composition to the low-grade sedimentary rocks of the Oakdale formation east of the Wekepeke fault. The Paxton schists were closer to the Fitchburg granite intrusion, whose heat raised the schists to higher metamorphic grades than those of the Oakdale formation.

Glacial Lake Nashua

Massachusetts 2 crosses the Nashua River at the interchange for Devens and the towns of Shirley and Ayer. Here the Nashua River meanders through the Oxbow National Wildlife Refuge. The Nashua River flows northward from its watershed divide near Worcester, which impounded Glacial Lake Nashua. The southeast-flowing North Nashua River and the north-flowing Nashua River have exposed lake sediments over a 12-mile-wide area.

Extensive sand and gravel plains of Shirley, Ayer, and Devens indicate the glacier stood still for a while near Massachusetts 2 and the Nashua River. During this standstill, Glacial Lake Nashua deposited lake sediments that extend from Massachusetts 2 south for about 10 miles to between Clinton and Boylston Center.

Deposits of delta sands and gravels are up to 165 feet thick. Lake-bottom silt and clay deposits are rare because Glacial Lake Nashua filled rapidly with coarse sediment. As the water level dropped, lake terraces developed in the Nashua Valley. Leominster, with its prominent spire of St. Cecilia's Church, stands on kame plain sands and gravels of Glacial Lake Nashua.

Massachusetts 3 (The Southeast Expressway)
Quincy—Sagamore Bridge
49 miles

Massachusetts 3 and I-93 run together from downtown Boston to Quincy, where I-93 heads west, and Massachusetts 3, the Southeast Expressway, continues all the way to Cape Cod, passing through geologically interesting seascapes and semirural communities. Our route primarily crosses Precambrian rocks of the Avalon terrane and architecturally prized granites of Silurian age. Glacial material blankets the bedrock along the southern half of the route.

Quincy

Quincy is the hometown of two presidents of the United States, John Adams and John Quincy Adams. It is also home to numerous granite quarries. The South Shore Museum in Quincy and the Blue Hills Museum in Randolph feature the history of granite quarrying and fabrication. The South Shore Museum recounts interesting stories of Scandinavian, Italian, and other ethnic groups recruited from Europe for their skill in the extraction, fabrication, and transportation of granite.

Quincy straddles the Blue Hills fault—the boundary between the Boston Basin and the Quincy pluton of the Blue Hills. The Blue Hills was a volcano about 440 million years ago, and the magma chamber crystallized

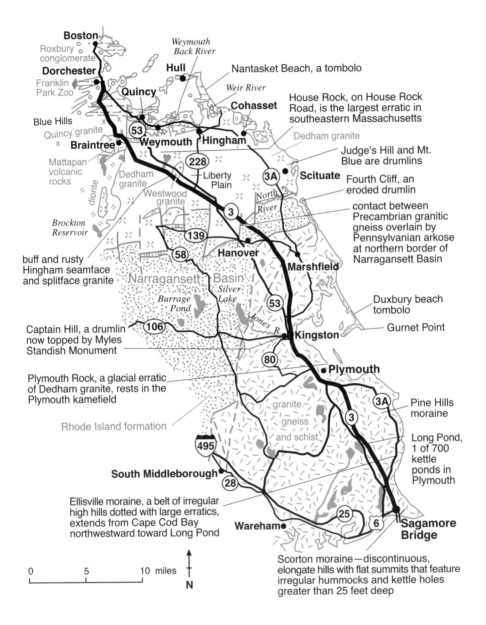

Boston

Roxbury conglomerate

Dorchester

Franklin Park Zoo

Weymouth Back River

Hull

Nantasket Beach, a tombolo

Weir River

Cohasset

House Rock, on House Rock Road, is the largest erratic in southeastern Massachusetts

Quincy

Blue Hills

Quincy granite

Braintree

Weymouth

Hingham

Dedham granite

Mattapan volcanic rocks

diorite

Dedham granite

Westwood granite

Liberty Plain

North River

Judge's Hill and Mt. Blue are drumlins

Scituate

Fourth Cliff, an eroded drumlin

contact between Precambrian granitic gneiss overlain by Pennsylvanian arkose at northern border of Narragansett Basin

Brockton Reservoir

buff and rusty Hingham seamface and splitface granite

Narragansett Basin

Silver Lake

Burrage Pond

Hanover

Marshfield

Duxbury beach tombolo

Captain Hill, a drumlin now topped by Myles Standish Monument

Jones R.

Kingston

Gurnet Point

Plymouth

Plymouth Rock, a glacial erratic of Dedham granite, rests in the Plymouth kamefield

Rhode Island formation

granite

gneiss

and schist

Pine Hills moraine

South Middleborough

Long Pond, 1 of 700 kettle ponds in Plymouth

Ellisville moraine, a belt of irregular high hills dotted with large erratics, extends from Cape Cod Bay northwestward toward Long Pond

Wareham

Sagamore Bridge

Scorton moraine—discontinuous, elongate hills with flat summits that feature irregular hummocks and kettle holes greater than 25 feet deep

0   5   10 miles

N

Geology along Massachusetts 3 between Quincy and Sagamore Bridge.
—Modified from Zen and others, 1983

as Quincy granite. Granites and associated volcanic rocks underlie most of the Blue Hills upland area. The Quincy granite is prized for its range of hues from light to extra dark and a golden variety called "gold leaf." The distinctively dark granite of St. Paul's Church in Boston and Minot's Lighthouse in Scituate came from Quincy.

One of the most attractive kinds of Quincy granite is the medium- to coarse-grained, gray to dark green variety that contains the minerals quartz, microperthite (a feldspar), riebeckite (a bluish amphibole), and aegerine (a greenish pyroxene). Its ability of take on a high polish and its massive character, free of hidden internal flaws and breaks, made it an integral part of the economic development of early New England. A reddish granite of the same composition probably acquired its hue when steeped in its own magmatic juices, or hot fluids, as the pluton crystallized.

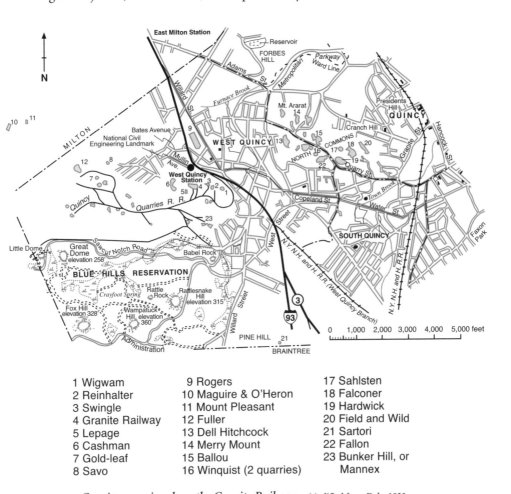

1 Wigwam	9 Rogers	17 Sahlsten
2 Reinhalter	10 Maguire & O'Heron	18 Falconer
3 Swingle	11 Mount Pleasant	19 Hardwick
4 Granite Railway	12 Fuller	20 Field and Wild
5 Lepage	13 Dell Hitchcock	21 Sartori
6 Cashman	14 Merry Mount	22 Fallon
7 Gold-leaf	15 Ballou	23 Bunker Hill, or
8 Savo	16 Winquist (2 quarries)	Mannex

Granite quarries along the Granite Railway. —Modified from Dale, 1923

Quincy granite in Dell Hitchcock Quarry, circa 1923. Note the gently dipping, horizontal joints that increase in thickness with depth. —T. Nelson Dale photo, U.S. Geological Survey

Dark green ball of polished Quincy granite from the Wigwam Quarry is 76 inches in diameter and weighs 22,000 pounds. Photo was taken in 1923 at the fabricating plant at the Wigwam Quarry. Ball is now displayed at the traffic circle on Massachusetts 3 on the north side of Fore River Bridge at Quincy Point, near the Quincy shipyard. —T. Nelson Dale photo, U. S. Geological Survey

You can still see evidence of extensive quarrying in the hills in and around Quincy. Heavy chains, derrick cables, and a variety of tools are anchored to the rock in some of the quarries. On a large monolithic granite base at the southern end of the Reinhalter Quarry, you can see bolts in a concrete matrix that once supported a derrick. It lifted granite blocks out of the deep quarry.

Granite Railway

The Granite Railway, a National Civil Engineering Landmark, was established in 1826 by Gridley Bryant to transport to the sea for shipping splendid architectural-quality granite from the Bunker Hill, or Mannex Quarry as it was called earlier. The old railroad mainly follows the route of Massachusetts 3 and I-93 to a stone pier on the Neponset River. Schooners transported the granite to docks at the foot of Breeds Hill in Charlestown. This rail, the first commercial rail in the country, transported granite used in the construction of the Massachusetts State Prison in Boston's North End and the Bunker Hill Monument in Charlestown. Specially designed wagons, pulled by horses, carried the blocks along the iron rails.

The Granite Railway. The outside blocks of granite have bolted metal strips for rails and the inner blocks have grooved fixtures for the stabilizing cable. —Taylor Heewon Khym photo

Polished Quincy granite obelisks flank the entrance to the narrow-gauge cog railway. You can see the railway at the headquarters of the Quincy Quarries Historic Site adjacent to Massachusetts 3 on Mullin Avenue in West Quincy.

Boston Has Its Faults

Boston has its faults—geological, that is! From Dorchester to Braintree, a number of narrow, northeast-trending faulted anticlines and synclines in late Precambrian sedimentary rocks are overturned to the southeast. These structures are separated from one another by high-angle reverse faults, probably developed in late Paleozoic time during the Alleghanian collision, about 275 million years ago. A reverse fault is similar to a thrust fault—rock on one side of the fault is pushed over the top of rock on the other side—but its angle exceeds 45 degrees. The reverse faults, in turn, are offset by northerly trending faults, commonly associated with basalt dikes about 200 million years old—early Jurassic time.

Massachusetts 3 and I-93 cross over three faults bounding the folded sedimentary rocks: the Mount Hope fault at Tenean Beach just north of the Neponset River, the Neponset fault at the Neponset River crossing, and the Blue Hills fault near West Squantum Street in Milton. You can only see these faults farther inland and on some Boston Harbor islands. The Blue Hills, a Metropolitan District Commission Park and Recreation Area, is the largest fault block.

The regionally important Ponkapoag fault, a normal fault of late Paleozoic age, forms the southern margin of the Norfolk Basin and the northern margin of the Sharon Upland. The fault truncates the Norfolk Basin just west of where Massachusetts 3 separates from I-93. As it continues on its east-northeasterly course, the fault forms the southern boundary of the Boston Basin in Hingham and Hull. The Ponkapoag fault has been active during two separate periods. The first was normal fault movement during Devonian time. The second activity was southeast-directed thrust faulting during Alleghanian folding. The Braintree slate on the north side of the fault and the Dedham granite on the south side moved at both times.

Massachusetts 3 crosses the Ponkapoag fault just north of Union Street at exit 17 in Braintree. You can see the Ponkapoag fault in the cliff along the railroad tracks on Commercial Street at Weymouth Landing. At another site along the northbound lane of Massachusetts 128 just west of Massachusetts 37 in Braintree, you can see broken and rusty, weathered Quincy granite in the fault zone.

Braintree Slate and Weymouth Limestone

The Weymouth formation of early Cambrian age and the Braintree slate of middle Cambrian age are well known around the world because of their shelly fossils and trilobites, described recently by Edward Landing of the New York Geological Survey. The Weymouth formation contains animal shells that pre-date trilobites. It includes gastropods and other mollusks, brachiopods, and tracks of animals. The Braintree argillite contains the trilobite *Paradoxides harlani,* which reached 1 foot in length. Large trilobites have been found near the Fore River shipyard. The Weston Observatory at Boston College exhibits some of these specimens.

Glacial Lake Bouvé

Glacial Lake Bouvé occupied an area that includes the present townships of Hingham, Weymouth, and Braintree, the eastern half of Quincy, and adjoining parts of Randolph, Holbrook, and Rockland. A ridge of bedrock and till formed the southern shore of the lake. From its western margin in the Blue Hills, the ancient lake shoreline looped around to the east, 1 mile north of Prospect Hill in southern Hingham. Glacial Lake Bouvé was named for the famous geologist Thomas T. Bouvé.

Geology along Massachusetts 3 near Quincy. Shaded area shows extent of Glacial Lake Bouvé. —Modified from Crosby and Grabeau, 1900

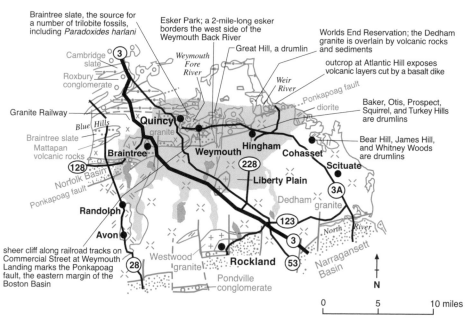

It seems likely that meltwater flowing from the Narragansett Bay–Buzzards Bay lobe created Glacial Lake Bouvé. Swift streams flowing from the ice cap over Boston and the Blue Hills carried sands and gravel into the lake, forming delta plains near Massachusetts 228 in Hingham. Liberty Plain is at 140 feet elevation and Glad Tidings Plain is at 68 feet elevation. A lower plain, which has an elevation of 50 feet, formed in Hingham Center when the lake had partially drained as the glacier receded. A still lower plain in North Weymouth, at Quincy Point, and in parts of Hingham formed the floor of the lake as it drained into Hingham Harbor. The present basin of Glacial Lake Bouvé drains north into Boston Bay through the Weymouth Fore River estuary, the Weymouth Back River, and the Weir River.

Drumlins and Tombolos

Along the coast from Hull to Duxbury, tombolos—bars of gravel and sand—connect islands to other landmasses. Nantasket Beach in Hull is a series of drumlins tied together by tombolos. Duxbury Beach, the long, narrow stretch of sand that separates Duxbury Bay from Cape Cod Bay, is a tombolo that connects the mainland to the islands of Gurnet Point and Saquish Head.

Numerous hills in Hingham, Cohasset, and Scituate are drumlins. Otis Hill in Hingham is one-half mile long, one-third mile wide, and 120 feet high. Like other drumlins in southeastern Massachusetts, Otis Hill is elongated in a southeasterly direction, the same direction that the glacial ice sheet moved. Fourth Cliff, the end of the long spit at the mouth of the North River in Marshfield, is an eroded drumlin.

Glacial Erratics

Erratics, large boulders transported by glacial ice, litter southeastern Massachusetts. House Rock on House Rock Road in Weymouth is the largest erratic in the area at 37 feet high, 42 feet wide, and 3,500 tons. The most famous erratic is probably Plymouth Rock, a boulder of Dedham granite set down in the Plymouth kamefield at the edge of Cape Cod Bay.

Hingham Seamface and Splitface Granite

An important architectural granite is quarried at Plymouth Quarries in Hingham and Weymouth on Massachusetts 53. The approximately 600-million-year-old pink to buff granite occurs as stocks or small batholiths that intruded the Dedham granite south of Boston. Quarrymen named this fine- to medium-grained granite "seamface and splitface" because the vertical, closely spaced, smooth joints are a rusty color, whereas the freshly

Bouvé glacial erratic at George Washington Boulevard and Rockland Street, Hingham, en route to Nantasket Beach. The coarse-grained rock may be Westwood granite. —Taylor Heewon Khym photo

split surfaces that break at 90 degrees to the joints are pink to buff. Builders arrange blocks of freshly split surfaces with those of rusty joint surfaces so that they form a mosaic of color in the walls of buildings. The Episcopal churches in Hingham and Cohasset are Hingham granite. The Carroll School of Management on the Boston College campus has strategically placed blocks of the steel gray, striped Chelmsford granite within a mosaic of Hingham seamface and splitface granite.

Narragansett Basin

Near Hanover, Massachusetts 3 crosses onto the 2-mile-broad neck of the Narragansett coal basin of Pennsylvanian age. The coal basin extends to the sea south of Scituate and north of Plymouth Bay. These dark, coal-bearing sedimentary rocks overlie the eroded surface of late Precambrian granite. One of the few places you can see the contact between the Dedham granite and basal conglomerate of the coal basin is in Hanover. Arkose with thin, reddish brown quartzite resting on an erosional surface of the Pre-cambrian granite is well exposed 300 feet west of Church Hill Cemetery, 2 miles northeast of Hanover and north of Massachusetts 53.

Glacial Lake Taunton and the Jones River Spillway

The ice margin of the receding Buzzards Bay lobe furnished an abundant supply of meltwater to Glacial Lake Taunton. Massachusetts 3 crosses the eastern part of the former lake, where thick sand and gravel deposits representing kame deltas overlay varved clay beds up to 20 feet thick. Foreset beds and stream-deposited topset beds are exposed along the high ground in western Kingston, Plympton, and Bridgewater. The Jones River, in the lowland of the former Glacial Lake Taunton, emerges from Silver Lake.

The Cape Cod Bay lobe of the glacier formed the eastern shore of the lake and prevented the waters from escaping to the east through the Jones River lowland. Instead, water exited to the southwest through the Taunton River near the city of Fall River. Once the Cape Cod Bay lobe melted farther north, the lake drained eastward through the Jones River spillway, which was 15 feet lower in elevation than the Fall River spillway.

Massachusetts 3 crosses the Jones River spillway in Kingston. At the gaging station on Elm Street, water abruptly plunges 15 to 20 feet from the placid level of the Jones River marshland to the flowing brook below the station. When the Cape Cod lobe melted northward from this spot, the lake must have drained with a velocity sufficient to move large boulders.

Looking upstream from the Jones River gaging station on Elm Street, Kingston, at the site of the spillway for Glacial Lake Taunton. —Taylor Heewon Khym photo

Glacial Deposits from Plymouth to Sagamore Bridge

From Plymouth to the Sagamore Bridge over Cape Cod Canal, Massachusetts 3 crosses glacial deposits that contain a large aquifer. The groundwater basin is one of the largest aquifers in Massachusetts and one of the most precious natural resource legacies from glacial times. The blanket of sand and gravel from 40 to 160 feet thick has a calculated potential yield of 300 gallons per minute per well, with a total water storage capacity of 540 billion gallons. Myles Standish State Forest lies in the middle of this groundwater basin.

The main landforms between Plymouth and the Sagamore Bridge are, from north to south, Monks Hill moraine, the hummocky Plymouth kamefield, Pine Hills ground moraine, Ellisville moraine, Wareham outwash plain, and Scorton moraine. The northwest-trending Ellisville moraine, a 0.5-mile-wide belt of irregular high hills dotted with thousands of erratics, is the largest of these features. It is bordered on the south by the Wareham pitted plain, almost entirely west of Ellisville, and on the north by an irregular boundary where the moraine abruptly rises above ground moraine and sandy deposits.

The breadth of the Sandwich moraine at the Cape Cod canal is exposed along Massachusetts 6W between the Sagamore Bridge on the east and the Bourne Bridge on the west. See map on page 83.

Massachusetts 12 and Massachusetts 13
Webster—Townsend
50 miles

Our route follows river drainages for much of the way. A drainage divide near Worcester separates the north-flowing Nashua River from the south-flowing French and Blackstone Rivers. This divide blocked meltwater in glacial times, impounding a large lake in the Nashua Valley.

The Avalon, Nashoba, and Merrimack terranes converge near Webster, where a bulging anticline presses the Avalon rocks into the narrow Nashoba terrane. A mere ten-minute walk will take you from the Avalon terrane to the Merrimack terrane. Between North Oxford and Clinton, Massachusetts 12 parallels the Clinton-Newbury fault zone. Between Clinton and the New Hampshire line, Massachusetts 12 and Massachusetts 13 cross rocks of the Nashua belt of the Merrimack terrane.

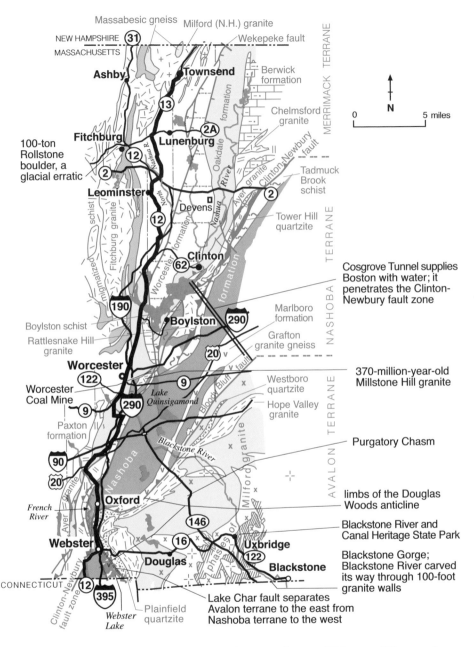

Massabesic gneiss   Milford (N.H.) granite

NEW HAMPSHIRE ⑶①   Wekepeke fault
MASSACHUSETTS

Ashby   Townsend   Berwick
                                 formation
          ⑬   Chelmsford
                                 granite
Fitchburg   Lunenburg ②Ⓐ

⑫
②
                                 Tadmuck
Leominster   Brook
                                 schist
          ⑫   Devens   ②
                                 Tower Hill
                                 quartzite

100-ton
Rollstone
boulder, a
glacial erratic

            Clinton
          ⑥②
                                 Cosgrove Tunnel supplies
⑲⓪   Boston with water; it
                                 penetrates the Clinton-
                                 Newbury fault zone
            Boylston ⑤②⓪   Marlboro
                                 formation
Boylston schist   Grafton
Rattlesnake Hill   granite gneiss
granite   ②⓪ v
Worcester   v fault   370-million-year-old
⑫②   ⑨   v   Millstone Hill granite
Worcester   Westboro
Coal Mine   quartzite
⑨   ②⑨⓪ Lake   Hope Valley   Purgatory Chasm
Paxton   Quinsigamond   granite
formation

⑨⓪   limbs of the Douglas
②⓪   Oxford   Woods anticline
French   Blackstone River and
River   Canal Heritage State Park
Webster   ⑭⑥
         ⑯   Uxbridge   Blackstone Gorge;
Douglas   ⑫②   Blackstone River carved
CONNECTICUT   Blackstone   its way through 100-foot
⑫   granite walls
⑶⑨⑤
      Plainfield   Lake Char fault separates
Webster quartzite   Avalon terrane to the east from
Lake   Nashoba terrane to the west

MERRIMACK TERRANE

NASHOBA TERRANE

AVALON TERRANE

Geology along Massachusetts 12 and Massachusetts 13 between Webster and Townsend.
—Modified from Zen and others, 1983

French and Blackstone Rivers

The Blackstone River originates in Worcester and flows southeast into Narragansett Bay at Providence. In the past, the Blackstone River was one of the most heavily industrialized rivers in the country. Near the Rhode Island border, the Blackstone River flows through a gorge with 100-foot walls of Milford granite. The fast-flowing river eroded potholes in the bedrock.

The French River is a tributary to the Thames River and Long Island Sound. The French River drainage basin has many acres of lakes and ponds, including Lake Webster, or Lake Char, one of the largest natural lakes in the state. Native Americans call it Lake Chargoggagoggmanchauggagoggchaubunagungamaugg, which is reported to mean, "You fish on your side, we fish on our side, and nobody fishes in the middle."

Pothole filled with rocks along the Blackstone River. River currents swirl the rocks around, drilling the pothole and rounding the rocks. —W. C. Alden photo, U.S. Geological Survey, 1906

Pothole in water-worn rock in the bed of the Blackstone River below the dam west of Blackstone. The elegant glacial geologist W. C. Alden provides a sense of scale. —W. C. Alden photo, U.S. Geological Survey, 1906

Lake Char Fault and the Nashoba Terrane

The Lake Char fault is Connecticut's name for the Bloody Bluff fault, the fault that separates the Avalon terrane from the Nashoba terrane. Despite geologists' predilection for long, amusing terms with history-based roots, we usually use the short version of the Lake Char name for the fault. The Clinton-Newbury and the Lake Char fault zones nearly converge but are not coincident near Webster. More faults cut rock here per square mile than in any other part of Massachusetts except for near the Clinton-Newbury fault zone farther north between Worcester and Lowell.

The Lake Char fault appears to have been a thrust fault early in its history. Then rapid uplift of the Nashoba block in Mississippian time created normal fault movement, which was also common during Mesozoic time.

Movements along the Clinton-Newbury and the Lake Char faults in late Permian to Triassic time shaved off rocks from the west and east sides of the Nashoba terrane. Rock formations of the Avalon terrane underlie the eastern shore of Lake Webster (Char); rocks of the Nashoba terrane underlie most of the lake and part of the western shore. Rocks in Webster and Dudley, just across the French River from downtown Webster, belong to the Merrimack terrane.

Just as the Bloody Bluff fault uses the Connecticut name down here, so too do the rocks of the Nashoba terrane. The Quinebaug formation is the equivalent of the Marlboro formation, and the Tatnic Hill is the equivalent of the Nashoba formation.

Douglas Woods Anticline

The northwest bulge of the Avalon terrane at the west end of Douglas Woods represents a northwest-plunging anticlinal structure with a core of Precambrian granite gneiss. The gneiss is Hope Valley alaskite, a phase of the Milford pluton, and intrudes still-older Plainfield quartzite. The anticline probably formed through buoyant uplift of the relatively light granitic gneiss along normal faults. The Precambrian rocks of the core of the anticline have displaced the entire Quinebaug and part of the Tatnic Hill formations of the Nashoba terrane. I-395 skirts the maximum curve of the fold at the leading edge of the Douglas Woods highlands, between exit 2 near Webster's East Village and exit 4 in Oxford.

A zone of highly sheared rock—mylonite—follows the shape of the anticline, but it's unclear whether it formed earlier or later than the fold. It may be that the mylonite was generated when the gneiss of the Douglas Woods anticline was compressed against the Nashoba terrane rocks, possibly when the Meguma terrane collided with the Avalon terrane. This same collision produced the great arcuate bend of the Avalon terrane against the

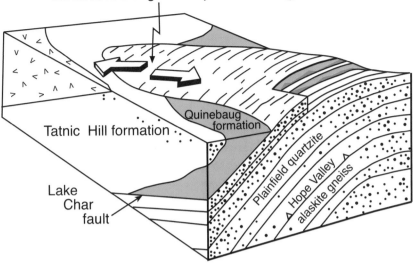

movement along fault is parallel to hinge of fold

Tatnic Hill formation

Quinebaug formation

Lake Char fault

Plainfield quartzite

Hope Valley alaskite gneiss

Northwest-plunging Douglas Woods anticline may have formed as bouyant Hope Valley alaskite and Plainfield quartzite rose upward. Arrows show directions of relative movements above and below the Lake Char fault. —Modified after Goldstein, 1982

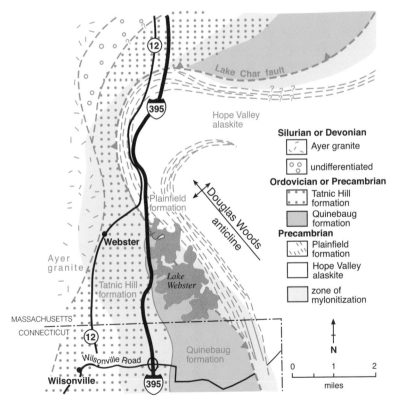

Lake Char fault

Hope Valley alaskite

Silurian or Devonian

Ayer granite

undifferentiated

Ordovician or Precambrian

Tatnic Hill formation

Quinebaug formation

Precambrian

Plainfield formation

Hope Valley alaskite

zone of mylonitization

Plainfield formation

Douglas Woods anticline

Webster

Ayer granite

Lake Webster

Tatnic Hill formation

MASSACHUSETTS
CONNECTICUT

Wilsonville Road

Wilsonville

Quinebaug formation

N

0   1   2

miles

Geologic map of the Douglas Woods anticline showing rock formations involved and mylonitic shear zones produced by deformation. Arrows show the fold axis and the direction it plunges into the earth. —Goldstein, 1982

Nashoba terrane. Near the Connecticut-Massachusetts line, the Hope Valley alaskite and Plainfield quartzite range from partly mylonitized to
intensely mylonitized over an east-west distance of about 4 miles. On the
northeast flank of the anticline, the zone of mylonitization is nearly 3 miles
broad.

A number of features of the anticline are visible along and near I-395.
Just south of the border in Thompson, Connecticut, you can examine
mylonitized and unmylonitized rocks of the sillimanite-bearing Tatnic
Hill (Nashoba formation) along Wilsonville Road near the I-395 ramps.
In Webster, the Lake Char fault passes through exit 2 of I-395 and separates unmylonitized Plainfield quartzite to the east from mylonitized rusty-
weathering gneiss to the west. About 0.6 mile north of exit 4 at Oxford on
the north boundary of the Webster quadrangle, a roadcut exposes folds
in the Tatnic Hill formation on the north flank of the Douglas Woods
anticline.

Purgatory Chasm

Purgatory Chasm, in Purgatory Chasm State Park off of Massachusetts
146, is a one-quarter-mile-long gorge with vertical walls of massive Milford
gneiss. The gorge is 50 feet wide and the walls are 10 to 70 feet tall. It may
be a keystone fault—a break in the rock that forms at the crest of a fold.

Nashua Valley and Glacial Lake Nashua

Our route traverses the length of the the Nashua Valley, the most prominent drainage area in Massachusetts east of the Connecticut Valley. The
headwaters of the Nashua River are near South Bay of Wachusett Reservoir, and it flows north to Nashua, New Hampshire, where it empties into
the Merrimack River. Massachusetts 12 crosses five major tributaries—from
south to north, the Quinapoxet, Stillwater, North Nashua, Squannacook,
and Nissitissit Rivers.

The Nashua Valley follows the regional northeast trend of the easily
eroded, low-grade metamorphic phyllites and schists of the Nashua belt.
The tributaries all flow southeast, the same direction in which the ice sheet
advanced across the Worcester Upland en route to Long Island Sound, as
they follow valleys carved by meltwater.

A series of glacial lakes occupied the Nashua Valley during the melting
of the ice sheet. The drainage divide that separates the Blackstone Basin to
the south from the Nashua Basin to the north impounded glacial meltwater. The lake progressively increased in size, filling the area between the
divide and the receding ice. Successive stages of the lake covered a combined length of about 35 miles, though open water did not extend the full

length at any one time. The southeast-flowing tributaries from the Worcester Upland carried fragments of plutonic and high-grade metamorphic rocks of the Wachusett Mountain belt into the basin. The history of Glacial Lake Nashua not only dominates the glacial history of the Nashua River valley but is one of the most well-documented stories in the geological evolution of central New England.

Geologists William C. Alden, Richard Jahns, and Carl Koteff produced classic studies on Glacial Lake Nashua and developed methodologies for deciphering glacial history. Dick Jahns and his assistant, Bill Muehlberger,

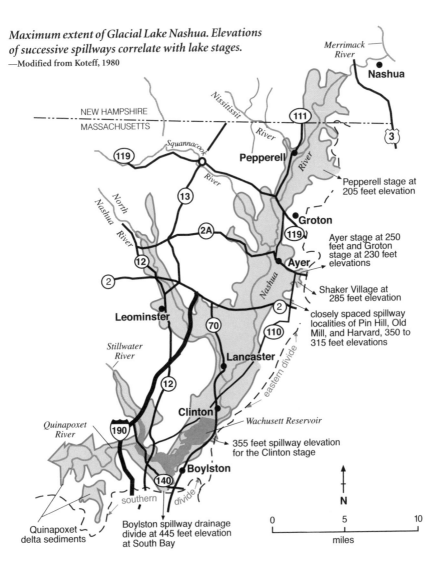

Maximum extent of Glacial Lake Nashua. Elevations of successive spillways correlate with lake stages.
—Modified from Koteff, 1980

were eager to get a firsthand look at the sands and gravels, so they dug several hundred 3-foot holes in the lake sands. Their work revealed the evolution of a lake whose source of meltwater kept on the move. The ice continued to recede northward, exposing progressively lower spillways. In all, Glacial Lake Nashua had six major stages, each corresponding to the most recently exposed spillway. All spillways were on the eastern and southern sides of the lake, through passes in the drainage divide.

In the following discussion, I give the present-day elevations of the spillways. Keep in mind that the land rebounded after the weight of the ice disappeared. The amount of uplift increased to the north.

Boylston Stage. The Boylston stage, the first stage of Glacial Lake Nashua, began when the ice melted north of the Blackstone-Nashua Rivers drainage divide. The divide impounded water between it and the ice front. When the water reached an elevation of 445 feet, it overflowed through two low spots in the divide on either side of a hill of gray phyllite. This double outlet, or spillway, is near South Bay of the Wachusett Reservoir, the west channel near the Boylston–West Boylston line and the east channel near the intersection of Massachusetts 70 and Massachusetts 140. The overflow drained through Lake Quinsigamond and the Blackstone Valley to Long Island Sound. The southern part of the Wachusett Reservoir covers most of the lake deposits of the Boylston stage.

Initially, meltwater occupied most of the Stillwater and lower reaches of the Quinapoxet River valleys near where they merge to become the Nashua River. Sediment filled the Boylston lake stage so rapidly that the open water did not remain long.

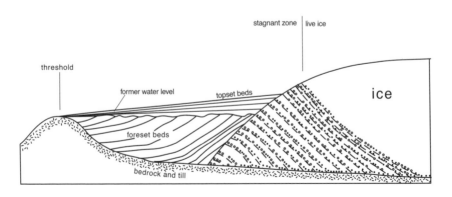

Sand and gravel in foreset and topset beds eventually filled the Boylston stage of Glacial Lake Nashua. —Modified from Koteff, 1974

Foreset and bottomset beds of sand plain 3 miles southwest of Clinton, near the Sterling–West Boylston–Boylston line. These sediments were deposited during the Boylston stage of Glacial Lake Nashua. —W. C. Alden photo, U.S. Geological Survey, circa 1925

Clinton Stage. When the ice had melted back north of the Carville Basin on the Boylston-Clinton line, it exposed a lower outlet at an elevation of 355 feet on the south side of Rattlesnake Hill. As the ice continued to melt northward during the Clinton stage, meltwater progressively filled both the Nashua Valley, which widened to the north, and the North Nashua Valley. The Clinton stage had more open water and was more long-lived than any other stage. Where the two rivers joined, the lake was 8 miles wide and extended northward to about the present location of Massachusetts 2 between Devens and Leominster. Numerous kame delta deposits, thicker than 130 feet, developed in front of the tongue of ice in the valley and between the valley wall and the glacier. Delta deposits 11 miles north of the Rattlesnake Hill spillway in Leominster attest to the breadth and longevity of the Clinton glacial stage.

Short-lived Stages. As the ice front continued to melt northward along the northward-sloping east wall of the Nashua Valley, progressively lower, granite-floored spillways were uncovered at elevations of 350, 325, 315, and 285 feet. These closely spaced spillways of Pin Hill, Old Mill, Harvard, and Shaker Village controlled the lowering water levels of Glacial Lake Nashua.

Ayer Stage. The elevation of the eastern valley wall dropped considerably near Ayer to an elevation of 250 feet. The lake drained east via Spectacle

Pond in Ayer to Stony Brook in Westford. Surface runoff from the higher elevations to the west as well as meltwater from ice in the North Nashua Valley eroded the soft deposits of the Clinton stage. Erosion produced extensive terraces in Clinton and Lancaster.

Groton Stage. The lake in the Groton stage was essentially the same as the Ayer stage, only slightly larger. The spillway, developed on the sandy plain at the Ayer gap at Devenscrest 1 mile east of the town of Ayer, was at 230 feet elevation.

Pepperell Stage. The Pepperell stage began as the ice front melted north of the Nissitissit River. The spillway developed in two steps. First, meltwater spilled over the Groton delta deposits in northeastern Groton at an elevation of 220 feet, near present-day Kemp Street. Second, meltwater flowed eastward through a small gap cut into outwash in southwestern Dunstable at an elevation of 205 feet. Swamp deposits at Unkety Brook near Groton Street occupy the gap now. The Dunstable spillway controlled a broad expanse of water for a long time. This lake stage persisted until the ice melted north of a late-stage outlet at an elevation of 195 feet. Glacial Lake Nashua drained completely into Glacial Lake Merrimack after the ice front reached southern New Hampshire.

Merrimack Terrane

The 432-million-year-old Ayer granite, with its distinctive large crystals of feldspar, is well represented along Massachusetts 12 between Webster and the Worcester Airport but then is absent for 22 miles until it appears sparingly near Clinton. It reappears abundantly in the Rockingham belt of northeastern Massachusetts.

From Worcester north, Massachusetts 12 enters a wedge of low-grade metamorphic rocks of the Nashua belt, the Oakdale and Worcester formations. The Nashua belt sits between the Rockingham and Wachusett Mountain belts.

The 370-million-year-old Millstone granite pluton underlies the high ground in eastern Worcester, west of Lake Quinsigamond and the University of Massachusetts Medical Complex, as well as several other health-related facilities. The 5-mile-by-1.7-mile Millstone Hill outcrop belt, the site of several quarries, is near the crest of Belmont Street-Massachusetts 9 at Bell Pond or Chandler Hill. This gray biotite and muscovite, middle to late Devonian granite trends north-south, is younger than the Silurian Oakdale formation, and has been quarried as a construction and architectural stone for many decades.

Between Worcester and Leominster, Massachusetts 12 parallels the Wachusett Mountain belt to the west. The belt consists mainly of Paxton

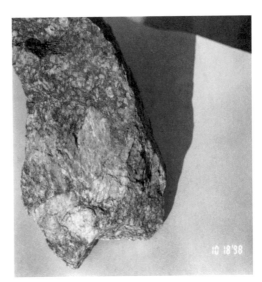

*Ayer granite contains
two generations of
large feldspar crystals.
The small ones are 1/2
to 3/4 inch long and
the big ones are several
inches long. Specimen
from Cosgrove Tunnel
in Clinton.*

and Littleton schists and large intrusions of Fitchburg granite. The Massabesic complex of Precambrian gneisses and Devonian and Pennsylvanian granite occupies a large area of the belt north of Lunenburg near Townsend and the New Hampshire line. The heat of Fitchburg granite intrusion raised the schists to higher metamorphic grades than the rocks farther east in the Nashua belt. I discuss the Fitchburg granite in the guide for Massachusetts 2 from Fitchburg to the Connecticut River.

Worcester Coal Mine

The Coal Mine Brook formation occurs in the Merrimack terrane near Worcester just west of the Clinton-Newbury fault zone. A small amount of coal was mined at one time at the coal mine along Coal Mine Brook on Plantation Street. Plant fossils in the formation correlate with the coal-bearing strata of the Narragansett Basin in southeastern Massachusetts. If Pennsylvanian coal swamps were widespread over the Merrimack terrane, they have largely eroded except for within small fault blocks.

Clinton-Newbury Fault Zone

The north end of Lake Quinsigamond follows the Clinton-Newbury fault zone in eastern Worcester. I first encountered the fault zone during the construction of the deep bedrock tunnel, the Wachusett-Marlboro Tunnel, that brings water from Washusett Reservoir southeast to Southborough, a distance of 8 miles. A complex series of closely spaced, inactive faults cuts the

1.5-mile-broad fault zone. The tunnel, later renamed the Cosgrove Tunnel, is 260 to 400 feet below the surface. Stations pump water at both ends.

The main branch of the Clinton-Newbury fault zone crosses Rattlesnake Hill, 4 miles west of West Berlin and south of Willow Road. Steeply dipping to the northwest, the fault is exposed along the lower part of the east margin of the massive Rattlesnake Hill granite pluton. You cannot see the exposure from the road.

Wekepeke Normal Fault and Massabesic Gneiss Uplift

The Wekepeke normal fault, most recently active in Mesozoic time, forms the western margin of the Nashua belt. This fault system extends at least 50 miles through Massachusetts and continues into New Hampshire. The Oakdale formation of the Rockingham belt to the east dropped down along east-dipping normal faults of the branching Wekepeke system.

The Massabesic gneiss complex consists of 623-million-year-old Massabesic granite gneiss, 402-million-year-old Fitchburg granite, and 275-million-year-old Milford (New Hampshire) granite. The Massabesic granite gneiss has Avalonian fingerprints—it chemically and mineralogically resembles Avalonian basement rocks.

The complex, at the northeast end of the Fitchburg pluton, is bounded on the west and east by the Campbell Hill and the Wekepeke normal faults, respectively. Essentially an uplifted fault block within the Wachusett Mountain belt, the structure extends from near Townsend to well beyond Manchester, New Hampshire. The Massabesic gneiss moved upward relative to the Oakdale sediments to the east of the Wekepeke fault. Massive intrusions of the Devonian and Permian granites rendered the area buoyant, lifting up the Massabesic gneiss along high-angle faults.

Massachusetts 24
Randolph—Fall River
37 miles

Massachusetts 24 crosses the Norfolk Basin, the Sharon Upland, the Narragansett Basin, and the Fall River Upland between Randolph and Fall River. The two Pennsylvanian basins alternate with upland areas of older igneous rocks, which resist erosion better than the weaker basin sediments. The northern end of the route begins in the Norfolk Basin at the southern margin of the Blue Hills volcano.

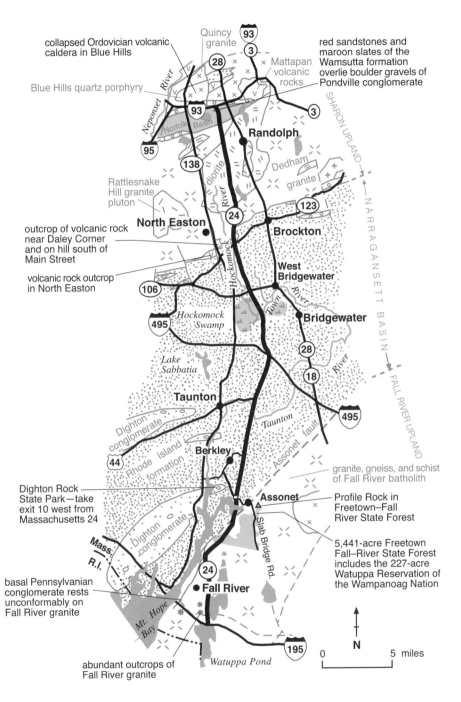

collapsed Ordovician volcanic caldera in Blue Hills

Blue Hills quartz porphyry

Quincy granite

Mattapan volcanic rocks

red sandstones and maroon slates of the Wamsutta formation overlie boulder gravels of Pondville conglomerate

Rattlesnake Hill granite pluton

Dedham granite

outcrop of volcanic rock near Daley Corner and on hill south of Main Street

volcanic rock outcrop in North Easton

Hockomock Swamp

Lake Sabbatia

Dighton Rock State Park—take exit 10 west from Massachusetts 24

granite, gneiss, and schist of Fall River batholith

Profile Rock in Freetown–Fall River State Forest

5,441-acre Freetown Fall–River State Forest includes the 227-acre Watuppa Reservation of the Wampanoag Nation

basal Pennsylvanian conglomerate rests unconformably on Fall River granite

abundant outcrops of Fall River granite

Neponset River

Norfolk Basin

Randolph

SHARON UPLAND

NARRAGANSETT BASIN

FALL RIVER UPLAND

diorite

River

North Easton

Brockton

West Bridgewater

Hockomock

Town

River

Bridgewater

River

Taunton

Dighton conglomerate

Rhode Island formation

Berkley

Taunton

Assonet fault

Assonet

Slab Bridge Rd.

Dighton conglomerate

Mass. R.I.

Fall River

Mt. Hope Bay

Watuppa Pond

N

0   5 miles

Geology along Massachusetts 24 between Randolph and Fall River.
—Modified from Zen and others, 1983

Blue Hills Volcano

The Blue Hills volcano, a large caldera that blew its top about 440 million years ago, sits just north of I-93, north of Randolph. The position of the Blue Hills is important to the shape of the Norfolk Basin. The hard volcanic and plutonic rocks of the Blue Hills block were shoved southward against the soft rocks of the Norfolk Basin during the Alleghanian mountain building event in late Permian time. The Norfolk syncline is overturned toward the south, indicating that late-stage tectonic shoving came from the north.

You can see the contact between the Norfolk Basin rocks and the Blue Hills rocks in the northwest cloverleaf connecting I-93 with Massachusetts 28 south at North Randolph. The nearly vertical to northerly dip of the contact indicates that the Pennsylvanian-age beds are the northern limb of the overturned syncline.

Red Rocks of the Norfolk Basin

The Norfolk Basin, mainly a synclinal structure, is a narrow, 2.5-mile-wide strip of land between the Blue Hills and the Sharon Upland. Terrestrial sediments, the Pondville conglomerate and Wamsutta redbeds, fill the Norfolk Basin and are in turn covered with Pleistocene glacial deposits.

The on- and off-ramps to Massachusetts 24 from I-93 follow deep cuts in coarsely granular and massive red sandstones and finely granular red slates of the Wamsutta redbeds. The sandstone consists of quartz and feldspar—not all that different from granite. The sand probably moved only a short distance from the weathered outcrops of the underlying Westwood and Dedham granites. Indeed, some of it may have moved hardly at all. Similar rocks of the same age in Oklahoma are called granite wash.

Crossbedding, ripple marks, scour and fill structures, graded beds, and mud cracks reflect deposition in a floodplain. The tilted beds are the northern limb of the Norfolk syncline. The Giant conglomerate, part of the Pondville formation, is magnificently exposed at and near the junction of Massachusetts 28 with I-93.

Sharon Upland

The southern contact of the Norfolk Basin with the northern margin of the 8-mile-wide band of Sharon Upland is about 1.4 miles south of I-93 on Massachusetts 24. The upland consists of dark diorite with lesser amounts of pale Dedham granite. The granite arrived as molten magma about 610 million years ago during late Precambrian time. The southern margin of the diorite is near Waldo Lake, just east of Massachusetts 24. The Dedham granite and the associated volcanic rocks are nicely exposed in parts of North Easton. They are north of and beneath the folded Pennsylvanian sedimentary strata near Daley Corner and the campus of Stonehill College.

Volcanic rocks are spatially associated with the intrusive Dedham granite. The numerous outcrops of volcanic rocks near North Easton, which lie along the northern margin of the Narragansett Basin, are dark greenish lavas that show flow banding. These rocks were probably rhyolites before hot water and steam circulating through the hot volcanic pile altered them. Their layers stand at steep angles because they are folded. These volcanic rocks lie beneath the Pennsylvanian strata and may either be about 600 million years old, the same age as the Mattapan volcanic rocks in the Boston Basin, or slightly older than the coal strata—geologists do not yet know.

Close view of volcanic rock at outcrop in North Easton.

View looking northeast to the broad outcrop of volcanic rocks 1,500 feet west of Massachusetts 138 on the south side of Main Street, North Easton.

Associated granite dikes may be 595 million years old, about the same age as the Westwood granite, a widespread granite closely related to the Dedham granite.

Taunton River Basin

The Taunton River drains the second-largest watershed in Massachusetts. The river drops only 20 feet in its 40-mile length, and a broad expanse of wetlands, including the 6,000-acre Hockomock Swamp, takes advantage of the flat gradient. The tide carries salt water about 12 miles upstream of Mount Hope Bay. The tide affects the water level to about the Winnetuxet River in Halifax Township. The drainage basin supports a great diversity of plants and animals, including the upland sandpiper, a bird that is in decline in its eastern range. Alewives and shad live in the river.

Narragansett Coal Basin

The Narragansett coal basin, by far the largest coal basin in New England, occupies almost 1,000 square miles, much of it low-lying, swampy forests. A sequence, at least as thick as 10,000 feet, of nonmarine sedimentary rocks underlies these extensive lowlands.

Between Brockton and Assonet, Massachusetts 24 crosses 28 miles of Pennsylvanian formations in the Narragansett basin. Rock outcrops are scarce, except where three large synclines have brought hills of the Dighton conglomerate to the surface. The synclines look like elongated ovals oriented northeast-southwest on a geological map. One is at North Dighton, another west of Taunton and north of Rehoboth, and another at Dighton Rock State Park and Somerset. The synclines tend to form hills here because the durable conglomerate was compressed into the trough of the fold and became even more resistant to erosion. The center of the fold contains very few open fractures that might permit water to seep in and weather the rock.

Glacial Lake Taunton

Glacial Lake Taunton flooded large parts of the Narragansett Basin. The lake, composed of two not entirely contemporaneous bodies of water, extended at least 17 miles from east to west and 14 miles southwest from Hockomock Swamp. It flooded parts of Halifax, Middleborough, Lakeville, Norton, Dighton, Rehoboth, and Somerset Townships. The Assonet fault defines the southeastern boundary between the Glacial Lake Taunton lowland and the higher ground farther southeast. The spillway for its high-water stage was at Fall River, along what is now the Taunton River. When the Cape Cod Bay ice lobe, the east bank of Glacial Lake Taunton, receded north of the Jones River, it exposed a lower spillway into Cape Cod Bay. The lake drained east and left the Fall River spillway high and dry.

Dighton Rock State Park

Dighton Rock is an 11-foot-by-5-foot chunk of layered sandstone of the Dighton conglomerate. The rock is covered with petroglyphic inscriptions attributed in part to the Portuguese explorer Miguel Corte Real, who passed through Massachusetts in 1511. The rock, a glacial erratic, originally rested along the bank of the Taunton River. It is now housed in the museum at Dighton Rock State Park.

Fall River Upland

Massachusetts 24 crosses the Assonet fault, the southeastern margin of the Narragansett Basin, near Assonet. To the southeast of the fault, the Fall River batholith of late Precambrian age surfaces. The granite of the Fall River batholith resembles the late Precambrian Dedham and Westwood granites near Boston.

Profile Rock is an 80-foot-high outcrop of Fall River granite in Freetown–Fall River State Forest. The main hill, which rises tens of feet higher than Profile Rock, is a glacially eroded and polished knob of granite. It has an apron of talus at its base. A climb to the top of the hill provides a broad

Daguerreotype of Dighton Rock taken by Captain Seth Eastman in 1853 at low tide in the Taunton River. Chalk highlights distinctive inscriptions. Historians attribute some of these to Portuguese explorer Miguel Corte Real, who passed through the region in 1511. —Daguerreotype courtesy of Manuel Luciano da Silva

View looking west at Profile Rock—the profile of giant's face capped by a pointed headdress. A highly reflective joint surface near the top of the headdress may in time render the rock unstable.

view of the Narragansett Basin with its prominent hills of Dighton conglomerate.

Granite of the Fall River batholith is abundantly exposed along Massachusetts 24 in Fall River and in a deep roadcut in Tiverton, Rhode Island, a short distance south of Fall River. The high ground of the Fall River batholith offers a magnificent view west across Narragansett Bay and its islands all the way to the highlands of the Providence Plantations.

Massachusetts 128
Gloucester—Peabody
20 miles

Cape Ann, a rocky headland swept by wind and waves, is one of the most scenic areas of the state. Massachusetts 128 between Gloucester and Peabody follows the craggy backbone of this rocky and knobby peninsula. With picturesque harbors and spectacular headlands jutting eastward into the Atlantic, Cape Ann attracts vacationers, artists, birdwatchers, geologists, and historians.

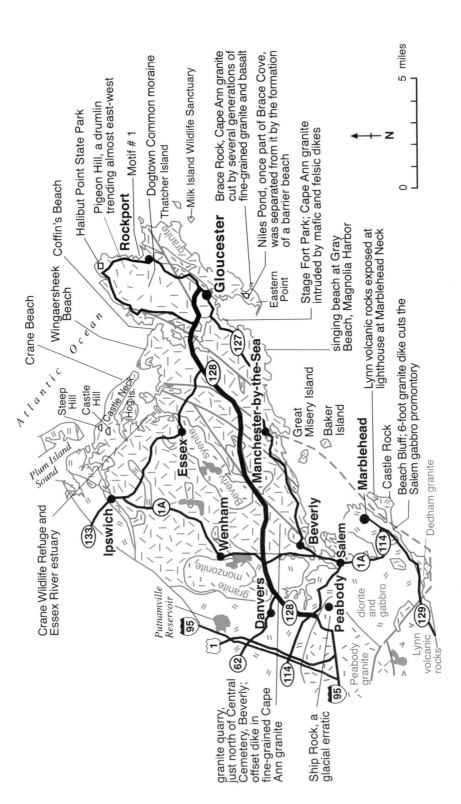

Crane Beach

Wingaersheek Beach

Coffin's Beach

Halibut Point State Park

Pigeon Hill, a drumlin trending almost east-west

Rockport — Motif # 1

Dogtown Common moraine

Thatcher Island

Milk Island Wildlife Sanctuary

Gloucester

Brace Rock, Cape Ann granite cut by several generations of fine-grained granite and basalt

Niles Pond, once part of Brace Cove, was separated from it by the formation of a barrier beach

Eastern Point

Stage Fort Park; Cape Ann granite intruded by mafic and felsic dikes

singing beach at Gray Beach, Magnolia Harbor

Lynn volcanic rocks exposed at lighthouse at Marblehead Neck

Marblehead

Castle Rock

Beach Bluff; 6-foot granite dike cuts the Salem gabbro promontory

Dedham granite

granite

127

128

Manchester-by-the-Sea

Great Misery Island

Baker Island

Beverly

Salem

Peabody

1A

114

129

Lynn volcanic rocks

diorite and gabbro

Peabody granite

Danvers

128

95

1

62

114

95

Ship Rock, a glacial erratic

granite quarry, just north of Central Cemetery, Beverly; offset dike in fine-grained Cape Ann granite

granite monzonite

Putnamville Reservoir

Wenham

1A

Essex

Beverly syenite

Castle Neck Hog Is.

Steep Hill

Castle Hill

Plum Island Sound

Crane Wildlife Refuge and Essex River estuary

Ipswich

133

Atlantic Ocean

N

0   5 miles

Bedrock geology along and near Massachusetts 128 between Gloucester and Peabody. —Modified from Zen and others, 1983

Motif No. 1, a famous landmark and seascape painted by generations of artists in Rockport Harbor. The pier in the foreground and the riprap breakwater in the distance are Cape Ann granite.

The famous *Gloucester Fisherman*, cast in bronze by Leonard Craske, commemorates generations of seafaring families. Rudyard Kipling, with his classic *Captains Courageous*, and the salty spinner-of-yarns James B. Connolly are among the artists and writers whose work has spread the fame of Cape Ann.

The name *Cape Ann* is used in an extended sense for the rocky and boulder-strewn, 160-square-mile area between Rockport, Danvers, Ipswich, and Manchester-by-the-Sea. Granites and syenites of the Cape Ann complex of late Ordovician or early Silurian time form most of the peninsula. The complex is a major alkaline intrusion in the Avalon terrane. The younger Peabody granite at 370 million years of age and related plutons intruded the Cape Ann complex and the dark gabbro plutons of the Peabody area.

Sandy Beaches and Glacial Features

Sandy barrier beaches, along the shoreline northwest of Gloucester and Rockport, have encroached into the marshlands behind the beaches. Wingaersheek and Coffin's Beaches in Gloucester and Crane Beach in Ipswich are magnificent recreational areas as well as classic localities for the study of glacial and beach features. Sand from Crane Beach surrounds two east-southeast oriented drumlins, Castle Hill and Steep Hill. The drumlins and beach form barriers on three sides and, together with the drumlins' associated kame terrace, they abut the evolving salt marsh behind the beach.

Marine and estuarine silty clay underlies Long Island in Crane Wildlife Refuge and the Richard T. Crane Jr. Memorial Reservation. Castle Hill and Hog Island are drumlins. Movie director Nicholas Hytner used 177-foot-tall Hog Island as a seaside landscape in *The Crucible*, the film version of the play about the Salem witch trials.

Wildlife thrives on these barrier islands and marshlands. Plum Island and the Parker River National Wildlife Refuge are well known for the great variety of birds that nest there or migrate through. The Crane Wildlife Refuge is part of the estuary system that includes part of Castle Neck, five islands, and the salt marsh along the Castle Neck River.

Glacial marine deposits of fine sand and gravel lie inland behind the saltwater marshland amid drumlins and bedrock hills covered with till. Nearly horizontal topset beds cover dipping foreset beds laid down at or near the edge of the melting glacier. In southern Ipswich and northern Hamilton, waves cut cliffs, formed benches on drumlins, and washed away much of the associated kame deposits. Well-developed eskers form a series of northwest-trending, steep ridges near Cutler Road off of Massachusetts 1A in Hamilton.

Pocket and Singing Beaches

Between Gloucester and Beverly, delightful pocket beaches—tiny, secluded beaches—sit between the rocky headlands. Interspersed among the pocket beaches are rocky "singing" beaches. A chorus—some folks might inelegantly call it a rock group—of granite, syenite, rhyolite, basalt, and cobbles of quartz clatter up and down the beach, creating music in tune with the melody of waves. These boulders become rounded and smaller with each song. Listen for them at Manchester-by-the-Sea and at Gray Beach near Magnolia Harbor.

Boulder Till and Boulder Trains

Evidence of the continental ice sheet on Cape Ann is far from subtle. The ice scattered extremely large boulders throughout the Cape Ann area in till, boulder trains, and as solitary erratics. Most of the surficial material that covers the bedrock of Gloucester and Rockport consists of till capped by a thin, hard-packed boulder-rich gravel left by the melting glacier.

The 12-square-mile Dogtown Common area, a public park in the heavily wooded, rugged hills north of Gloucester and west of Rockport, has a network of well-marked trails. Huge granitic boulders of a boulder train fill the area. Most of these boulders are 4 to 5 feet in diameter and many are up to 10 to 15 feet in height.

You can still see cellar foundations of Dogtown Common, a settlement that once existed among these boulders. During settlement times, before

A glacier transported Ship Rock, a glacial erratic, to this spot near Peabody. This 2,200-ton boulder is Peabody granite, but it differs texturally from the Peabody granite upon which it perches. —Photo courtesy of Peabody-Essex Museum

The boulder moraine of Dogtown Common, northeast of Riverdale Village in Gloucester.

Roger Babson's aphorism, "HELP MOTHER," decorates a boulder along the Babson Trail in Dogtown Common. How many parents have dragged their reluctant offspring to this trail in the hope of implanting Babson's sentiments in their souls?

the availability of powerful earth-moving equipment, these boulder trains posed nearly insuperable obstacles. In Rockport, southeast of Beach Grove Cemetery, a moraine of uniformly large boulders forms a high ridge 50 feet broad and 900 feet long.

Those in search of Horatio Alger–like inspiration and guidance can do no better than to walk the Roger Babson Trail on the southeast slope of Dogtown Common and follow the advice of the founder of the world-famous Babson Institute, now Babson University. Deeply carved 1-foot-high letters in boulders spaced strategically along the trail read: "WORK," "STUDY," "BE ON TIME," "SPIRITUAL POWER," "KINDNESS," and "HELP MOTHER."

Rift Granites

Cape Ann features two beautiful granites of two different ages. The older and more plentiful one, the Cape Ann granite, crystallized about 450 million years ago. The younger rock, the Peabody granite, crystallized about 370 million years ago. Plutons of Peabody granite occur in Wenham, Peabody, and Reading. Both granites are alkalic—that is, rich in potassium and sodium—which suggests they originated from rifting, or extensional tectonics. Geologists call these rocks rift granites.

Quarries

Granite quarries pockmark the upland area of Cape Ann between Rockport and Lanesville, the northern part of Dogtown Common. These quarries have produced magnificent hornblende granite for architectural purposes. The Cape Ann granite at Rockport is similar in age, mineralogy, massive character, and beauty, both rough and polished, to the famous

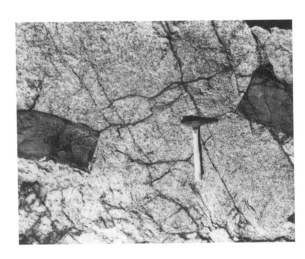

Broken and offset andesite dike in fine-grained Cape Ann granite in a quarry in Beverly. The dike intruded a fracture as the granite began to crystallize, but tectonic stresses broke the dike. —Priestly Toulmin III photo, U.S. Geological Survey

Quincy granite. Because of its consistency of texture, hardness, and strength, the Cape Ann granite is a favorite architectural stone. It was used in such structures as Boston's Custom House and the Longfellow Bridge over the Charles River between Cambridge and Boston. Cape Ann paving stones decorate the streets of such cities as Boston, New York, Philadelphia, Washington, D.C., and New Orleans.

In the early 1900s, granite quarrying was a thriving industry in Rockport. The granite from Rockport resembles the Quincy granite, but an experienced quarry worker can recognize the difference. You can see demonstrations of granite splitting at Babson Farm granite quarry in Halibut Point State Park. The granite cliffs of Halibut Point, Rockport, are 50 feet high and form a landmark for sailing ships rounding the northern tip of the cape. The Johnson Quarry, on Johnson Road off the end of Pigeon Hill Road, is the only quarry still operating in Rockport and has operated continuously since 1898.

Mineral and Rock Paradise

The bedrock of Essex County captured the attention of famous pioneering geologists of the late nineteenth century and the twentieth century. Henry S. Washington, Benjamin K. Emerson, Charles H. Clapp, John Henry Sears, and Priestly Toulmin III noticed the unusual alkaline minerals and petrology. Alkalic granites and syenites are silica-poor rocks that contain more sodium and/or potassium than is needed to form common feldspar minerals rich in silica. As a result, sodium- and potassium-rich minerals form.

The Cape Ann complex contains patches of Beverly syenite. Salem Neck, where the Beverly syenite intrudes into the edge of the Salem gabbro plu-

Cape Ann granite was quarried in Halibut Point State Park.

ton, contains a great mass of essexite, a black, mafic rock with visible crystals of alkaline minerals. John Sears named a light blue variety of essexite—salemite. A syenite containing the sodium-rich and silica-poor mineral nepheline is exposed for 8 miles along the shore between Salem Neck and Gales Point in Manchester-by-the-Sea.

Fault Blocks

The Cape Ann region has a number of northeast-trending faults. Most faults in Rockport indicate left-lateral movement on the basis of offset of basalt dikes. A fault in Beverly has been offset dextrally—the block to the north has been offset to the right relative to the block to the south.

Rift granites and fault blocks suggest that extensional tectonic forces were working to split this part of the Avalon terrane apart. A series of northeast-trending blocks bounded by faults clusters near the Avalon-Nashoba terrane boundary near Danvers and Middleton. These narrow fault blocks include the Lynnfield ultramafic deposit that intruded along brittle faults near the western Avalon margin. We know these forces were active as recently as Cretaceous time—a geologist encountered a 118-million-year-old rhyolite dike along one of the faults while drilling a 6,000-foot-deep borehole.

You can see the Lynn volcanic rocks near the lighthouse at Marblehead Neck. Red, well-cemented volcanic agglomerate contains blocks and fragments of layered tuff. Castle Rock consists of intensely jointed, welded tuff.

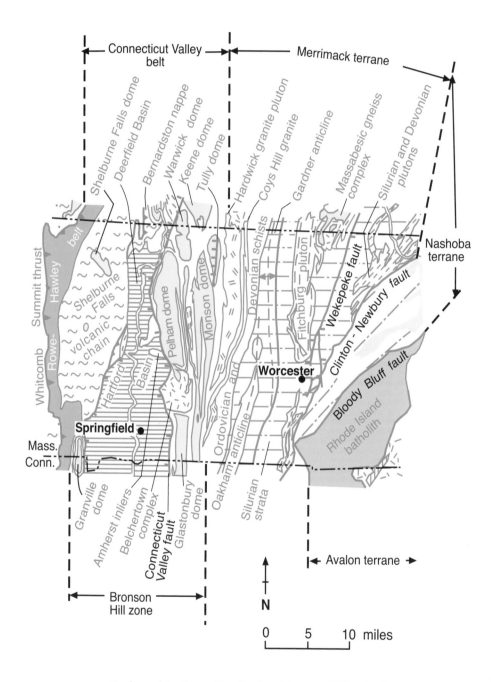

Geology of the Central Lowland and Bronson Hill Upland.
—Modified from Zen and others, 1983

Central Lowland and Bronson Hill Upland

As you enjoy a trip through the Connecticut River valley or across the Bronson Hill Upland, remind yourself that this is where supercontinents collided, creating huge mountains that divided eastern and western Massachusetts. The narrow, linear, north-trending bands of rocks you see on a geologic map reflect the great east-west forces that compressed the earth's crust in Paleozoic time. The collision created giant folds in the rocks. Later, the region began rifting apart as hot magma welled up from below as the Pangaean supercontinent began to break in two.

The eastern edge of the Bronson Hill Upland, marked by the eastern margin of the giant folds, is 335 feet in elevation at the Fitchburg airport but rises rather steeply toward the west, beyond Athol and Ware, to elevations of 1,600 feet at Mount Grace and about 1,400 feet in the Pelham dome. The Millers and Chicopee Rivers drain the western part of the Bronson Hill Upland and flow into the Connecticut River.

COLLIDING CONTINENTS

Western Massachusetts, including the Bronson Hill volcanic island chain, is part of the eastern margin of the late Precambrian Laurentian continent—early North America. Eastern Massachusetts consists of small blocks of former continents that I believe were all part of the Precambrian supercontinent Gondwana—what is now West Africa and South America. Central Massachusetts is essentially where Gondwana and Laurentia collided between 425 and 370 million years ago in late Silurian and Devonian time, forming mountains as grand as the Alps in Europe. The collision, the Acadian mountain building event, affected much of central New England.

The comparison with the Alps gives you an idea of scale and also illustrates a similar mountain building event. The Alps are the product of a continental collision; they formed when Europe collided with Africa as recently as 15 million years ago. Studies of the relatively young Alps help geologists understand what happened in Massachusetts. The alps of

Massachusetts had nearly 400 million years to erode, so we see only the deep interior of these former mountains.

Nappes, gigantic folds that are pushed up and out of the continental crust by intense pressure, were first recognized in the Alps and are prevalent in central Massachusetts. As the Acadian mountain building event continued, the Laurentian and Gondwanan continents pressed progressively closer together and squeezed rock formations up and out of the collision zone like toothpaste from a tube. At the same time, rocks deep within the earth's crust melted, generating magma. Considerable amounts of granite, including the Hardwick and Coys Hill granites of Devonian age and associated dark mafic rocks, formed in the great Acadian collision between Laurentia and pieces of Gondwana.

Somewhere in central Massachusetts, a collision boundary divides the Laurentian and the Gondwanan pieces. It would be difficult for me to take you to a single place, point to two kinds of rocks, and tell you with a straight face that those on one side belong to Laurentia and those on the other belong to Gondwana. In a continental collision, blocks, slivers, and slabs of one continent break off and penetrate the opposing continent, resulting in a great mingling of rock formations on both sides of the collision boundary. Ultramafic rocks associated with the Coys Hill granite may be pieces of the Iapetus Ocean floor that were swallowed up in the collision.

Central Massachusetts is a major zone of tectonic divergence—that is, it features tectonic structures such as nappes with thrust faults that were pushed

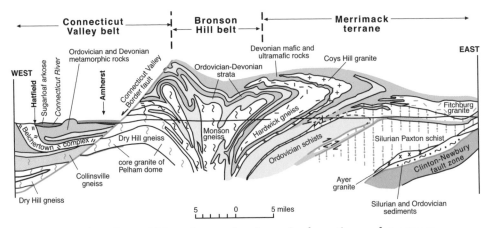

Cross section of central Massachusetts showing major formations and structures. Giant folds are schematically shown as they may have looked prior to erosion.
—Modified from Zen and others, 1983

in opposite directions at the same time. At depth in the earth's crust, the collision squeezed the rocks intensely, creating the divergence near the surface.

A north-trending axial zone that cuts central Massachusetts resembles the Insubric Line of the Alps—the line at the heart of the collision zone between the European and African continents. Geologists have measured an anomolous ridge of high gravity along the axis in Massachusetts, suggesting that magma from the earth's mantle has welled up and underplated this central zone. The boundary *at depth* between the colliding continental masses may be very close to the present location of the Connecticut Valley. Near the land *surface,* I believe the Bronson Hill domes are near the center of the collision surface between the Gondwanan and Laurentian terranes.

Domes of the Bronson Hill Volcanic Belt

The Bronson Hill volcanic belt developed in late Ordovician time between 454 and 442 million years ago along the margin of Laurentia as a volcanic island chain, a lot like today's Aleutian Islands or Japan. The Bronson Hill volcanic rocks were laid down on a basement of late Precambrian to middle Ordovician gneissic basement rocks, now exposed in the core of domes and blanketed by middle Ordovician to early Devonian schists and quartzites.

The Monson gneiss of late Precambrian to Ordovician age is the dominant core rock of the Bronson Hill belt. The Monson gneiss, a band of variable thickness from Connecticut to New Hampshire, consists of stratified and layered gneiss with biotite and plagioclase, amphibole schist, and gneiss with large crystals of microcline. It is called an augen gneiss (*augen* is German for "eyes") because these large buff to pink crystals of microcline feldspar look like eyes. The Monson gneiss also contains lenses of ultramafic oceanic crust called peridotite.

The older stratified cover rocks that rim the uplifted domes of gneiss are the Ammonoosuc and Partridge formations of middle Ordovician time and the Clough quartzite of Silurian time. The Ammonoosuc formation ranges from dark, mafic volcanic rocks in the lower part to light-colored, felsic rocks in the upper part. These strata erupted from the Bronson Hill volcanic chain during Ordovician time, but the question of when these rocks were first deformed is not yet resolved to every geologist's satisfaction.

The rock layers of the overlying Partridge formation contain mafic and felsic volcanic rocks that interlayer with black, metamorphosed, sulfidic schists—formerly shales. A thin package of rocks of Silurian age overlie these Ordovician rocks in central Massachusetts. The distinctive Clough formation consists of quartzite, conglomerate, calcareous quartzite, and schist. The overlying Fitch formation contains sulfidic schists and some lime silicates. These sedimentary rocks were deposited unconformably on the eastern margin of Laurentia following the Taconic event and well before the collision with the Gondwanan microcontinents.

When the Laurentian and Gondwanan plates were squeezed together in the Acadian collision in late Silurian to late Devonian time, the basement gneiss escaped the tremendous pressure by slipping upward and outward. The older, low-density gneiss floated upward like a beach ball submerged in water. The nearly upright Pelham and Monson gneiss domes, so elongate they look nearly like anticlines, are near the center of greatest uplift, tightest compression, and other structures created by the collision.

The Pelham dome contains some 600-million-year-old rocks that geologists have identified as part of the Avalon terrane of eastern Massachusetts. How did they get to central Massachusetts? Geologists believe that the west-dipping Avalon plate sank deep beneath the Nashoba plate in a subduction zone. Geophysical evidence indicates that the subducted tip of the Avalon plate extends to a considerable depth beneath central New England, where it meets the east-dipping Laurentian margin. If this is true, the Avalonian rocks in the Pelham dome may have been brought to the surface as the dome rose during Devonian time. These 600-million-year-old igneous rocks in the Pelham dome mark the westernmost occurrence of rocks with Gondwanan fingerprints—namely, an age similar to the major rock formations of the Avalon terrane.

MERRIMACK TERRANE

Many of the rocks of central Massachusetts are part of the Merrimack terrane, one of four Gondwanan terranes in Massachusetts. The Merrimack terrane consists of six northeast- to north-trending belts. From west to east, they are the Ware, Gardner, Wachusett Mountain, Southbridge, Nashua, and Rockingham belts. The belts include a mixture of metamorphosed sedimentary, volcanic, and plutonic rocks of Ordovician, Silurian, and early Devonian age, with a few rocks of Pennsylvanian age. Distinct structural features such as a giant fold or a similar metamorphic history delineate each belt.

Spectacular collisional structures characterize the eastern and western margins of the Merrimack terrane. The western margin is the collision zone between Laurentia and Gondwana; the eastern margin, marked by the Clinton-Newbury fault, is the subduction boundary between the Nashoba terrane and the Merrimack terrane. On the geologic map you can see an enormous concentration of faults along the eastern margin of the Merrimack zone. Both the Merrimack terrane and the Nashoba terrane to the east represent small continental blocks that formed along the fringe of Gondwana.

Silurian and Devonian Schists

The Merrimack terrane contains the westernmost edge of a massive breadth of west-dipping rock of Silurian age. The main rocks, the Paxton schist and its less-metamorphosed equivalents, the Oakdale and Eliot formations of the Nashua and Rockingham belts, occur throughout the terrane. They formed the original crust of the Merrimack microcontinent. These rocks plunged beneath the Laurentian continental crust during the Acadian collision in late Silurian to Devonian time. In places, the Silurian strata eroded prior to the deposition of sediments in Devonian time. The main way to distinguish the Silurian schists and gneisses from the Devonian is that the latter, the Littleton schists, preserve primary sedimentary structures, such as graded beds, even in zones of high-grade metamorphism.

Ware Belt

Gray-weathering schists of Silurian and Devonian age in tight folds dominate the eastern half of the north-northeast trending Ware belt. The west-dipping schists are the east limb of a great fold that was squeezed out of the collision zone between the continental plates. The western half of the Ware belt consists of the 4-mile-wide Hardwick pluton. The distinctive Coys Hill granite occupies the middle of the belt along with some mafic and ultramafic plutons near the Mass Pike. The southern end of the Hardwick pluton is cut off along the eastern margin of the Monson gneiss. Major faulting accompanied the domal uplift of the Monson gneiss in Devonian time. The Coys Hill granite also pinches out 6 miles from the Connecticut line, probably because of faulting.

The Hardwick pluton of early Devonian age has two distinctive granites: a light-colored, muscovite-biotite granite, commonly with large crystals and low in quartz, and a banded, fine-grained black granite gneiss, sometimes with large crystals. The black granite obtains its color from very coarse and abundant biotite and magnetite.

The Coys Hill granite of early Devonian age is unforgettable because of its large crystals of microcline feldspar. French geologists named similar granites in the Alps *dents de cheval,* or "horse-tooth" granites. Twin crystals of feldspar are commonly 2 to 4 inches square and 3/4 inch thick. In many localities, the closely spaced feldspars nearly crowd out the coarse biotite and garnet groundmass. Some of the feldspars have granulated borders due to brittle fracturing, and in some places the feldspar crystals are elliptical and look like eyes.

Close-up view of characteristic coarse-grained, rectangular as well as rounded feldspars of the Coys Hill granite, Rock House cliffs, West Brookfield.

Gardner Belt

The Gardner belt of the Merrimack terrane is an anticline in the westernmost edge of an astounding breadth of strata of Silurian age that underlie substantial parts of the Merrimack belt all the way east to the Atlantic Ocean near the New Hampshire line. The rocks of the Gardner belt, like the Ware belt, are part of the east limb of upside-down rocks squeezed out of the collision zone between the continental plates. The Gardner belt consists of a thick sequence of limy quartz and feldspar sandstone members of the Paxton schist that appear abruptly below the Littleton schists.

Wachusett Mountain and Southbridge Belts

The Wachusett Mountain belt is the easternmost, far-traveled Devonian nappe, or giant fold. It was squeezed up and out of the collision zone and moved at least 15 miles to the east over the previously formed folds of the Gardner and Ware belts. The crystalline rocks of the Wachusett Mountain belt lie on top of a large block of Paxton schists that form the Southbridge belt. Together, the two belts form a nested pair of synclines. The upper Wachusett Mountain belt was thrust over the Southbridge, and then both belts were folded at the same time. To understand this, picture the Wachusett Mountain belt shaped like a canoe resting on the Southbridge belt of Paxton schist, with the south prow of the canoe plunging northward. The Wachusett Mountain belt is absent south of Worcester, possibly eroded away.

The Fitchburg granite intruded the rocks of the Wachusett Mountain belt 402 million years ago in early Devonian time. This was before they were thrust up and to the east from the depths of the Acadian collision zone. In cross section the granite pluton looks like a squid swimming east over a bit of Devonian Littleton schist and, below that, a great thickness of the Silurian Paxton schist.

Nashua and Rockingham Belts

The Silurian and Devonian sedimentary rocks of the Nashua belt are essentially devoid of igneous intrusions. The low-grade metasedimentary rocks contain well-developed sedimentary structures. The most easterly belt of the Merrimack terrane, the Rockingham belt, occupies a wedge whose southern point is just north of Clinton in north-central Massachusetts. The Clinton-Newbury fault zone defines the eastern boundary of the Rockingham belt. The Nashua and Rockingham belts are discussed in more detail in the Eastern Seaboard chapter.

MESOZOIC RIFT BASINS

The supercontinent Pangaea was finally assembled by the end of Paleozoic time, about 250 million years ago. The assembly did not last long, geologically speaking, as Pangaea began to rift apart about 200 million years ago in Mesozoic time. We know rifting was under way in earnest in eastern Massachusetts because the 300-foot-wide Medford gabbro dike intruded about 190 million years ago in Jurassic time. Rift valleys formed throughout New England as the northern landmasses of North America, Greenland, and Europe broke away from the southern continents of South America and Africa as the Atlantic Ocean began to open between them. The North Atlantic basin opened in late Jurassic time and the South Atlantic basin in Cretaceous time. The ocean floor continues to rift apart to this day.

The Connecticut Valley rift basin complex, including the Deerfield and Hartford Basins in Massachusetts, has a 50-million-year history of faulting, sediment deposition, and lava eruption. Crystalline strata of Paleozoic age, such as those exposed on Mount Warner west of Amherst, underlie the valley at various depths and were extensively faulted. By early Jurassic time, deep-reaching faults tapped magma that squirted up to the surface as dikes and poured out broadly on the surface as the Holyoke, Deerfield, and

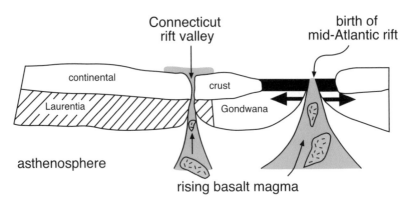

Pangaea begins rifting apart near the suture zone of the Laurentian continent with the microcontinental blocks of Gondwana. —Modified from Hoffman, 1991

Hampden basalts. The dikes fed magma to fissure flows that once covered much of southern New England. Basalt dikes at North Spencer, Holden, and Tyngsboro, which intrude older crystalline rocks far to the east of the Connecticut Valley, served as feeder dikes to much more widespread flood basalts. The amount of basalt flows and related volcanic rock increases southward toward Long Island Sound.

The 80-mile-long Hartford Basin of Massachusetts and Connecticut reaches a maximum breadth of 25 miles near the Massachusetts-Connecticut border. The basin extends from New Haven to Northampton, just north of Mount Holyoke and Mount Tom. The Deerfield Basin, a much smaller feature just north of the Hartford Basin, is 25 miles long and 7 miles broad. It extends from just north of the Holyoke Range to Bernardston. A cross-valley uplift, or fault block, of older crystalline rocks of Paleozoic age divides the basins near Amherst.

The Deerfield and Hartford Basins slid down against the Connecticut Valley border fault, which bounds their eastern margins. The effects of such faulting are dramatically illustrated by lava ridges shaped like boomerangs with both horns cut off at the border fault. The fault dropped a great block of basin strata of late Triassic and Jurassic age thousands of feet down. The older, harder rocks to the east moved up relative to the valley strata.

As rifting progressed into Jurassic time, rotated fault-block mountains, similar to mountain ranges of the Basin and Range in Nevada, shed sediments into the adjacent basins. Swiftly flowing rivers deposited gravel and sand in alluvial fans along the base of the valley border faults. The now-eroded Pelham dome, the Belchertown pluton, and the Glastonbury dome in the eastern highlands and metamorphic schists of the Berkshire Hills to the west served as sediment sources. The rivers transported coarse stream deposits and floodplain mudstones some distance from the alluvial fans. Sediments and volcanic rocks accumulated in the rift basin for about 50 million years.

Dinosaur Footprints

The Mesozoic sedimentary rocks contain dinosaur footprints, fish remains, insects, and other fossils. Before geologists discovered age-determining fossils in the Connecticut Valley red rocks, they thought these rocks might be of Devonian age, as are the redbeds of the Catskill Mountains of New York. The first clue to the presence of Jurassic wildlife in the valley was the discovery of dinosaur tracks in the mudstones of the Shuttle Meadow formation of the Hartford Basin in 1802. Later, other distinctive fish, insect, and reptilian fossils of Jurassic age were found.

Dinosaur footprints. A boy named Pliny Moody discovered the prints in Jurassic strata while plowing a field in South Hadley in 1802. —Edward Hitchcock drawing, 1841

Life after Jurassic Time

By the end of Cretaceous time, about 65 million years ago, the entire region had been eroded nearly to sea level, and a blanket of sediment rested unconformably on top of the older crystalline rocks. The geologic record for the Connecticut Valley is sparse until 23,000 to 22,000 years ago, when the last ice sheet left its mark. Prior to the ice age, the Connecticut River flowed through the basin formed during continental rifting in Mesozoic time. The river, like a slow-spinning buzz saw, cut down through the basalt and other ridges in response to gradual uplift of the land. This downcutting was probably accomplished by about the beginning of Miocene time, 24 million years ago.

The pre-Pleistocene landscape of the Connecticut Valley was in many respects much the same as the present except that the valley was deeper and broader. The topography controlled the Pleistocene ice and the location of glacial lakes that formed from the melting glacier.

GLACIAL GEOLOGY

Most of the glacial features visible in the Connecticut Valley and surrounding uplands were left by the Connecticut Valley–Bronson Hill lobe of the Wisconsinan ice sheet, the last glacial advance. Numerous drumlins, large glacial erratics, and scattered rock outcrops characterize the upland areas. Stratified glacial sediments, postglacial floodplains, and swamp deposits fill valleys.

The western part of the lobe occupied the Connecticut Valley and receded more slowly than the eastern part of the lobe that sat on the uplands. The difference in melting rates produced complex relationships among drumlins, lake bottom deposits, and delta sands and gravels. For instance, streams washed sediment from the melting uplands over the top of the ice lobe in the valley.

Glacial Lake Hitchcock

As the glacier receded, large meltwater lakes filled basins. The largest in Massachusetts, Glacial Lake Hitchcock, filled the Connecticut Valley. This lake, named for pioneer geologist Edward Hitchcock of Amherst College, extended some 220 miles from the sediment dam at Rocky Hill, Connecticut, to its northern shore at St. Johnsbury, Vermont.

Glacial Lake Hitchcock began as a puddle on the north side of a 1-mile-wide series of delta sediments that plugged the Connecticut Valley lowland about a dozen miles south of Hartford, Connecticut, at Rocky Hill. The lake increased in length as the ice melted northward. A spillway, the New Britain Channel, existed to the west of Rocky Hill and permitted Glacial Lake Hitchcock to drain slowly, thus preserving for many years the fragile sand and gravel deposits of the dam.

At maximum size, Glacial Lake Hitchcock was about 20 miles wide, with islands of basalt ridges. Prior to the lake's formation, the Holyoke Range and East Mountain, ridges of basalt, stood up as nunataks, or bedrock islands within the ice. Glacial meltwater channels along the ice margins on both sides of the ridge cut into the bedrock. Once the ice melted, waterfalls formed where meltwater streams cutting through the bedrock were suddenly at a higher elevation than the surrounding lowland. Wave-cut cliffs and benches attest to lake levels more than 300 feet above the valley bottom.

Sediments at the bottom of a glacial lake often have couplets, or varves, of dark clay layers alternating with light silty layers. Geologists believe that varves are a seasonal phenomenon. The coarser-grained sediments—silts and sands—are deposited in summer when warm weather triggers melting of sediment-laden ice. This layer grades up into a layer of finer sediments—clay—deposited during winter when ice covers the lake, and fine particles

*Glacial Lake
Hitchcock.*
—Modified from
Larsen and Koteff,
1988; Lougee, 1957

and organic matter settle to the bottom. You can imagine that if you counted every varve in the lake bottom deposits, you would know the age of the lake.

A tunnel excavated in 1946 on the Smith College campus in Northampton uncovered glacial varves. The first field trip of my career was through this excavation site, led by Marshall Schalk of Smith College, where I saw the alternating summer and winter layers of sediment precipitated from the glacial meltwater. Using my pocketknife, I carved out a specimen of those varves for Weston Observatory Museum. Today, the specimen is as hard as a brick.

Glacial varves from lake bottom clays excavated from the campus of Smith College in 1946 in Northampton.

An ingenious geologist, Ernst Antevs, studied the varves from one end of Glacial Lake Hitchcock to the other in 1927. Believing that each varve represented a yearly cycle, Antevs counted these very fine-grained layers from the south end of the lake northward and came up with 4,000 years for the duration of the lake. Counting varves was a previously undiscovered, untested, and even simplistic field method of Antevs's own devising. At the time many geologists did not accept his work, but after radiometric age dating was developed and performed on carbon-based materials in the lake sediments, Antevs's ballpark figure proved remarkably close.

Based on radiometric dating, geologists determined that the southern ice margin stood in central Connecticut until about 15,000 to 14,000 years ago. The lake then began to form as the ice front receded north. The ice margin had receded to the Chicopee delta of Glacial Lake Hitchcock near Springfield, Massachusetts, by about 14,000 years ago. Glacial Lake Hitchcock to the south of the Holyoke Range drained about 12,700 years ago, when the water breached the dam at Rocky Hill. The entire lake drained when water breached the sediment dam at the Holyoke Range about 12,400 years ago, yielding a total life span of 3,600 years. The life span of Glacial Lake Hitchcock was longer than that of any other glacial lake in southern New England.

Delta Deposits in the Lake

The Connecticut Valley not only records the presence of Glacial Lake Hitchcock but also the formation of enormous deltas of stratified sand and gravel laid down on lake bottom sediments. Meltwater rivers and streams from the highlands east and west of the valley carried great volumes of sediment into the valley from time to time. Braided streams deposited alluvial fans above lake level, while other streams fed the lake with sediments. Mastodons and woolly mammoths inhabited the lakeshores and were hunted and eaten by early Americans.

Two great deltas and a few smaller deltas extended into Glacial Lake Hitchcock in Massachusetts, giving rise to present-day sand plains. These include the great deltas of the glacial Chicopee and Westfield Rivers and smaller deltas such as the Sunderland delta near Deerfield.

Glaciated Upland

As the continental ice sheet moved southward over Massachusetts, it scraped away weathered bedrock and earlier glacial deposits, leaving relatively unaltered bedrock throughout the upland. The valleys, commonly the location of deeply leached fault zones, were literally scooped out by the glacial bulldozer.

The Wisconsinan ice sheet transported weathered bedrock and whatever residual glacial materials were left on the land's surface by previous Pleistocene ice sheets. As it melted, the ice sheet shed its load, a 10- to 15-foot-thick blanket of till. This heterogeneous mixture of clay, sand, gravel, and boulders now drapes the glacially polished and scratched bedrock. The matrix of till is typically a compact clay with tiny grains of crystalline rocks, some of which may be weathered.

A notable feature of some glacial deposits in the Bronson Hill Upland is a deep iron-manganese staining in the upper, permeable zones. This stain is especially prevalent in the eastern Ware and Gardner belts of the Merrimack terrane that are underlain by Ordovician to Devonian schists. Leaching of iron and manganese from fragments of rusty-weathering sulfidic schists produces this color.

Don't mistake the iron staining on the surface for the typically reddish basal layer of glacial sediment. These sediments were derived from the red rocks of the Connecticut Valley and transported to the uplands.

Throughout the upland, stratified sands and gravels are mainly confined to river and stream valleys. Runoff from the melting ice cascaded into valleys, where water deposited stratified sediments in ponds and lakes. Small glacial lakes formed in the uplands from time to time. Some valleys

Rum Rock, a coarse pegmatite, is one of largest glacial erratics in the central Massachusetts region. It is west of Massachusetts 32, just inside the Barre Township boundary. —W. C. Alden photo, U.S. Geological Survey

contain 30 to 50 feet of unstratified till, and a few are filled with as much as 90 to 100 feet of glacial drift, much of which may be till.

At least 275 drumlins exist in central Massachusetts. These streamlined hills may reach a thickness of 100 feet. Some drumlins have bedrock cores; others contain a core of older till from the Illinoian ice sheet. The Wisconsinan glaciation scraped away much of the Illinoian till and buried what little remains, but the older till is sometimes exposed where the glacier smoothed off the tops of hills. The Illinoian till is dark gray and compact, with only a few pebbles and cobbles, and is generally unoxidized. The younger Wisconsinan till is typically oxidized olive gray to olive brown, is sandier than the older till, and contains more pebbles and cobbles. It is not compacted. The older drumlin till was deposited, compacted, and weathered before the latest episode of glaciation began.

Postglacial Rebounding

After the ice melted from the region and glacial lakes drained, the earth's crust rebounded. Since Glacial Lake Hitchcock drained, the Connecticut River has cut 160 feet down into the lake deposits near Springfield and has eroded laterally over 2 miles on each side of the present river. This uplift and resulting downcutting through the glacial deposits exposed the lake bottom varved clays that permitted Ernst Antevs to work out his calendar for the life span of Glacial Lake Hitchcock.

Originally horizontal planes associated with glacial lakes' water levels—planar features such as wave-cut terraces, cliffs, and shorelines—are now tilted upward 4 to 5 feet per mile to the north. John Peper showed that a glacial lake delta north of Palmer at 650 feet in elevation is on the same glacial lake water plane as the South Monson spillway at 615 feet in elevation. A similar uplift rate of about 5 feet per mile in deltas of Glacial Lake Hitchcock corroborates Peper's uplift rate.

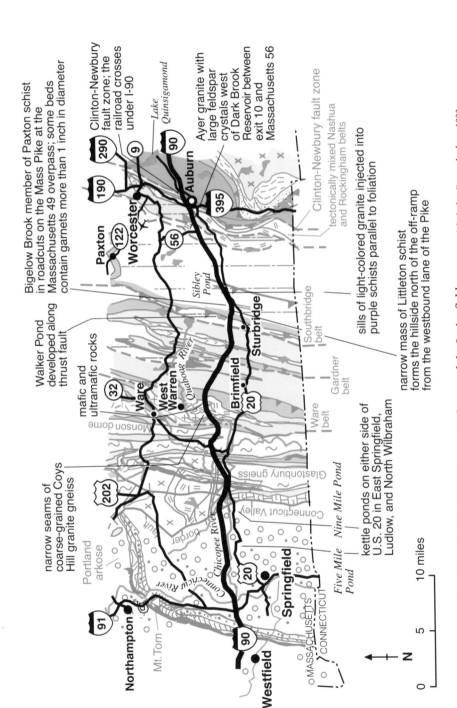

Bigelow Brook member of Paxton schist in roadcuts on the Mass Pike at the Massachusetts 49 overpass; some beds contain garnets more than 1 inch in diameter

Clinton-Newbury fault zone; the railroad crosses under I-90

Lake Quinsigamond

Ayer granite with large feldspar crystals west of Dark Brook Reservoir between exit 10 and Massachusetts 56

Clinton-Newbury fault zone

tectonically mixed Nashua and Rockingham belts

sills of light-colored granite injected into purple schists parallel to foliation

narrow mass of Littleton schist forms the hillside north of the off-ramp from the westbound lane of the Pike

Walker Pond developed along thrust fault

mafic and ultramafic rocks

Ware

West Warren

Quaboag River

Brimfield

Sturbridge

Southbridge belt

Gardner belt

Ware belt

narrow seams of coarse-grained Coys Hill granite gneiss

Monson dome

Glastonbury gneiss

Connecticut Valley border fault

Nine Mile Pond

Chicopee River

kettle ponds on either side of U.S. 20 in East Springfield, Ludlow, and North Wilbraham

Portland arkose

Five Mile Pond

Connecticut River

Mt. Tom

Springfield

Northampton

Westfield

MASSACHUSETTS
CONNECTICUT

N

0   5   10 miles

Paxton

Worcester

Auburn

Sibley Pond

Geology along the Mass Pike between Auburn and the Springfield area. —Modified after Zen and others, 1983

ROADGUIDES TO THE CENTRAL LOWLAND AND BRONSON HILL UPLAND

Interstate 90 (Mass Pike)
Auburn—Springfield Area
46 miles

The Mass Pike crosses a fascinating array of narrow bands of rocks between Auburn and the Connecticut Valley border fault. Roadcuts abound because construction workers blasted this major thoroughfare straight through the bedrock obstacles in their way. The Mass Pike traverses the collision zone between the Laurentian continent and the amalgamated pieces of Gondwana before dropping down into the Connecticut River valley along the Chicopee River.

Auburn

Auburn was settled in 1714. It was originally named Ward for the acclaimed Revolutionary War officer Artemas Ward. But the name was changed in 1837 to end the confusion between Ward and Ware Townships, and perhaps also in response to the British poet Oliver Goldsmith's phrase, "Sweet Auburn, loveliest village of the plain" from "The Deserted Village," published in 1770. By any name, it is a lovely place to begin or end a trip.

Auburn sits on the Clinton-Newbury fault zone, a spectacular structure that bounds the eastern margin of the Merrimack terrane. The Nashoba microcontinent collided with the eastern edge of the Merrimack terrane during an early stage in the assembly of the Pangaean supercontinent. Later, the collision boundary became an oblique strike-slip fault. Faults are concentrated along the eastern margin of the Merrimack zone near Auburn. The Mass Pike crosses the Clinton-Newbury fault at the Providence & Worcester Railroad overpass just east of I-290. Here the fault separates the Tadmuck Brook schist and Nashoba formation of the Nashoba terrane from the Boylston schist in the Merrimack terrane.

Nashua Belt

The 5-mile area between the intersection of the Mass Pike with U.S. 20 in Auburn Township and with Massachusetts 56 in Oxford Township consists mainly of narrow fault blocks in the Nashua belt. This belt consists primarily of lightly metamorphosed siltstones, slates, and phyllites of the Oakdale formation and the overlying Worcester formation. West of Dark

Brook Reservoir, between exit 10 in Auburn Township and Massachusetts 56, a pluton of Ayer granite intrudes the Oakdale sediments along the western border of the Nashua belt. The granite may have intruded along a fault between the Oakdale formation and the Paxton schist, which is thought to be merely a higher metamorphic equivalent of the Oakdale formation. The 432-million-year-old Ayer granite of Silurian age resembles a leopard skin with big gray spots—large feldspar crystals.

Southbridge Belt

The alert roadside observer who has driven Massachusetts 2 across northern Massachusetts will remember that the Wachusett Mountain belt lay just west of the Nashua belt. Along the Mass Pike, the Southbridge belt is west of the highly faulted Nashua and Rockingham belts. A colossal collision between the Merrimack microcontinent and Laurentia pushed the rocks of the Wachusett Mountain belt eastward over the top of the Southbridge belt. The entire package of rock was folded together and plunged down to the north. Erosion has removed the Wachusett Mountain mass south of Worcester.

Highly contorted, boudinaged, and brecciated granite dikes in the dark Southbridge member of the Paxton schist. Looking north along the westbound lane of the Mass Pike just east of Northside Road overpass in Charlton. —Mary Margaret Cooney photo

The Paxton formation, named for the town of Paxton, where hundreds of stone walls have been constructed of this slabby schist, makes up most of the Southbridge belt and is one of the most broadly distributed formations in the state. It extends without major interruption along the Mass Pike from just west of Massachusetts 56 in Oxford Township to 1 mile west of the junction with I-84 in Sturbridge Township.

The Mass Pike crosses the Southbridge and Bigelow Brook members of the schist. The Southbridge member, interbedded quartz- and feldspar-rich rocks and lime silicate beds, covers a broad area east of Massachusetts 169 in Charlton City. The Bigelow Brook member of the Paxton schist is west of Massachusetts 169. It has been thrust over the Southbridge member along the Black Pond fault. I-90 crosses the fault at the Brookfield Road underpass. The Bigelow Brook member resembles the Southbridge member but contains garnets larger than 1 inch in diameter. The Southbridge member in southern Massachusetts is free of garnet-bearing schist beds.

The Bigelow Brook member of the Paxton schist contains sills of gray to white Devonian granite that parallel the foliation of the enclosing schist. Geologists call these intrusions *lit-par-lit* intrusions, French for "bed-by-bed." You can see one from the Massachusetts 49 bridge over the Mass Pike.

View to the north of boudinaged granite in Paxton schist. Outcrop in the median, opposite the west end of the service station along the eastbound lane of the Mass Pike and west of Depot Road overpass in Charlton Township. Note fault blocks and relative movement indicated by arrows. —Mary Margaret Cooney photo

Granite sills intrude the Bigelow Brook member of the Paxton schist along the low-angle, west-dipping foliation. Looking southeast from Massachusetts 49 bridge over the Mass Pike (Mile Marker 79.2) at the Charlton town line. —Mary Margaret Cooney photo

The prominent and linear, northeast-trending valley of Walker Pond at Wells State Park follows the west-dipping Hamilton Brook thrust fault. The fault busted up the rocks and then the ice age glaciers excavated them, forming a conspicuous valley. A belt of diorite intrudes a narrow mass of Littleton schist that forms the hillsides east and west of Walker Pond. The Littleton schist overlies the Bigelow Brook member.

Sturbridge Graphite Mines

In 1633 on a scouting expedition from Plymouth, John Oldham and Samual Hall encountered Indians whose faces were blackened by graphite from a place called Black Hill, which may be the same as Leadmine Hill, or Mountain, in southwestern Sturbridge Township. In 1638, John Winthrop Jr. purchased the tract of land and attempted to mine the graphite. He called it the Tantiusque Lead Mine (*lead* is the old word for graphite). People sporadically attempted to mine the graphite from 1640 to 1904 with little success, but shafts and tunnels honeycomb the area.

Graphite is highly metamorphosed coal, which is in turn cooked peat. We do not know for sure when the peat was originally deposited or when the coal formed. Coal in Massachusetts is primarily Pennsylvanian in age, but the graphite at Sturbridge is present in rocks of probably Ordovician age, possibly squirted into fractures during deformation.

To see the abandoned workings, take I-84 south to the second inter-change. Follow the frontage road farther south to Vinton Road and turn west. Turn northwest onto Leadmine Road just west of Leadmine Brook. Leadmine Hill is west of Leadmine Pond.

Gardner Belt

The Mass Pike crosses from the Southbridge belt onto the Gardner belt 1 mile west of the I-90 junction with I-84. The Gardner belt is a westward-inclined and westward-dipping giant fold of Paxton schists that has been complexly folded with an equally deformed batch of Partridge schists of Ordovician age. Along the Mass Pike in Sturbridge and Brimfield Townships, look south for sweeping views of hogback ridges and narrow valleys along the trend of the near-vertical foliation and layering of the metamorphosed, folded strata.

View to north of dark gray Partridge formation. Small granitic sills, possibly of Silurian age, parallel the foliation, and a thicker dike of granite, possibly of Devonian age, cross-cuts the foliation and earlier sills at a gentle angle. Along the Mass Pike 0.75 mile southeast of the Warren-Brimfield town line. —Mary Margaret Cooney photo

Ware Belt

The West Warren complex, a northeast-trending belt of mafic and ultramafic rocks near Warren, traverses nearly the breadth of Massachusetts. These rocks were squeezed up from the seafloor as the Gondwanan and Laurentian continents collided and swallowed the Iapetus Ocean in the Acadian collision of Devonian time. The West Warren complex of Devonian age is closely associated with the narrow strip of Coys Hill granite, which also runs nearly the width of the state, pinching out 6 miles short of the Connecticut line. The complex contains diorite, norite, and other more mafic rocks. Peridotite, a greenish black rock from the earth's mantle, appears in the hills near West Warren and northern Brimfield. Good outcrops exist near Reed Street and Brook Road just north of I-90.

Along the Mass Pike, the Coys Hill pluton is about 0.5 mile wide but thickens to about 1 mile wide near Massachusetts 9 in West Brookfield. It contains large, horse-tooth crystals of microcline and some garnet and sillimanite. The Coys Hill granite intrudes rusty-weathering, sulfidic schists of Silurian and Devonian age. Warren, north of the Mass Pike and rich in rusty schists, was called *Quabaug*, meaning "red-water place," by the Native Americans. Leaching of the easily weathered schists, and possibly other iron-bearing rocks, produces reddish seeps of groundwater.

Bronson Hill Domes

The Bronson Hill belt of crystalline rocks, the easternmost rocks of the Laurentian continent, is a series of gneiss domes mantled by schists and quartzite. The boundary between the schists of the Ware belt and the Monson gneiss of the Bronson Hill dome is just west of the Quaboag River. West Brimfield nearly straddles the boundary between Laurentia on the west and pieces of Gondwana on the east.

The Monson gneiss of Ordovician age forms the core of the Bronson Hill belt. The Glastonbury dome lies to the west of the Monson gneiss dome. The Great Hill syncline, a fold in volcanic rocks of Ordovician age and some Silurian quartzite, continues south between the Monson and Glastonbury domes to Great Hill on the Connecticut line. Between Palmer and Ludlow, both the Mass Pike and U.S. 20 are confined to a steep-sided valley formed of the Monson and Glastonbury gneiss domes draped with Ordovician schists.

The Mass Pike passes just south of the Belchertown plutonic complex of Devonian age, which bulges westward into the Connecticut Valley southeast of the Holyoke Range. That complex is essentially a domal uplift that deformed the northern end of the Glastonbury dome as well as the southern end of the Pelham dome.

Connecticut Valley Border Fault

The Mass Pike crosses the Connecticut Valley border fault in Ludlow Township, 1.6 miles west of the Mass Pike bridge over the Chicopee River, just east of the river's right-angle turn to the west. The border fault forms the eastern boundary of a rift basin of Mesozoic age. Rocks of the Bronson Hill belt to the west of the normal fault dropped thousands of feet down relative to rocks on the east and are now buried by sediments that filled the deep basin. Despite the fill, the boundary between the uplands and the Connecticut River valley is pronounced. The Connecticut Valley border fault follows the steep front of Minechoag Mountain to the north and Sunset Ridge to the south in Wilbraham Township.

Glacial deposits cover most of the bedrock along the Mass Pike from the border fault to west of the Connecticut River. Rocks of the Portland formation, reddish brown arkose and sandstone deposited in the early Jurassic rift basin, surface along the Chicopee River. You can see them

Close view of dipping redbeds of the Portland arkose. Under the south abutment of the Massachusetts 21 bridge over the Chicopee River near Indian Orchard in Springfield.
—Mary Margaret Cooney photo

under the south abutment of the Massachusetts 21 bridge. This locality also features coarse cobble conglomerate and numerous potholes in the Chicopee River bed. A few scattered outcrops of Portland arkose occur along the Mass Pike between the Connecticut River and the West Springfield exit.

Chicopee River and Its Glacial Delta

Three meandering rivers—the Swift, Ware, and Quaboag—originate in the mountainous upland of central Massachusetts and join at a village appropriately named Three Rivers. Together they form the Chicopee River, an important river in glacial times. The glacial Chicopee River deposited the vast Chicopee delta in the eastern part of Glacial Lake Hitchcock, which filled the Connecticut Valley as the ice sheet melted. Between Wilbraham and the Connecticut Valley you will travel over the Chicopee delta deposits that were washed down from the highlands into the lake.

The Chicopee delta still maintains a generally triangular shape with the apex pointing up the Chicopee River to the metamorphic highlands of Wilbraham and Palmer Townships. Because of postglacial rebounding of the earth's crust, the Chicopee River cut deeply into the sediments of the delta. Ludlow, at the apex of the delta, was incorporated separately from Springfield in part because of the difficulty of crossing the deeply incised river. The numerous ponds on either side of U.S. 20 in East Springfield, Ludlow, and North Wilbraham are kettles.

Interstate 91
Connecticut River Valley
55 miles

A great rift that would become the future Connecticut Valley split the northern part of the Pangaea supercontinent almost 200 million years ago. Triassic and Jurassic sedimentary rocks and basalts filled the rift basin, and dinosaurs roamed the region. Glacial Lake Hitchcock occupied the valley during the melting of the last ice sheet, and the modern Connecticut River follows a course cut through the rift basin rocks by its preglacial predecessor. You can see remnants of rifting and glacial history as you travel along the valley corridor.

Glacial Lake Hitchcock

Sediments of Glacial Lake Hitchcock cover the lowlands along I-91. This enormous lake extended from a sediment dam at Rocky Hill, Connecticut,

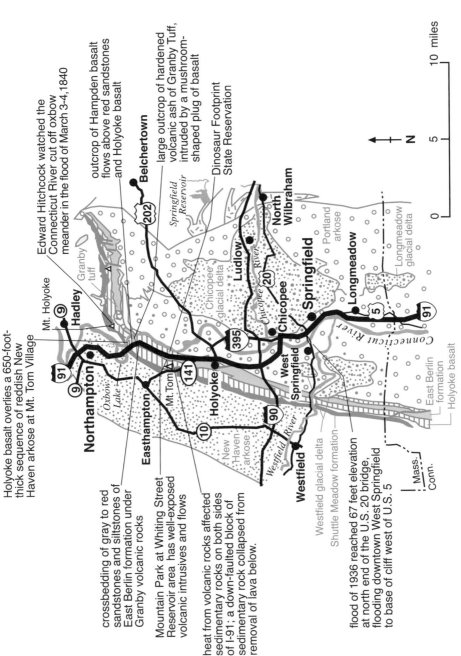

Holyoke basalt overlies a 650-foot-thick sequence of reddish New Haven arkose at Mt. Tom Village

Edward Hitchcock watched the Connecticut River cut off oxbow meander in the flood of March 3-4, 1840

outcrop of Hampden basalt flows above red sandstones and Holyoke basalt

large outcrop of hardened volcanic ash of Granby Tuff, intruded by a mushroom-shaped plug of basalt

Dinosaur Footprint State Reservation

crossbedding of gray to red sandstones and siltstones of East Berlin formation under Granby volcanic rocks

Mountain Park at Whiting Street Reservoir area has well-exposed volcanic intrusives and flows

heat from volcanic rocks affected sedimentary rocks on both sides of I-91; a down-faulted block of sedimentary rock collapsed from removal of lava below.

flood of 1936 reached 67 feet elevation at north end of the U.S. 20 bridge, flooding downtown West Springfield to base of cliff west of U.S. 5

Belchertown

202

Springfield Reservoir

Granby tuff

Mt. Holyoke

9

Hadley

Northampton

91

9

Oxbow Lake

Easthampton

Mt. Tom

141

10

Holyoke

90

Westfield

Westfield River

New Haven arkose

Westfield glacial delta

Shuttle Meadow formation

West Springfield

395

Chicopee

Chicopee River

Ludlow

20

North Wilbraham

Springfield

Chicopee glacial delta

Portland arkose

Longmeadow

5

91

Connecticut River

East Berlin formation

Holyoke basalt

Longmeadow glacial delta

Mass. Conn.

N

0   5   10 miles

Geology of the Hartford Basin in the Connecticut Valley between Longmeadow and Northampton. —Modified from Zen and others, 1983

to St. Johnsbury, Vermont. It began to form in central Connecticut about 15,600 years ago as the ice sheet receded.

Between the Connecticut line and the I-91 bridge over the river in Chicopee, I-91 runs on gravel fill that raises the highway above the level of the floodplain. Since Glacial Lake Hitchcock drained catastrophically about 12,700 years ago, the Connecticut River has cut 160 feet vertically into the lake deposits of the Springfield area and has eroded sideways over 2 miles on each side of the present river. The earth's crust rebounded after the weight of the ice was gone and triggered the downcutting of the river. I-91 runs along the base of a 30- to 40-foot-high sandy terrace of the postglacial Connecticut River.

Note the steep slopes held up by retaining walls just south of the I-91 bridge across the Connecticut River in Springfield. During pile-driving operations for bridge construction, varved clay deposits slumped, requiring stabilizing procedures. Varved clays are notably slippery and unstable.

Two major deltas and one small delta formed in Glacial Lake Hitchcock near Springfield. The smaller one, the Longmeadow glacial delta, formed along the Massachusetts border on the east side of the valley. The southern margin of the ice sheet lay northeast of Bass Pond, about 2.5 miles east of Springfield College. A south-flowing meltwater stream deposited outwash down a narrow valley between the ice tongue in the center of the Connecticut Valley and the hills of till in eastern Springfield and East Longmeadow. The stream emptied into Glacial Lake Hitchcock south of the border and formed the delta. When the ice sheet melted north of the Chicopee River, the meltwater abandoned the Longmeadow course and delta and followed the river.

The Chicopee delta still maintains a generally triangular or deltoid shape, with the apex pointing upriver to the metamorphic highlands of central Massachusetts. A large volume of sediment poured from this region into Glacial Lake Hitchcock. Because of regional uplift, the Chicopee River has cut deeply into the delta sediments.

The swift, east-flowing Westfield River formed a delta on the west side of the Connecticut Valley. Glacial meltwater carried coarse- to fine-grained sands and gravels from the Berkshires. The Westfield River deposited its load as delta sediments in the western arm of Glacial Lake Hitchcock, west of Provin and East Mountains, ridges of basalt lava.

Dunes

The most extensive dunes in New England lie on the eroded sand-plain deposits of Glacial Lake Hitchcock. In postglacial times, the Connecticut River cut down through soft lake sediments, and winds blowing from the north produced transverse (east-west), longitudinal (north-south), and parabolic sand dunes. In Longmeadow west of I-91 and east of the river, and in Agawam west of the river, large longitudinal to curved dune fields formed on a lower river terrace cut in lake bottom sediments. These dunes are closely associated with channels cut by the river during floods. The dune fields are 2 miles long and 650 feet wide.

Complex transverse and parabolic dunes formed on the upper river terrace in Longmeadow south of the Springfield line and Porter Lake, and southwest of Franconia Golf Course. Dunes also formed in western Chicopee from south of the Springfield line to north of the I-91 bridge across the Connecticut River. The dunes on the upper terraces may be up to 50 feet high.

Hartford Basin

The Hartford Basin, part of the continental rift basin of Jurassic age, extends from New Haven, Connecticut, to Northampton. It is 15 to 25 miles wide and 80 miles long, and it filled with a thick sequence of sedimentary and volcanic rocks. A fault block of Paleozoic basement rocks separates the Hartford Basin from the Deerfield Basin farther north.

The first sedimentary rock to fill the valley—the 6,500-foot-thick, coarse- to fine-grained New Haven arkose of Triassic age—was deposited in alluvial fans. This conglomeratic sandstone consists, in part, of weathered feldspars from deeply weathered granitic rocks in the nearby mountains. Coarse sediments, now red and brown, were most likely various shades of brown and gray when deposited but reddened after burial as iron-rich groundwater circulated through the rock and deposited iron minerals such as hematite.

The red mudstones, sandstones, and conglomerates of the New Haven arkose contain soil layers cemented by calcium carbonate, known as caliche, which contain variously shaped white calcite veins, including vertical tubelike forms. The presence of caliche soils tells us that the climate was semiarid, receiving only 4 to 20 inches of annual rainfall.

The Shuttle Meadow formation consists of red lakebeds and is overlain by fine-grained gray mudstone, sandstone, and black shale of the 550-foot-thick East Berlin formation. Lakes and streams that lapped onto the alluvial fans along the eastern escarpment deposited these sediments, some of which extend west beyond the faulted and eroded margin of the Hartford Basin.

The top of the sediment stack, the Portland formation, consists of lake sediments covered with 3,900 feet of mainly braided river sands and mudstones deposited in floodplains. All along the eastern highland, river-deposited fans interfingered and merged one into the other.

youngest	Portland formation (early Jurassic)
↑	Granby tuff (early Jurassic)
	Hampden basalt (early Jurassic)
	East Berlin formation (early Jurassic)
	Holyoke basalt (early Jurassic)
	Shuttle Meadow formation (early Jurassic)
	Hitchcock volcanic rocks (early Jurassic)
oldest	New Haven arkose (early Jurassic and late Triassic)

Sequence and ages of sedimentary and volcanic rocks of the Hartford Basin.

Volcanic Rocks

Basalt magma erupted from the rift faults within and outside of the valley during early Jurassic time, spreading into large lava flows and squeezing up along fractures as vertical dikes. The Hartford Basin experienced three outpourings of lava and volcanic rocks—the Hitchcock volcanic rocks, the Holyoke basalt, and the Hampden basalt with its associated Granby tuff, all of early Jurassic age.

Two flows of Holyoke basalt—a 320-foot-thick flow and a 255-foot-thick flow—form a north-trending mountain ridge from Hartford, Connecticut, to Easthampton Township, where it turns east. The ridge includes the rugged profiles of Provin Mountain, East Mountain, and Mount Tom along the west side of the valley. The Connecticut River cuts through the basalt ridge at the village of Mount Tom in Easthampton. From this big bend in the basalt ridge, the Holyoke Range runs easterly to where the Connecticut Valley border fault chops it off abruptly.

The view of Mount Tom in the Holyoke Range, looking southeast from Northampton, reveals a very steep western profile and a gentler southeastern slope. The gentle slope is parallel to the original surface of the basalt flow. The basalt cracked into great columns at right angles to the layering of the lavas. The fractures shape the steep cliff face on the west side of the range.

I-91 crosses through this sequence of basalt between Whiting Street Reservoir and the Mount Tom water gap. Numerous outcrops are visible from the highway. Near the reservoir, sedimentary rocks of the East Berlin formation outcrop on both sides of I-91. Mountain Park at the Whiting Street Reservoir has well-exposed volcanic rocks and lava flows.

As I-91 curves to the northwest to pass through the water gap, roadcuts expose the volcanic and sedimentary sequence. If you're headed north, you'll pass through the sequence from younger to older rocks. You'll see the Granby tuff, Hampden basalt, East Berlin formation, and Holyoke basalt. At the Mount Tom village, Holyoke basalt overlies a 650-foot-thick sequence of New Haven arkose.

Dinosaur Footprints

A boy named Pliny Moody found the first dinosaur tracks known in North America while plowing on the family farm at South Hadley in 1802. Some of them were of truly biblical proportions, so people called them the tracks of "Noah's Raven." The scientific and possible religious implications of these tracks fascinated the Reverend Edward Hitchcock, a professor of geology at Amherst College. He collected more than 21,000 of them in the early nineteenth century. No trip through the Connecticut Valley would be complete without examining Hitchcock's collection at the Pratt Museum of Natural History on the campus of Amherst College.

Although only two fossil skeletons have been documented from the Connecticut Valley, the study of footprints leads geologists to believe that the Jurassic dinosaurs of the basin fell into two distinct groups. A bird-hipped vegetarian frequently visited the lakeshores and drying lake floors.

Casts in sandstone of tracks of the dinosaur, Coelophysis, *a fast runner about 8 to 10 feet long. The little bumps are raindrop imprints.*
—From outdoor museum of Weston Observatory

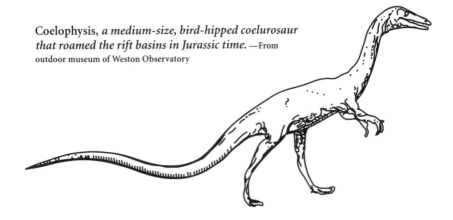

Coelophysis, *a medium-size, bird-hipped coelurosaur that roamed the rift basins in Jurassic time.* —From outdoor museum of Weston Observatory

Small to large lizard-hipped dinosaurs occupied another ecological niche on drier ground.

The Dinosaur Footprint State Reservation is on the east side of U.S. 5 along the Connecticut River at Smiths Ferry. The thin-bedded red shale of the Portland formation was long famous here because large dinosaur tracks and plant fragments were well exposed. John Ostrom of Yale tracked nineteen pairs of *Eubrontes* whose 15-inch-long prints headed west from the Connecticut River.

Mount Tom Water Gap

By the end of Cretaceous time, about 65 million years ago, the entire region had been eroded nearly to sea level. The land was uplifted perhaps 15 million years ago during Miocene time. As the land rose, the Connecticut River cut gently through the sedimentary rock layers. The uplift of land was slow enough that when the river reached the ridge of Holyoke basalt, it kept cutting down through the rock instead of changing its course.

The Roaming Connecticut River

In Hadley and Northampton Townships, the floodplain of the Connecticut River is up to 3 miles wide. Since townships along the river have long placed their common boundaries in the middle of the river, meandering changes in the river channel result in township borders shifting from flood to flood. The accompanying gain and loss of land depends on the artistic whim of the river. In the vicinity of Hadley, Hatfield, and Northampton, the river has meandered over 2.5 miles from a straight southerly course to within a few hundred yards of the western margin of the basin.

Oxbow in the Connecticut River and Northampton as seen from Mount Holyoke.
—Sketch by Ora Hitchcock, wife of Edward Hitchcock; published in Hitchcock, 1841

In the flood of 1936, the Connecticut River was on the brink of taking a shorter course right through the village of Hadley. The river stayed its course, but Hadley remains strategically located for a quick trip downriver sometime in the future.

In anticipation of a similar sudden change of course by the Connecticut River in flood stage, Professor Edward Hitchcock took students to the top of Mount Holyoke on March 3 and 4, 1840. They observed the fickle river abandon its meandering course along the oxbow south of Northampton and forge a direct course south through the water gap in the Holyoke Range. The river left behind Oxbow Lake, which you can see along I-91 and from lookouts on Mount Holyoke and Mount Tom.

Mount Warner and the Amherst Inliers

Mount Warner and other hills near Amherst are composed of crystalline rocks of Paleozoic age—as is the Bronson Hill Upland to the east. These rocks, part of a large fold that trends northwest, are in an uplifted fault block. The block separates the Hartford Basin to the south from the Deerfield Basin to the north. These crystalline rocks probably underlie at great depth the sedimentary rocks of the basins.

View looking southwest from Mount Holyoke. Lava ridge of Mount Tom at far left. Connecticut River at lower right. —Carol Skehan photo

A view from Mitch's Marina on Massachusetts 47 on the east side of the Connecticut River in Hadley looking west-southwest across the floodplain during flood on April 4, 1998. Twelve feet of the light post remained above flood stage in this secluded cove near the main course of the river. Rows of trees indicate low-water position of riverbank.

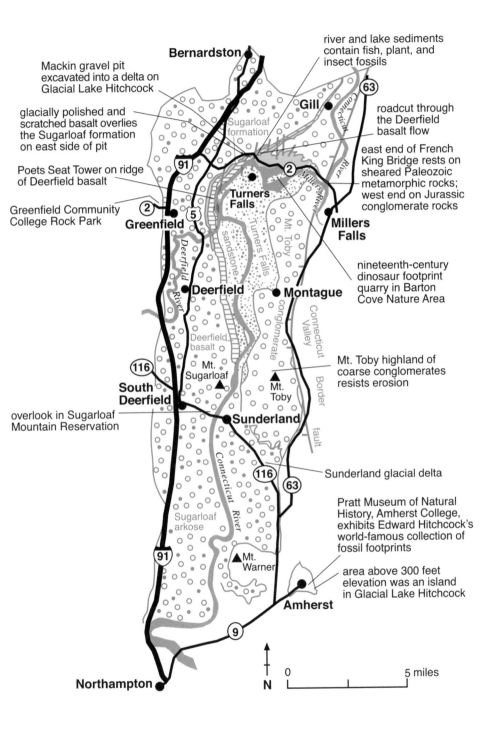

Geology of the Deerfield Basin in the Connecticut Valley between Northampton and Bernardston. —Modified from Zen and others, 1983

Deerfield Basin

The Deerfield Basin has much less volcanic activity and fewer deep-seated faults than the Hartford Basin. The basin's one lava flow, the 180-foot-thick Deerfield basalt, is sandwiched between coarse- to fine-grained sedimentary strata. Rocky Mountain—an appropriately named, thick, east-dipping, black ridge of Deerfield basalt—is in the Pocumtuck Range in Greenfield. Poets Seat Tower, on the crest of the lava flow between Turners Falls and Greenfield, boasts a spectacular view of the valley and surrounding uplands.

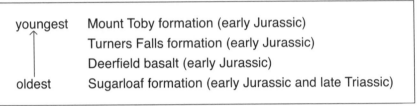

youngest   Mount Toby formation (early Jurassic)

Turners Falls formation (early Jurassic)

Deerfield basalt (early Jurassic)

oldest   Sugarloaf formation (early Jurassic and late Triassic)

Sequence and ages of sedimentary and volcanic rocks of the Deerfield and Northfield Basins.

The oldest sedimentary rock in the basin is the Sugarloaf formation, a sequence of reddish arkose and some conglomerate, sandstone, siltstone, and shale of late Triassic and early Jurassic age. The arkose is a coarsely granular rock composed of weathered feldspar and sand derived from eroded granite. The Deerfield basalt overlies the Sugarloaf formation and is in turn overlain by the Turners Falls sandstone, primarily an arkose with conglomerate and shale.

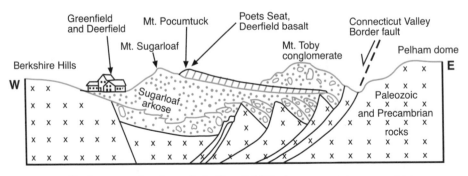

Generalized cross section through the Deerfield Basin. —Modified from Wise and Belt, 1991

The Mount Toby formation tops the sequence. It is an alluvial fan complex composed of pebbles of granitic gneiss. It resists erosion because of a cement of albite feldspar. It also contains black shale lakebeds and breccias that formed from ancient landslides.

Sugarloaf Mountain

The outcrops around the overlook in Sugarloaf Mountain State Reservation in South Deerfield along Massachusetts 116 are Sugarloaf arkose. The lookout provides a magnificent view to the south of Mount Tom, Mount Holyoke, and the Seven Sisters—seven wavy ridges of Holyoke basalt that define the northern margin of the Hartford Basin. The anticlinal neck of Paleozoic rocks that divides the Deerfield and Hartford Basins may be visible to the southeast, about halfway to the Holyoke Range. To the northeast you can see Mount Toby.

Sunderland Delta and Hanging Valley

Braided meltwater streams formed the Sunderland delta of Glacial Lake Hitchcock. Coarse sediments in the delta are productive aquifers that recharge readily because of the highly permeable overlying sediments and

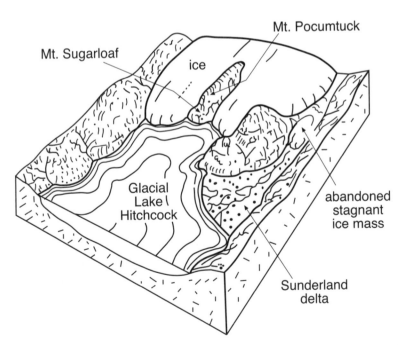

Mount Sugarloaf and the Pocumtuck Range emerge from beneath the receding ice front.
—Modified from Brigham-Grette, 1991

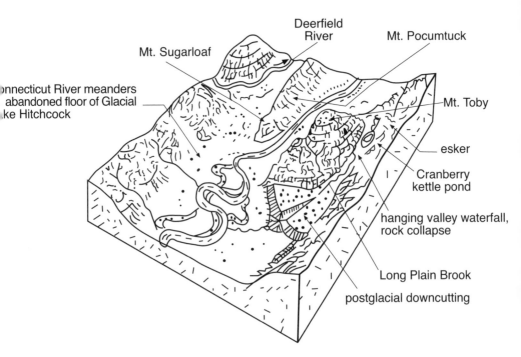

Glacial Lake Hitchcock drained from the Deerfield Basin about 12,400 years ago. Long Plain Brook eroded through the Sunderland delta, depositing a small fan at its foot. Stream sediments clogged the valley west of Mount Sugarloaf, diverting the Deerfield River to the north. —Modified from Brigham-Grette, 1991

soils. After the lake drained, Long Plain Brook eroded through the delta, depositing a small fan at its foot.

At the head of Long Plain Brook, you can see a poorly developed hanging valley with a waterfall and a rock collapse along the east side of Mount Toby. In the waning stages of the ice age, a residual, stranded glacier sat in the narrow valley between the uplands to the east and the east slope of Mount Toby. An east-flowing stream from Mount Toby emptied into the valley, dumping its load of sediment over the top of the ice block. When that ice block melted, the stream cascaded over the steep valley wall, giving rise to the modest hanging valley and waterfall just south of Cranberry Pond, the kettle created by the buried ice block. The eastward dip of the arkosic sandstone beds along the valley wall renders the rock unstable, causing a collapse.

Turners Falls

South of Massachusetts 2 in Turners Falls, fine-grained lake deposits of the Turners Falls formation contain fossils of fish, clams, and plants such as horsetails, rushes, and ferns. These fossiliferous rocks are well exposed along the rocky banks of the Connecticut River and its backwaters at Turners Falls. Sandstone of the Turners Falls formation contains burrows and abundant traces left by walking and swimming arthropods. Impressions of freshwater clams exist in mudstones and sandstones of the nearby Sugarloaf arkose.

Because the late Triassic rocks lack fish fossils, the abundance and diversity of fish fossils in early Jurassic strata (more than fifty species) raise

Four genera of fossil fishes in the sedimentary rocks of the Connecticut Valley.
—From McDonald, 1982

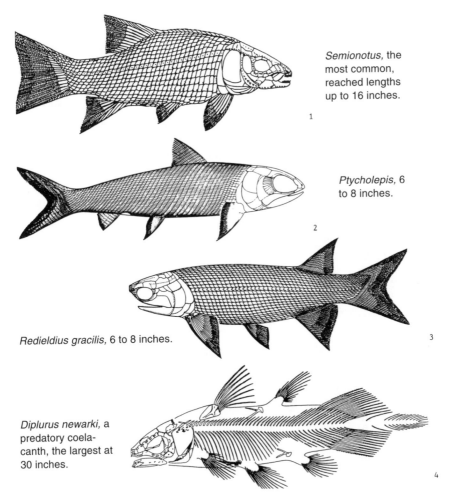

Semionotus, the most common, reached lengths up to 16 inches.

1

Ptycholepis, 6 to 8 inches.

2

Redieldius gracilis, 6 to 8 inches.

3

Diplurus newarki, a predatory coelacanth, the largest at 30 inches.

4

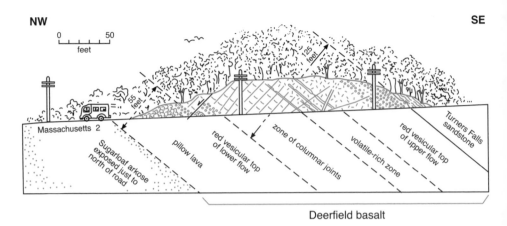

NW

SE

0   50
feet

125 feet

55 feet

Massachusetts 2

Sugarloaf arkose exposed just to north of road

pillow lava

red vesicular top of lower flow

zone of columnar joints

volatile-rich zone

red vesicular top of upper flow

Turners Falls sandstone

Deerfield basalt

Schematic cross section of Jurassic lavas and associated strata along Massachusetts 2 in Greenfield just west of the Connecticut River. —Modified from Wise, Hubert, and Belt, 1992

questions concerning the rate of evolution and dispersal of fish. One species, *Ptycholepis,* was common in European marine strata and probably swam to the lakes from the nearby, newly forming Atlantic Ocean.

A lengthy roadcut along Massachusetts 2 west of the Connecticut River exposes strata of the Deerfield Basin. This cross section includes the fossil-fish-bearing Turners Falls sandstone as well as the thick volcanic layers that form the lava ridge between Turners Falls and Greensfield. Parking places and picnic spots on Massachusetts 2 near Riverside and near the lava ridge make exploration of the rock formations easy. You can see three distinct layers of Deerfield basalt in the roadcut. Pillows of basalt created by extrusion of lava into water form the bottom layer, basalt with columnar joints occupies the middle layer, and reddish vesicular basalt tops the lava flow.

The Mackin gravel pit southeast of the split of Massachusetts 2 and Massachusetts 2A in Greenfield Township is part of a small delta of Glacial Lake Hitchcock. You can see pillow basalts overlying the Sugarloaf formation at the edge of the pit. The glacier polished and scratched the basalt.

Greenfield Community College Rock Park

The Greenfield Community College Rock Park, on a knoll behind the Science Wing of the college, contains 132 specimens from the long and rich geologic history of the Connecticut Valley and from older geological formations in the adjacent uplands. To build the college, workers wisely excavated varved clays of Glacial Lake Hitchcock and replaced them with more stable delta gravels.

A specimen of Turners Falls sandstone at the rock park contains unusual armored mud balls. Mud balls form when a chunk of mud from a streambank falls into the water. The current tumbles and rounds the chunk, which eventually rolls along the streambed, picking up the pebbles that form its armor. The mud ball must be quickly buried in sediment to be preserved. Only about ten known localities of mud balls exist in the world. You can see more mud balls at outcrops of sandstone below Turners Falls Dam.

Bernardston Nappe and Northfield Outlier

Folded Ordovician, Silurian, and Devonian metasedimentary strata similar to those of the Bronson Hill domes underlie the 6 miles between Greenfield and the Vermont border. These rocks are part of a large fold—the Bernardston nappe.

The Northfield outlier is a small fault block along the Connecticut Valley border fault, 1 mile north of the Deerfield Basin and 1 mile south of the Vermont border. Like the bigger basins to the south, it contains early Jurassic sandstones, siltstones, and shale.

Armored mud balls enclosed in maroon Turners Falls sandstone of the Connecticut Valley, one of only about ten known occurences in the world. Specimen is in the Greenfield Community College Rock Park. —Richard Little photo

Massachusetts 2
Fitchburg—Connecticut River
50 miles

Massachusetts 2 winds up and over the Bronson Hill Upland of central Massachusetts. The eastern quarter of the route is in the Nashua River basin and the western three-quarters is in the Millers River drainage basin. The hilly terrain exposes intermediate- to high-grade metamorphic rocks and granites with complex mineral assemblages and wondrous geologic histories.

Fitchburg Granite

The Fitchburg plutonic complex, a 5-mile-wide, north-trending belt of granite and granite gneiss of early Devonian age, is in the middle of the 12-mile-broad Wachusett Mountain belt. The granite intruded the rocks of the Wachusett Mountain belt 402 million years ago, some distance west of its present location. The whole package was thrust up and to the east from the depths of the Acadian collision zone. In cross section the granite pluton looks like a squid swimming east over a bit of Devonian Littleton schist.

Think of the Fitchburg granite as a long jelly roll wrapped in a blanket of gray-weathering schists. The top of the jelly roll has eroded, exposing both the granite and patches of the schist. Swarms of schist are present between South Street on the Fitchburg-Leominster line and Westminster. The package of rocks forming the jelly roll, in turn, rests on and is surrounded by more metasedimentary rocks—the Paxton schists. The jelly roll structure goes by a more elegant name—the Southbridge syncline.

The Fitchburg granite was once quarried at Rollstone Hill. This biotite and muscovite granite typically contains tourmaline, sometimes as sprays of radiating crystals, and small amounts of garnet, magnetite, apatite, and zircon. Pegmatites in the granite have splays of tourmaline where they contact inclusions of schist.

Near Mill Pond Restaurant at the junction of Massachusetts 2A and Princeton Road on the west side of Fitchburg, gray, foliated granite outcrops contain inclusions of gneiss and schist of Silurian to Devonian age.

Several phases of the Fitchburg complex are exposed at and near Wachusett Mountain State Reservation in Westminster and Princeton Townships, west of Massachusetts 140. Plutonic masses of Fitchburg granite and pegmatite surface in areas to the southeast and east of the reservation. A stack of large, broadly recumbent folds of Paxton schist and Littleton schist crop out at lower elevations and as inclusions in the plutonic complex. Horizontal sills of granite up to 150 feet thick intrude the schists.

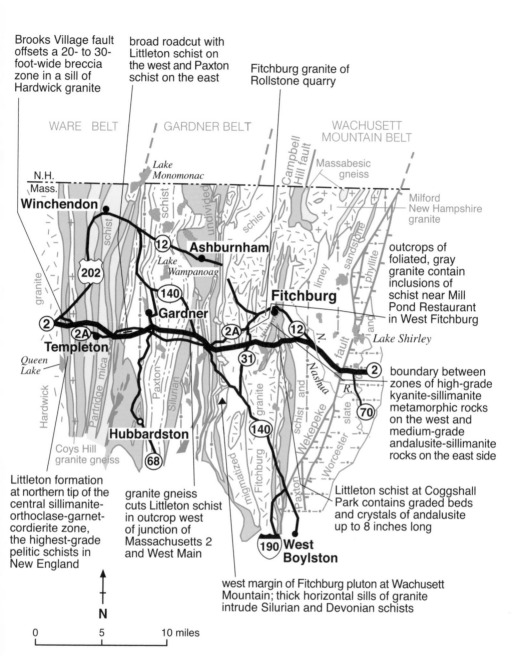

Brooks Village fault offsets a 20- to 30-foot-wide breccia zone in a sill of Hardwick granite

broad roadcut with Littleton schist on the west and Paxton schist on the east

Fitchburg granite of Rollstone quarry

WARE BELT   GARDNER BELT   WACHUSETT MOUNTAIN BELT

Lake Monomonac

N.H.
Mass.

Winchendon

Massabesic gneiss

Milford New Hampshire granite

schist

granite

undivided

schist

Campbell Hill fault

202

Lake Wampanoag

12   Ashburnham

140

Gardner

Fitchburg

limey

sandstone

phyllite

outcrops of foliated, gray granite contain inclusions of schist near Mill Pond Restaurant in West Fitchburg

2   2A

Templeton

Queen Lake

Hardwick

Partridge mica

Paxton

Silurian

2A

31

granite

12

Nashua

schist and

Lake Shirley

Wekepeke

slate

fault and

Worcester

2

70

boundary between zones of high-grade kyanite-sillimanite metamorphic rocks on the west and medium-grade andalusite-sillimanite rocks on the east side

Coys Hill granite gneiss

Hubbardston

68

migmatized

Fitchburg

Paxton

Littleton schist at Coggshall Park contains graded beds and crystals of andalusite up to 8 inches long

Littleton formation at northern tip of the central sillimanite-orthoclase-garnet-cordierite zone, the highest-grade pelitic schists in New England

granite gneiss cuts Littleton schist in outcrop west of junction of Massachusetts 2 and West Main

140

190   West Boylston

west margin of Fitchburg pluton at Wachusett Mountain; thick horizontal sills of granite intrude Silurian and Devonian schists

N

0   5   10 miles

Geology along Massachusetts 2 between Fitchburg and Templeton.
—Modified from Zen and others, 1983

foliated gneiss is exposed over a breadth of one-half mile straddling the crest of Wachusett Mountain and underlies much of the eastern shore of Wachusett Lake. Paxton schist underlies the western shore. Foliated biotite and muscovite granitic gneiss with inclusions of Littleton schist underlie the eastern slope of Wachusett Mountain. Massachusetts 31 (East Princeton Road) northeast of Princeton village and Sterling Road east of Merriam Road cross an ellipical pluton of Fitchburg granite.

Rollstone Boulder

A glacier deposited the 100-ton Rollstone Boulder on the crest of Rollstone Hill, just south of the center of Fitchburg. The glacial erratic is Kinsman quartz-monzonite, a distinctive granitic rock that outcrops more than 15 miles to the north in New Hampshire. To facilitate quarrying of the Fitchburg granite that forms Rollstone Hill, workers removed the erratic by splitting it into sections. They reassembled it on Main Street in downtown Fitchburg.

High-Grade Schists

Between Fitchburg and Templeton Townships, we cross high-grade metamorphic rocks. The schists along the east side of the Fitchburg granite are in the garnet zone of metamorphism, and those west of it are in the high-grade sillimanite-muscovite zone. The mineral assemblage of sillimanite and muscovite requires both high temperature and pressure to

Rollstone Boulder on Main Street in downtown Fitchburg. —Jacqueline Skehan Babineau

crystallize. These rocks must have formed deep within the earth to have reached the required conditions. The magma of the Fitchburg granite likely produced some of the heat of metamorphism.

An interesting mineral assemblage is present in outcrops of schists at Coggshall Park in the southern part of Fitchburg off Mount Elam Road near Massachusetts 2. These Littleton schists contain garnet, staurolite, sillimanite, and andalusite. Sillimanite crystals rim andalusite crystals up to 8 inches long. Graded beds of the original sedimentary rock are still recognizable.

Near the Gardner-Templeton town line, a broad roadcut exposes Littleton and Paxton schists. The Littleton schist, on the west end, contains sillimanite and fine-grained graphite. The Paxton schist is on the east end of the roadcut.

Gardner Belt

Between Westminster and Gardner, Massachusetts 2 traverses the Gardner belt. This belt includes both Paxton and Littleton schists that dip east and west on opposite limbs of the Gardner anticline. The highway crosses a Mesozoic normal fault, which marks the east margin of the belt, just east of the intersection with Massachusetts 140 between Wyman Pond and Round Meadow Pond.

On the west side of the junction between Massachusetts 2 and West Main Street in eastern Gardner, granite gneiss cuts through an outcrop of Littleton schist. Massachusetts 2 crosses the west margin of the Gardner belt just west of the Gardner-Templeton township boundary.

Ware Belt

At Templeton, our route crosses the east margin of the Ware belt, the westernmost belt of the Merrimack terrane. Tight folds in the gray-weathering schists characterize the belt's eastern half.

You can see some of the highest-grade schists in New England on the east side of Queen Lake, southeast of the village of Phillipston on Barre Road at the junction with Massachusetts 101. These Littleton schists contain the minerals sillimanite, orthoclase, garnet, and cordierite.

The Coys Hill granite of early Devonian age intrudes the sillimanite schists of the Ware belt. Massachusetts 2 crosses the narrow, ¼- to ½-mile-wide sliver of granite near the junction with Massachusetts 2A, midway between Templeton and Phillipston Four Corners. Large crystals of feldspar form twin crystals, commonly 2 to 4 inches square and ¾ inch thick.

Massachusetts 2 crosses the eastern margin of the 4-mile-wide Hardwick granite on the western side of Templeton. This light-colored granite of Devonian age resembles the Fitchburg granite.

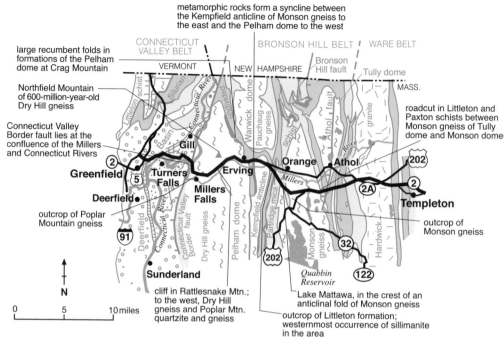

metamorphic rocks form a syncline between
the Kempfield anticline of Monson gneiss to
the east and the Pelham dome to the west

large recumbent folds in
formations of the Pelham
dome at Crag Mountain

Northfield Mountain
of 600-million-year-old
Dry Hill gneiss

Connecticut Valley
Border fault lies at the
confluence of the Millers
and Connecticut Rivers

outcrop of Poplar
Mountain gneiss

N

0   5   10 miles

roadcut in Littleton and
Paxton schists between
Monson gneiss of Tully
dome and Monson dome

outcrop of
Monson gneiss

cliff in Rattlesnake Mtn.;
to the west, Dry Hill
gneiss and Poplar Mtn.
quartzite and gneiss

Lake Mattawa, in the crest of an
anticlinal fold of Monson gneiss

outcrop of Littleton formation;
westernmost occurrence of sillimanite
in the area

*Geology along Massachusetts 2 between Templeton and the
Connecticut River.* —Modified from Zen and others, 1983

Bronson Hill Belt

In Athol Township, we encounter rocks of the Laurentian continent—
the Bronson Hill belt. When Gondwana collided with Laurentia in the
Acadian mountain building event in late Silurian through middle Devonian time, the gneisses of the continental crust of Laurentia rose to the
surface, squeezed out by intense pressures. Along our route, we pass the
Tully dome of Monson gneiss, the Monson dome of Monson gneiss, the
Warwick dome of Pauchaug gneiss, and the Pelham dome of Dry Hill gneiss.

Massachusetts 2 crosses the 10-foot-wide Athol fault just east of the South
Athol Road underpass in Athol. This normal fault of Mesozoic age has an
offset of 1,300 feet and cuts the southernmost end of the Tully dome.

A band of Littleton and Paxton schists separates the Tully dome from
the Monson dome. An outcrop along Massachusetts 2 in west-central Athol
exposes both schists. You can see the 440- to 445-million-year-old Monson
gneiss in a roadcut on Massachusetts 2 west of Massachusetts 122. Lake
Mattawa sits on the crest of an anticlinal fold of Monson gneiss. To the east
of the fold lies a syncline that exposes Ammonoosuc volcanic rocks and
the overlying Partridge formation of middle Ordovician age.

South wall of an extensive roadcut on Massachusetts 2 in Athol Township, just east of South Athol Road underpass. The Athol normal fault zone of cemented fault breccia separates the Tully body of Monson gneiss (left) from the sillimanite schist of the Littleton formation (right).

Along Massachusetts 2 between the junction with Massachusetts 122 and Erving, you can see a sequence of Ammonoosuc volcanic rocks, Partridge formation, ultramafic pods, Clough quartzite, Fitch formation, and Littleton schist. These stratified rocks covered the domes prior to uplift and are now exposed in a syncline between the Monson dome and the Kempfield dome, or anticline. Near Wendell Depot, Massachusetts 2 crosses over the Kempfield anticline of Monson gneiss.

A thin band of metamorphic rocks, 1 mile east of Erving, separates the Monson gneiss of the Kempfield anticline from the main body of older gneiss in the Pelham dome. To see some of these rocks, find a small outcrop of Littleton schist with a strong mica lineation about 1.2 miles east of Erving along the road. Walk 300 feet uphill from that outcrop to a ridge, 1.5 miles south of Laurel Lake. Here, amphibolite of the Erving formation of Devonian age records primary volcanic texture. From this outcrop, head 200 feet east to a large outcrop of well-bedded Clough quartzite that is highly contorted in the hinge of a major fold.

The 600-million-year-old Dry Hill gneiss, the oldest rock of the Pelham dome, was probably deposited as a layered volcanic rock. Because of its age

Poplar Mountain gneiss, a nearly flat-lying, sheared, dark gray rock characterized by large, oblong to rounded crystals of feldspar.

and mineralogy, geologists think it may have been part of the Avalon terrane of eastern Massachusetts. The west-dipping Avalon plate sank deep beneath the Nashoba plate in a subduction zone, possibly in Silurian or early Devonian time. The subducted tip extended west at a considerable depth beneath central New England. The Avalonian rocks in the Pelham dome may have been brought to the surface as the dome rose during Devonian time.

Dry Hill gneiss forms the bedrock of Northfield Mountain, which rises just north of the Millers River and Massachusetts 2 near Farley. A cliff at Rattlesnake Mountain southwest of the village of Farley exposes the contact of the Dry Hill gneiss and the 585-million-year-old Poplar Mountain gneiss, also part of the Pelham dome. Large crystals of microcline feldspar in the Poplar Mountain gneiss are probably sheared fragments of a pegmatite. You can see them in a roadcut on Massachusetts 2, 1 mile east of Millers Falls on the south slope of Poplar Mountain.

Connecticut Valley Border Fault

The Connecticut Valley border fault is a west-dipping normal fault along the eastern margin of the Deerfield Basin. The late Triassic and early Jurassic rocks of the basin cover a large fault block that dropped thousands of feet down relative to the Bronson Hill block to the east of the fault. Massachusetts 2 crosses the fault on the east side of the Connecticut River at its confluence with the Millers River. The east end of the French King Bridge rests on a narrow belt of Paleozoic rocks and the west end on Jurassic rocks.

The Millers River has eroded a deep valley into the Bronson Hill belt because of the much lower elevation of the down-faulted Deerfield Basin. The river meanders through a series of cascades from about 650 feet elevation near Athol to about 200 feet elevation at its confluence with the Connecticut River.

Between the Connecticut River and Greenfield, Massachusetts 2 traverses the northern end of the Deerfield Basin. The Berkshires to the west and the Pelham dome to the east shed coarse sediments onto alluvial fans in the basin. Some of the clasts deposited early in the rifting process are from low-grade metamorphic rocks that may have eroded from the surface of the Pelham dome after it was uplifted in Devonian time. The low-grade metamorphic rocks are completely eroded now, and we see only the high-grade rocks of the dome's interior in the uplands.

Massachusetts 9
Worcester—Amherst
50 miles

If you love unspoiled, scenic New England countryside, don't miss this trip over the forested hills of the central Massachusetts upland. Old-time architects designed this road back in the times of horse and buggy. It twists and turns, finding the easiest route from village to village. Roadcuts are sparse because the road builders did not blast through hills as they did on the Massachusetts Turnpike, but instead routed the road around hills.

Glaciation shaped much of the landscape along Massachusetts 9. As the last continental glacier crawled southward over Massachusetts in Pleistocene time, it scraped away weathered bedrock and most of the pre-Wisconsinan glacial deposits except for the older, basal parts of drumlins. The higher elevations now consist of relatively unaltered bedrock. The glacial bulldozer scooped out broken rocks along deeply leached faults, further deepening the existing valleys. It streamlined drumlins, deposited eskers and kames, and spread a thin mantle of ground moraine that tested the fortitude of farmers. Numerous lakes dot the landscape, some of which are kettles and some of which are the shrunken remnants of glacial lakes that occupied the narrow, scooped-out valleys.

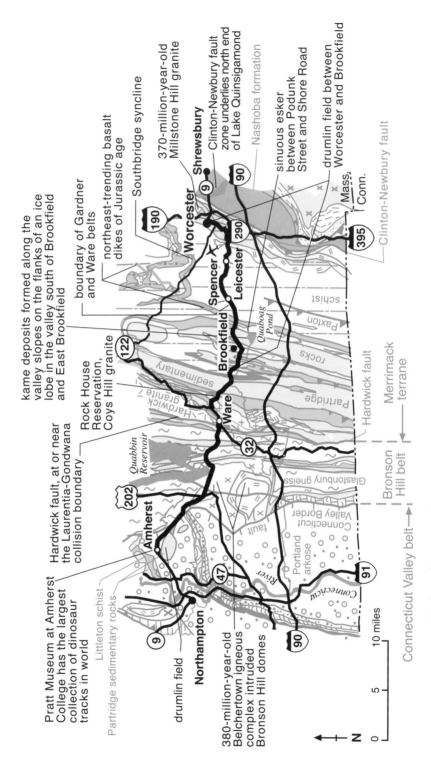

Pratt Museum at Amherst College has the largest collection of dinosaur tracks in world

kame deposits formed along the valley slopes on the flanks of an ice lobe in the valley south of Brookfield and East Brookfield

Hardwick fault, at or near the Laurentia-Gondwana collision boundary

Rock House Reservation, Coys Hill granite

boundary of Gardner and Ware belts

northeast-trending basalt dikes of Jurassic age

Southbridge syncline

370-million-year-old Millstone Hill granite

Clinton-Newbury fault zone underlies north end of Lake Quinsigamond

Nashoba formation

sinuous esker between Podunk Street and Shore Road

drumlin field between Worcester and Brookfield

Clinton-Newbury fault

Littleton schist

Partridge sedimentary rocks

drumlin field

380-million-year-old Belchertown igneous complex intruded Bronson Hill domes

Amherst

Northampton

Quabbin Reservoir

Ware

Hardwick granite

Partridge sedimentary

Brookfield

Spencer

Leicester

Worcester

Shrewsbury

Quaboag Pond

Paxton schist

Partridge rocks

Glastonbury gneiss

Connecticut Valley Border fault

Portland arkose

Connecticut River

Bronson Hill belt

Hardwick fault

Merrimack terrane

Mass. Conn.

Connecticut Valley belt

N

0   5   10 miles

Geology along Massachusetts 9 between Worcester and Amherst. —Modified from Zen and others, 1983

In places, the glacier melted back from the higher elevations while the valleys remained occupied by thick tongues of ice. Meltwater from the uplands flowed down ravines and deposited kames along the valley walls and on the sloping sides of the valley glacier.

Worcester

Worcester is on the east edge of what is sometimes called the central Massachusetts upland, a plateau dissected by erosion. A major drainage divide at Worcester separates the Blackstone River, which empties into Narragansett Bay, from the northeast-flowing Nashua River, which joins the Merrimack River in New Hampshire.

Faults of different ages coalesce in the eastern part of Worcester and in Shrewsbury Township, creating a number of fault blocks. At the north edge of the Blackstone River basin, the narrow Lake Quinsigamond valley follows the Clinton-Newbury fault zone, which divides the Nashoba terrane from the Merrimack terrane. Glacial ice scoured out the fault-broken rock of this valley. In a distance of slightly more than 2 miles west of Lake Quinsigamond, thin slivers of the Boylston formation, Ayer granite, Worcester phyllite, Millstone Hill granite, and Oakdale formation appear in rapid succession. The slaty and limy siltstone of the Oakdale formation underlies downtown Worcester. The east-dipping Wekepeke normal fault, just east of Lincoln Square near the junction of Massachusetts 9 with I-290, is responsible for the difference in elevation between the downtown lowland and high ground of resistant granite at Millstone Hill to the east.

Just west of Lake Quinsigamond, Pennsylvanian-age schists contain a fossiliferous meta-anthracite coal bed several feet thick. It occurs in a fault zone near Coal Mine Brook on Plantation Street. If swamps were extensive in the Merrimack terrane in Pennsylvanian time, their deposits have been largely eroded away. A fault block preserved this small remnant of Pennsylvanian history.

Glacial Lake Quinsigamond, considerably more extensive than the present Lake Quinsigamond, formed in the glacially scooped valley. A bedrock dam southeast of Worcester near Grafton impounded meltwater. Fluvial deposits of coarse cobble gravel up to 45 feet thick overlie a thin till layer in the upper slopes of the Quinsigamond and Poor Farm Brook valleys. These fluvial deposits grade down into pebble/cobble and sand beds that overlie foreset delta beds that dipped into the lake. Fine-grained lake bottom sediments up to 65 feet thick underlie the delta deposits. Kettles, such as Mud, Newton, Hall, and City Farm Ponds, formed at the north end of the Quinsigamond Valley.

In the southern part of Worcester at Hadwen Park on the west side of Kettle Brook, a splendid esker reaches a height of 30 feet. This narrow, well-developed, 2-mile-long esker trends southwest through the park. Geologist William C. Alden noted that the esker makes an "oxbow curve close to the hillslope on the west and then away from it, as if this slope had deflected the esker-forming stream." At the southern border of the park, at Clover and Webster Streets, the esker consists of unsorted coarse to fine gravel as well as boulders more than 1.5 feet in length.

Drumlin Field

A number of drumlins cluster in the central Massachusetts upland between Worcester and Brookfield Townships. The glacially streamlined hills have a pronounced southeasterly trend between Worcester and Spencer and a strong southerly orientation from Spencer to Brookfield. The ice shaped the drumlins in the direction it was traveling. Though ice movement in the Illinoian and Wisconsinan glaciations was generally to the southeast, it varied at the local level. In Cherry Valley 1.5 miles west of the Worcester-Leicester line, Chapel Street passes between two drumlins with slightly divergent long axes.

In southwestern Worcester, Fairlawn Hospital sits on the more easterly of two drumlins along Massachusetts 9 between Coes Reservoir and Massachusetts 122. Incidently, Loring J. Coes Jr., a research mineralogist at Norton Company in Worcester, is the namesake of Coes Reservoir. In 1953, he synthesized and described a very high-pressure form of silica, now called coesite. Subsequently, geologists have found the mineral coesite occuring naturally at meteorite impact sites.

Southbridge Syncline

Between the Worcester Polytechnic Institute and Spencer, Massachusetts 9 crosses nearly 12 miles of Paxton schist of Silurian age. The schist here is part of the Southbridge syncline, a gigantic fold structure in central Massachusetts produced by the collisions between the Gondwanan and Laurentian continental plates. About 2.5 miles north of Massachusetts 9 in Leicester Township, crystalline rocks of the Wachusett Mountain belt lie on top of the Paxton schist of the Southbridge syncline after being thrust from the west along a fault. Both belts of rock were then folded at the same time, creating a nested pair of synclines. The Wachusett Mountain belt is shaped like a canoe resting on the Paxton schist. The southernmost rocks of the Wachusett Mountain belt plunge northward under Asnebumskit Hill just 1.5 miles north of the Worcester Municipal Airport.

Gardner and Ware Belts

The west-dipping thrust-fault boundary between the Southbridge belt on the east and the Gardner belt on the west passes along the eastern margins of Lake Lashaway and Quaboag Pond in East Brookfield Township. The Gardner and Ware belts contain spectacular, highly contorted rocks of Ordovician, Silurian, and Devonian age—they developed enormous folds when the continental plates squeezed them up and out of the collision zone.

Kames and Eskers in the Quaboag River Valley

A sinuous esker winds southwest of East Brookfield between Podunk Street and Shore Road where the Brookfield River empties into Quaboag Pond. Kames occur along the valley slopes south of Brookfield and East Brookfield where the Quaboag and Brookfield Rivers converge in Quaboag Pond and as far south as the moraine at the head of Quacumquasit Pond. When the ice lobe occupied the Quaboag Valley farther to the west, kame terrace deposits formed on the southeast margin of the ice. They are on the northwest side of Massachusetts 67 in Warren.

Coys Hill and Hardwick Granites

Massachusetts 9 passes to the northeast of Coys Hill, the namesake of Coys Hill granite, a narrow sliver of rock that runs from New Hampshire almost to Connecticut. Coys Hill granite contains large twin crystals of

Rock House, an overhanging cliff of Coys Hill granite gneiss in Rock House Reservation in West Brookfield.

microcline feldspar—two crystals that grew together and share a common axis. You can see Coys Hill granite at Rock House Reservation, a public park on Massachusetts 9 in West Brookfield. Native Americans used the overhanging cliff of Coys Hill granite as a cold-weather hunting camp. The cliff and collapsed granite blocks offered protection from wind and snow, and the sun warmed the southeast exposure.

The town of Ware is near the southernmost tip of the Devonian Hardwick granite pluton. The Hardwick fault to the south cuts out the pluton, the neighboring Coys Hill granite, and the West Warren complex, which consists of Ragged Hill gneiss as well as mafic and ultramafic rocks. This fault zone runs near the eastern margin of the Bronson Hill domes. It is part of the fairly broad collision boundary between the Laurentian supercontinent and Gondwanan microcontinental fragments.

Bronson Hill Domes

Massachusetts 9 leaves the Ware belt and enters the Bronson Hill domes between Ware and Ware Center, crossing the Monson dome and skirting the southern edge of the Pelham dome. Here, the zone of domes is a mere 3.5 miles wide. Only 23 miles to the north, the zone is 14 miles wide. Part of the reason the belt is so narrow along Massachusetts 9 is because a large igneous intrusion at Belchertown fills the space. The Monson gneiss—biotite and plagioclase gneiss, amphibolite, and microcline gneiss—forms the cores of the Monson and Pelham domes. A narrow synclinal dimple of sulfidic mica schist of the Partridge formation separates these two linear ridges. The schist contains dark lenses of amphibolite.

Belchertown Igneous Complex

The Belchertown igneous complex is just south of Quabbin Reservoir. The igneous rocks crystallized 380 million years ago in middle Devonian time. The 70-square-mile complex looks like two rowboats side by side, cruising north toward the Pelham dome. A normal fault, probably of Mesozoic age, separates the two pieces, with the western block lifted up relative to the eastern block.

Geophysical gravity measurements, which analyze subsurface density of bedrock, reveal a saucer-shaped pluton about 1 mile thick except in the southeast corner where it has a funnel shape. The funnel may be where the magma rose to the surface. Many plutons are vertical cylinders that extend downward indefinitely, but the remarkable Beltchertown pluton is more like a sill, an intrusion that parallels the structure of the country rock.

This igneous body intrudes middle Ordovician Ammonoosuc volcanic rocks and Partridge metasedimentary rocks, as well as the early Devonian

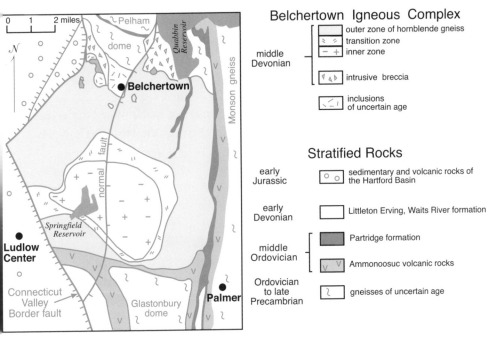

The Belchertown complex, a zoned pluton, shouldered aside rocks of the Bronson Hill domes. —Modified from Ashwal and others, 1979

Erving formation. Three Bronson Hill domes—the Monson, Pelham, and Glastonbury—and their associated cover rocks surround the pluton.

The igneous rocks of the Belchertown intrusion are zoned. The unaltered central core of the intrusion is granitic. It grades outward, through a transition zone, into hornblende gneiss whose foliation and lineation are the same as that in the surrounding rocks. This indicates that the complex and the surrounding rocks were deformed at the same time, probably at the end of the Acadian mountain building event, about 370 million years ago.

The town of Belchertown lies on the upthrown western side of the pluton, just west of the normal fault. Massachusetts 9 crosses the northern tip of the complex, including a small mass of breccia that formed during the Belchertown intrusion.

The Notch and Amherst

Between Belchertown and Amherst, Massachusetts 9 and the Connecticut Valley border fault pass through the 1,500-foot-wide Notch, within which nestle Arcadia Lake and Lake Holland. The east-trending basalt ridge of the Holyoke Range forms the west side of the Notch, and the Pelham dome forms the east side. The crystalline basement rocks on the west side of the fault dropped thousands of feet down relative to the Pelham dome on the east. The fault evolved in Mesozoic time during continental rifting.

The Pangean continent began to split apart, forming the Atlantic Ocean, about 200 million years ago. Smaller rifts formed along the edge of the continent, including here in the Connecticut Valley. Basalt welled to the surface and spilled out in flows, one of which formed the Holyoke Range to the south of Amherst.

Between the Notch and Amherst, Massachusetts 9 passes over sedimentary rocks that filled the rift basin. Red, pink, and gray sandstone and conglomerate of late Triassic age form a thin cover over the Ordovician and Silurian rocks that protrude through the basin strata south and north of Amherst. These crystalline rocks, called the Amherst inliers, are an uplifted fault block in the rift basin and divide the Deerfield Basin to the north from the Hartford Basin to the south. Mount Warner, to the west of Amherst, is one of the inliers. A drumlin field and stratified sediments of Glacial Lake Hitchcock cover the inliers to the south and north of Amherst and the Holyoke Range.

Amherst is home to Amherst College and the University of Massachusetts. The Pratt Museum at Amherst College houses a vast collection of dinosaur footprints collected in the early 1800s by Edward Hitchcock. Rocks deposited in the rift basin in Jurassic time preserved these footprints. In the 6 miles between Amherst and Northampton, Massachusetts 9 crosses rocks of the Hartford Basin that are covered by a thick veneer of lake sediments deposited in Glacial Lake Hitchcock. The area above 300 feet in Amherst was an island in this lake.

Massachusetts 32
South Monson—Royalston
50 miles

The western flank of the central uplands is one of Massachusetts's best-kept and most spectacular secrets. The southern two-thirds of our route is in the Chicopee River basin, the largest drainage basin in the state. The Swift, Ware, and Quaboag Rivers join in Palmer to form the Chicopee River, a principal tributary to the Connecticut River. The northern third of our route is in the Millers River drainage basin, also a tributary of the Connecticut River.

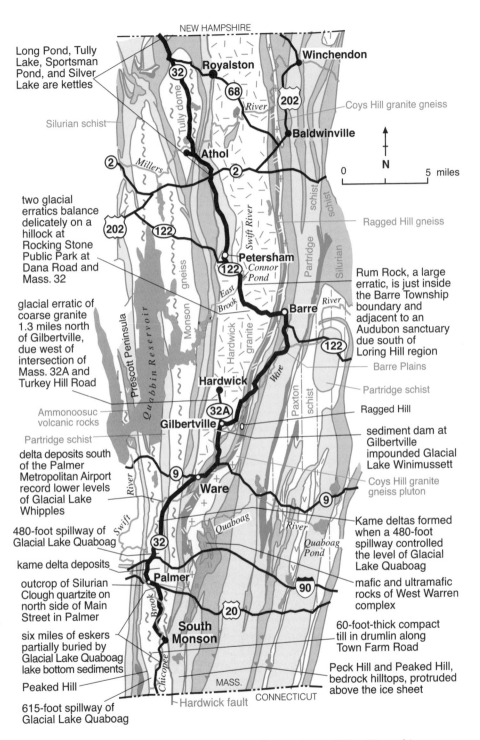

Long Pond, Tully Lake, Sportsman Pond, and Silver Lake are kettles

NEW HAMPSHIRE

Winchendon

Royalston

Silurian schist

Coys Hill granite gneiss

Baldwinville

Athol

Millers

N

0   5 miles

two glacial erratics balance delicately on a hillock at Rocking Stone Public Park at Dana Road and Mass. 32

Swift River

Ragged Hill gneiss

Petersham

Connor Pond

Rum Rock, a large erratic, is just inside the Barre Township boundary and adjacent to an Audubon sanctuary due south of Loring Hill region

glacial erratic of coarse granite 1.3 miles north of Gilbertville, due west of intersection of Mass. 32A and Turkey Hill Road

Barre

River

Barre Plains

Partridge schist

Ammonoosuc volcanic rocks

Hardwick

Ragged Hill

Gilbertville

sediment dam at Gilbertville impounded Glacial Lake Winimussett

Partridge schist

delta deposits south of the Palmer Metropolitan Airport record lower levels of Glacial Lake Whipples

Ware

Coys Hill granite gneiss pluton

480-foot spillway of Glacial Lake Quaboag

Quaboag   River

Kame deltas formed when a 480-foot spillway controlled the level of Glacial Lake Quaboag

kame delta deposits

outcrop of Silurian Clough quartzite on north side of Main Street in Palmer

Palmer

mafic and ultramafic rocks of West Warren complex

South Monson

60-foot-thick compact till in drumlin along Town Farm Road

six miles of eskers partially buried by Glacial Lake Quaboag lake bottom sediments

Peck Hill and Peaked Hill, bedrock hilltops, protruded above the ice sheet

Peaked Hill

MASS.
CONNECTICUT

Hardwick fault

615-foot spillway of Glacial Lake Quaboag

Geology along Massachusetts 32 between Connecticut and New Hampshire.
—Modified from Zen and others, 1983

This region of elevated bedrock has local relief of 300 to 500 feet. The hills and relatively narrow valley bottoms generally trend north, following the bands of bedrock. The Connecticut Valley ice lobe moved southeast across the central Massachusetts upland, scouring and exposing bedrock on the southeastern side of hills and depositing till on the summits.

Glacial Doings at the Border

Glacial sands and gravels cover the bedrock at the Connecticut line. A sand pit in the southern part of Monson, just east of Massachusetts 32 and south of Blanchard Road, exposes sandy lake bottom deposits, esker ridge gravels, and boulder-rich outwash deposited in contact with the ice.

Eskers, or ice-tunnel deposits, have the sinuous, meandering shape of a stream channel. An absolutely spectacular esker, partially buried by Glacial Lake Quaboag bottom deposits, follows the valley of Chicopee Brook for at least 6 miles. Swift-flowing water must have deposited the coarse sediments of the esker.

As the last ice sheet retreated to the north-northwest, it melted from the higher peaks first. Peck Hill, Bear Hill, and Peaked Hill near the Connecticut line protruded above the much-thinned glacier.

An exposed cross section of the boulder-rich esker that forms a long sinuous ridge along Chicopee Brook. On Massachusetts 32 at Oak Street in South Monson. —Mary Margaret Cooney photo

Glacial Lake Quaboag

When the ice sheet stood at the Connecticut state line, it contributed meltwater to the glacial Middle River of Connecticut. When the ice receded north of the state line to South Monson, it exposed a 615-foot drainage divide. Meltwater ponded north of the divide in the valley of Chicopee Brook. Overflow spilled south to the Middle River, depositing gravels near the state line.

As the ice thinned and receded, Glacial Lake Quaboag spread north over the top of the thinning margin of ice. Lake sediments were deposited directly on top of the ice that choked the lower reaches of the Chicopee Brook and Quaboag River valleys. When the ice melted, the sediments collapsed.

Extensive kame delta and terrace deposits on the hills north and west of the Quaboag River in eastern Palmer Township record the highest level of Glacial Lake Quaboag. The ice margin was east of the present business district of Palmer. The high-level stage of Glacial Lake Quaboag is now at about 660 feet elevation but drained south over the 615-foot gravel spillway south of South Monson. You may wonder how the lake level could be higher than the spillway. After the ice sheet retreated, the earth's crust

View to northeast from State Road in northern Monson across Quaboag River valley to kame deposits in sand and gravel pits. The kames are at 670 feet in elevation and consist of boulder and cobble gravel deposits near Palmer Reservoir and thick foreset beds south of it. —Mary Margaret Cooney photo

rebounded about 4 to 5 feet per mile to the north. Water planes that were horizontal are now tilted along this gradient. The glacial delta north of Palmer at 660 feet and a distance of about 11 miles north of the South Monson spillway is actually on the same glacial-lake water plane as the spillway at 615 feet in elevation.

The First Congregational Church of Monson, founded in 1762, sits on a river terrace cut into lake bottom deposits of Glacial Lake Quaboag. The terraces, which step down to Chicopee Brook to the east of Massachusetts 32, are mainly thick, sandy lake deposits. They contain ice-rafted boulders and some varved silt and clay.

As the ice front continued to recede, progressively lower spillways to the west controlled the water levels. A spillway at 605 feet sat high on the mountain north of Bald Peak 1.5 miles southwest of Palmer, and another spillway was at 555 feet at Glendale Church, about 3.4 miles southwest of Bald Peak. A spillway southeast of Palmer Center at an elevation of 480 feet controlled the level long enough for kame deltas to form northeast of Palmer Center and south of Forest Lake.

The First Congregational Church of Monson sits on an upper river terrace cut into lake bottom deposits of Glacial Lake Quaboag. —Mary Margaret Cooney photo

Monson Gneiss

Massachusetts 32, between the Connecticut border and Ware, passes right through the center of the dome of Monson gneiss. The Monson dome of the Bronson Hill belt runs north-south almost across the entire length of the state, and is so compressed it resembles an anticline. This rock was near the center of the collision between the Gondwanan and Laurentian continents. Rocks in the dome range from Precambrian to Ordovician in age. Stratified rocks, including the Ammonoosuc volcanic rocks of Ordovician age and the Clough quartzite of Silurian age, interlayer with and overlay the Monson gneiss.

The first ridge on the western skyline north of the Connecticut line consists of Ammonoosuc formation on the eastern, upright limb of the Bolton syncline. The Bolton syncline is a fold in the stratified Ordovician, Silurian, and Devonian rocks between the Monson dome and the Glastonbury dome

Well-bedded Clough quartzite contains reddish-weathering lime silicate layers. At Main Street (Massachusetts 20), six blocks west of Massachusetts 32 in Palmer. —Mary Margaret Cooney photo

to the west. When the foliage is gone in late fall and winter, Peaked Hill and the central high ridge to the west display resplendent white ledges of Clough quartzite glistening with muscovite. Mount Dumplin, a prominent craggy peak just north of Palmer, contains layered Ammonoosuc amphibolite schist with near-vertical foliation on the west limb of the Bolton syncline.

About 5 miles north of Mass Pike near Gibbs Crossing, Massachusetts 32 crosses a 0.5-mile-wide synclinal trough of Ammonoosuc volcanic rocks that rest on top of the Monson gneiss.

Glacial Lake Whipples

Forest Lake, near Massachusetts 32 in northern Palmer Township, is a remnant of Glacial Lake Whipples. Sand and gravel deposits east of Thorndike impounded the glacial lake in the Ware Valley. Unlike the deposits of Glacial Lake Quaboag, lake and river deposits of Glacial Lake Whipples, whose ice source was to the east along the Ware River, were not in contact with the ice front. The highest level of the lake was about 435 feet, measured at the contact between topset and foreset beds in the Whipples kame delta southwest of Forest Lake. Lower stages of the lake deposited a delta south of the Metropolitan Airport, west of the Ware River in the northernmost part of Palmer, and a kame delta northwest of the junction of the Penn Central and the Boston & Maine Railroads.

Muddy Brook

Muddy Brook flows south in a linear, north-south oriented valley, first following the trend of the Littleton formation but soon crossing over to follow the Partridge formation south to the Ware River near Ware. In the Muddy Brook valley, large blocks of ice broke off the receding ice margin and blocked the valley. Coarse piles of sediments added to the small ice dams. This process kept repeating itself as the ice continued to recede, giving rise to a series of progressively higher glacial ponds in the valley.

Ware and Gardner Belts

North of Massachusetts 9, Massachusetts 32 crosses rocks of the Ware belt. The Hardwick granite of early Devonian age is streaky, foliated, and biotite-rich. Horizontal and vertical joints make it suitable for architectural and paving uses.

The Hardwick pluton intruded Ordovician, Silurian, and early Devonian metasedimentary rocks of the Ware belt. The Hardwick magma incorporated and only partly digested numerous fragments of these rocks. Massachusetts 32 crosses these metasedimentary rocks in the rock-cored hills between 2.5 and 3.5 miles northeast of Gilbertville. In the hills just northeast of Gilbertville, a pollywog-shaped, foliated diorite body intrudes the Hardwick granite.

Just east of the thin metasedimentary strata along the eastern margin of the Hardwick pluton lies the very distinctive Coys Hill granite, a narrow intrusive body spotted like a leopard with big, round to rectangular, microcline crystals. It also contains sillimanite and garnet. The Coys Hill granite separates the Ware belt from the Gardner belt to the east. It occurs one-half mile west of Wheelright in a band that our route crosses twice.

Just west of the Coys Hill granite, another thin, but persistent, sill-like intrusion locally cross-cuts the metasedimentary strata of the Ware belt. The Ragged Hill gneiss, named after the prominent hill southeast of Gilbertville, is part of the West Warren complex that includes the mafic and ultramafic plutons in Warren Township along the Mass Pike.

The plate boundary between Laurentia and Gondwana is probably complexly deformed. If any one line or zone represents that plate boundary and the effects of that collision, in my opinion it may be represented by the Coys Hill granite, the Hardwick pluton, and the mafic and ultramafic plutons of the West Warren complex.

Glacial Lake Winimussett

Near Gilbertville glacial sediments dammed tiny glacial lakes along the receding margin of the ice sheet, giving rise to low-water stages of Glacial Lake Winimussett. As the lake deepened, deltas formed along the margin of the lake, and lake bottom sediments accumulated in the deeper and wider parts. At the high-water stage, open water covered an area three-quarters of a mile wide and more than 5 miles long. The lake occupied the valley of the north-flowing Winimusset Brook as well as part of the Ware River valley.

The Ware outwash, a 5-mile-long deposit, rests on top of the Glacial Lake Winimussett sediments. Braided streams from the receding ice margin deposited the thin outwash. In terraces along the banks of the Ware River between Ware and South Barre, you can see coarse gravels of outwash overlying kame delta deposits.

Quabbin Reservoir

The Windsor Dam, built in 1938, impounded the three branches of the Swift River, forming the Quabbin Reservoir. It is the principal source of potable water via pump stations and tunnels for Boston and forty-three other cities and towns. Quabbin Reservoir has been called the largest high-quality water impoundment in the world. Fifty-six thousand acres of publicly protected watershed surround the reservoir. Twenty-thousand acres in the Ware watershed are also protected, producing ample trophy-size game fish and providing habitat for a variety of wildlife, including the American bald eagle.

W. C. Alden studied the area before impoundment of the water and saw no evidence for a large glacial lake there. He interpreted the stratified sand and gravel he saw as glacial outwash deposited by meltwater rivers. Smaller, short-lived lakes may have occupied the valleys.

Melting along the approximately 20-mile-long margin of the ice sheet between Bassets Corner in Petersham Township and Quabbin Hill on the east side of Windsor Dam produced enormous volumes of glacial meltwater. This water escaped via the south-flowing East Branch of the Swift River and incised a channel through substantial deposits of moraines, kame terraces, and outwash, all now inundated by Quabbin Reservoir.

Glacial Erratics

A superabundance of boulders, the largest capable of straining the resourcefulness of even the most ingenious farmer, adorn the central Massachusetts uplands. The moving ice sheet picked up the rocks, then dropped them gently on the surface when the ice melted. A large glacial erratic of coarse granite sits 1.3 miles north of Gilbertville on Massachusetts 32A, due west of the junction with Turkey Hill Road. It is 89 feet in circumference. Rum Rock, a coarse pegmatite, is one of largest glacial erratics in the central Massachusets region. It rests east of Massachusetts 32, just inside the Barre Township boundary and adjacent to an Audubon sanctuary due south of Loring Hill.

Double Rocking Stone. A granite erratic perches on a lower granite erratic that in turn is delicately balanced on a rounded granite hillock. —Sketch from Hitchcock, 1841

An impish glacier arranged the erratics at Rocking Stone Public Park, at the junction of Dana Road and Massachusetts 32 in Barre Township. The ice set the first block of granite gneiss on the edge of a small cliff of Hardwick granite. This erratic and the bedrock make contact only near the center of the block. The ice then deposited a second erratic of granite gneiss directly on top of the first block—an artistic but potentially unstable arrangement. A landowner anchored these blocks to the bedrock so no unruly character could upset the balance.

Millers River

Massachusetts 32 crosses the drainage divide that separates the southwest-flowing rivers of the Chicopee River basin from the Millers River basin at the Petersham-Athol line. The Chicopee River basin is hilly, with upland relief ranging from 200 to 1,500 feet. The Millers River and its tributaries flow in steep, winding, narrow valleys. These meanders probably became incised within a plain during Cretaceous time or possibly later in Miocene time.

The till of the basin is particularly acidic because of the weathering of fragments of sulfidic schists. The till cannot buffer the effects of rainwater, especially acid rain, so the river and its aquatic biota are vulnerable to environmental degradation.

Athol sits on a great sand plain. A glacial lake probably formed north of the drainage divide near Lake Ellis. Numerous ponds and lakes in the Tully River drainage north of Athol are kettles within the extensive outwash deposits.

Tully Dome

The Tully body of Monson gneiss is a gigantic ellipsoid—in essence, a boudinaged mass of Monson gneiss that is completely separated from the main gneiss body. It is encased in a carapace of Ordovician and younger strata. It lies northeast of the main continuous outcrop of Monson gneiss and extends from south of Massachusetts 2 northward through Athol and Royalston Townships almost to the New Hampshire line.

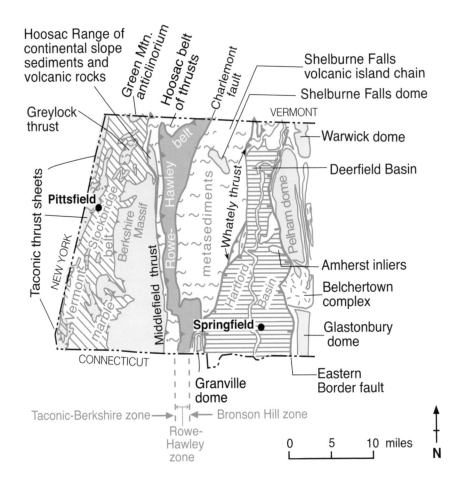

Hoosac Range of continental slope sediments and volcanic rocks

Green Mtn. anticlinorium

Hoosac belt of thrusts

Charlemont fault

Shelburne Falls volcanic island chain

Shelburne Falls dome

VERMONT

Warwick dome

Greylock thrust

Deerfield Basin

Taconic thrust sheets

Pittsfield

NEW YORK

Stockbridge belt

Berkshire Massif

Vermont marble

Rowe-Hawley belt

metasediments

Middlefield thrust

Whately thrust

Pelham dome

Hartford Basin

Amherst inliers

Belchertown complex

Springfield

Glastonbury dome

CONNECTICUT

Eastern Border fault

Granville dome

Taconic-Berkshire zone → ← Bronson Hill zone

Rowe-Hawley zone

0   5   10 miles

N

Geology of the Berkshires. —Modified from Zen and others, 1983

The Berkshires

When painters, musicians, poets, and recreationalists flock to the Berkshires, where exactly are they going? The term *Berkshires* can refer to three slightly different geographic regions. First, it refers in general to the entire area of western Massachusetts from the western edge of the Connecticut Valley to the New York border. Second, people commonly use the term to refer to the rugged terrain that rises rapidly west from the Connecticut Valley, including the Hoosac Range but not the valley of the Hoosic and Housatonic Rivers. This area is also called the *Berkshire Hills*. Third, the *Berkshires* refers to the narrow Vermont-Stockbridge Valley region—the scenic, fertile, marble-floored valleys of the Housatonic and Hoosic Rivers and the mountains west of the valleys. In this book, I will use *Berkshires* to mean the region as a whole as in the first usage, and *Berkshire Hills* for the second usage.

Geologists use yet another term, the *Berkshire Massif,* for the part of the Berkshire Hills that consists dominantly of thrust sheets of Grenville gneisses of middle Proterozoic age. These thrust sheets contain the oldest rock formations east of the Vermont-Stockbridge Valley and form the crest of the Berkshire Hills.

A century after the Pilgrims arrived at Plymouth, the distant Berkshires remained isolated from the east by the rugged mountains. During the French and Indian War in the mid-eighteenth century, Colonial settlement moved west only under the protection of forts and troops. Access to the Berkshires was chiefly by way of the south-flowing Connecticut, Housatonic, and Hudson Rivers. The first overland route into western Massachusetts was along the hostile Mohawk Trail. Another trail over the mountains in Becket was used as early as 1735. U.S. 20 follows this trail and was named Jacob's Ladder, a biblical comparison of the steep terrain with the steep ascent to heaven.

Farming in the uplands was difficult, though Scottish emigrants in Blandford managed to raise cattle and sheep on a limited scale. The lake bottom sediments of the Vermont-Stockbridge Valley provided fertile grounds for all kinds of farming.

View of the east face of Mount Greylock on the stormy skyline and Ragged Mountain (middle distance) *from West Summit Overlook on Massachusetts 2. Landslide scar is visible on the upper flank of Mount Greylock.*

Although mining and smelting of iron ore in the valleys was more limited in Massachusetts than farther south in Connecticut, it was an important resource during the Revolutionary War. Edward Hitchcock gave up recording the sites of iron ore in his nineteenth-century *Final Report on the Geology of Massachusetts* because nearly every town had a supply of bog iron. The availability of water power gave rise to mills, and the contruction of the 25,000-foot-long Hoosac railroad tunnel through Hoosac Mountain, completed in 1875 after twenty-four years of work, made the Stockbridge Valley accessible.

Since the early half of the twentieth century, the Vermont-Stockbridge Valley has been gradually transformed into an international center for the arts. Today, the Berkshires offer plenty of culture, scenery, and recreation. The Appalachian Trail traverses the region from north to south, and Mount Greylock, at 3,491 feet, is the highest mountain in Massachusetts. The Berkshires will not disappoint geology afficionados either! From whatever direction you approach western Massachusetts, wind gaps, water gaps, and mountain peaks provide magnificent views of the exquisite scenery and close-up views of the splendid geological features.

LAURENTIAN TERRANES

The rocks of the Berkshires were once the ancient margin of Laurentia, the early North American continent. Laurentian terranes include the Grenville gneisses of middle Proterozoic age that formed the basement of the continent; the Laurentian continental shelf, slope, and rise deposits along the shore of the Iapetus Ocean; and the remains of two offshore volcanic island chains—the Shelburne Falls volcanic chain of the eastern Berkshires and the Bronson Hill volcanic chain, which lies just east of the Connecticut Valley.

Plate tectonic movements in Ordovician time culminated in the Taconic mountain building event and joined the rocks of the continental margin and volcanic islands together. The Laurentian terranes formed the eastern margin of the ancient North American continent by middle Devonian time.

Taconic Mountain Building Event

In Ordovician and Silurian time, between 485 and 440 million years ago, the Shelburne Falls and Bronson Hill volcanic island chains slowly moved toward and finally collided with the eastern margin of Laurentia. The continental collision shoved rocks of the island chains, ocean bottom, and continental margin up and over rocks of the continental shelf.

If you peeled away each thin, east-dipping thrust sheet of Grenville gneiss and overlying deposits, one after another, and restored them to their original place before the continental collision, they would stretch for

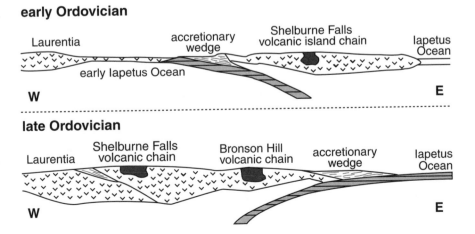

The Shelburne Falls and Bronson Hills volcanic island chains accrete to Laurentia during the Taconic mountain building event. —Modified from Karabinos and others, 1998

early Ordovician

Laurentia   accretionary   Shelburne Falls   Iapetus
                     wedge   volcanic island chain   Ocean

early Iapetus Ocean

W   E

late Ordovician

          Shelburne Falls   Bronson Hill   accretionary   Iapetus
Laurentia volcanic chain   volcanic chain   wedge   Ocean

W   E

many hundreds of miles to the east of the Berkshires. The collision squeezed these rock formations like an accordion and pushed thrust sheet after thrust sheet of Precambrian to middle Ordovician rocks immense distances westward over the top of the Laurentian continental shelf.

Thrust sheets composed of late Proterozoic to Ordovician sedimentary and volcanic rocks make up most of the eastern Berkshire Hills east of the Berkshire Massif. These thrust sheets dip steeply east to near vertical. The originally low-angle, east-dipping thrust surfaces have been compressed, folded, and squeezed up against a steep block of rocks, probably the resistant core of Grenville gneisses that underlies the Berkshire Hills to considerable depth. The breadth of outcrop of these thrust sheets ranges from 3 to 10 miles, not much greater than the aggregate thickness of the bundle of sheets.

Major faults along the western margin of the Green Mountains of southern Vermont and along the west side of the Berkshire Massif and Hoosac Range indicate that these thrust blocks have been stacked one on top of another, pushed from the east toward the west. Faults bound many of the formations in the Berkshire Hills. Fault finders can see such structures as the folded Whitcomb summit thrust and the folded Middlefield thrust that bound the eastern and western margins, respectively, of the Hoosac formation in the Hoosac Range.

Exactly when were the Grenville gneisses piled on top of one another in thrust sheets like a series of shingle blocks? Geologists are still investigating the timing of the Taconic mountain building event, but three sources of information support one theory that constrains the event to middle to late Ordovician time. First, radiometric age dating of metamorphic minerals in the thrusted rocks indicates the event began about 475 million years ago. Second, a certain graptolite, a diagnostic fossil found in sedimentary rocks about 455 million years old, is present in rocks beneath thrust faults in the Taconic Mountains. Since the fossil was preserved in rocks deposited just before thrusting began, the thrusting in the far western region of the Berkshires must have begun between 460 and 455 million years ago. Third, thrusting has not offset the Middlefield granite that intruded thrust rocks of the continental slope 447 million years ago, so the mountain building event was probably complete by the time of that intrusion.

Along the crest of the Berkshire Massif, geologists discovered another critical clue to determining the timing of the collision—Cambrian and Ordovician sedimentary rocks were deposited on the upturned and eroded edges of middle Proterozoic gneisses *before* the thrust faulting. The youngest unit present is the Walloomsac formation of middle Ordovician time. This relationship tells us that the thrust faulting of Grenville gneisses and overlying Paleozoic sediments was later than middle Ordovician time, the age of the Walloomsac beds.

This relationship also tells us that an earlier episode of mountain building deformed the gneiss at great depths in the earth before the Taconic event. This was the Grenville mountain building event about 1.1 to 1.2 billion years ago in middle Proterozoic time.

Thrust Sheets of the Taconic Mountains

The Taconic Mountains are the eroded remnants of thrust sheets that were pushed over the top of the continental shelf. The thrust sheets consist of sedimentary rocks that were deposited offshore on the continental slope. The continental slope sits between the continental shelf and the gentle incline that drops down to the abyssal plain of the ocean floor. The Taconic collision pushed continental slope rocks up from the deeper parts of the Laurentian margin onto the carbonate banks. The thrust sheets soon completely covered the continental shelf and dragged the shelf deposits of sandstone and limestone up into the thrust sheets as well. The leading edge of these thrust sheets, the Taconic thrust sheet, moved west toward the Hudson River. It now forms spectacular mountain masses such as Mount Greylock, the Taconic Range, and Mount Washington.

Sometimes we landlubbers forget that there is geologic "commotion in the ocean," as plate tectonics pioneer Bob Dietz used to say. Continental slope deposits that are thrust over the marble of the Stockbridge Valley far to the west of the crest of the Berkshire Hills are the same age as the continental slope rocks lying east of the crest. The far-traveled rocks to the west consist of the Taconic thrust sheets of Everett and Greylock schists.

The Everett and Greylock schists, sedimentary rocks originally on the continental slope and rise of Laurentia, were deposited at the same time and vicinity as the Hoosac formation in the Hoosac Range, but the Taconic mountain building event thrust them a great distance, possibly 50 miles or more, onto the western part of the continental shelf. The Everett schist consists mainly of pale green to greenish gray phyllite with tiny crystals of albite feldspar and chloritoid, a platy mineral that superficially resembles chlorite. The Greylock schist, similar in composition to the Everett schist, consists of green phyllite with minor beds of green quartzite, gray dolomitic phyllite, white albite-spotted phyllite, blue quartz schist and conglomerate, and salmon pink dolomitic marble.

Laurentian Continental Shelf

A continental shelf is the nearly flat-lying margin of a continent that sits between the shoreline and the noticeable change in slope—the continental

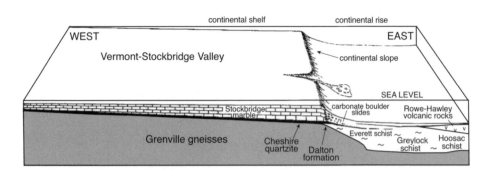

The continental shelf, slope, rise, and abyss of Laurentia in middle Ordovician time, before the Taconic mountain building event. Formations are in their original depositional environments. —Modified from Bird and Dewey, 1970

slope. Because the sea is fairly shallow over the shelf, marine deposits include beach sands and corals of carbonate banks. The early Paleozoic ocean, the Iapetus, covered the eastern continental shelf of Laurentia. Beach sands and carbonate banks that were deposited at the edge of the shallow ocean in Cambrian and Ordovician time have been metamorphosed into the white marble and quartzite that form the floor of the scenic Stockbridge Bowl and lowlands of the Housatonic Valley.

The quartzite and marble are connected to similar carbonate and sandy rocks in eastern New York. The ancient continental shelf extended west from near Great Barrington, Stockbridge, and North Adams in Massachusetts to near the Catskill and Helderberg Mountains in New York. This broad expanse of similar rock suggests that the ancient continental shelf was tens of miles broad.

Sedimentary strata that form the tops of the rugged Taconic Range, which rises up sharply along the western margin of the Stockbridge Valley, rest on *top* of the younger quartzite and marble deposits—a relationship created by thrust faults in the Taconic mountain building event. The marbles of the valley have also experienced a certain amount of westward thrusting and folding.

Dalton Formation and Cheshire Quartzite. The Dalton formation of late Proterozoic age contains pebble and cobble conglomerates with some carbonaceous schists, muscovite schists, glassy quartzite, and thin beds of tourmaline-bearing quartzite. A series of giant fault blocks that descended like stairsteps to the abyssal plain of the ocean basin broke the edge of the Laurentian shore of eastern North America during late Precambrian and early Cambrian time. The poorly sorted sediments of the Dalton formation accumulated in small basins and rifts on the tilted fault blocks. In places, the Dalton formation was deposited on the eroded, weathered surface of middle Proterozoic gneisses. This clay-rich weathered surface, called saprolite, is preserved at the base of the Dalton formation in Lenox.

The early Cambrian Cheshire formation, a glassy, fossiliferous quartzite of beach sand, overlies the Dalton formation near shore. Farther out to sea on the continental slope and rise, the Dalton formation graded up into the Hoosac sediments. Cambrian trilobites that scavenged their way across the Cambrian seafloor abound in similar rocks in northwestern Vermont but are not preserved in the Cheshire quartzite.

Stockbridge Marble and Walloomsac Formation. The Stockbridge marble—metamorphosed limestone beds—originally formed as carbonate banks on the continental shelf during early Cambrian to early Ordovician time, and were probably similar to those of the modern Bahamas. These carbonate banks were laid down on top of the Cheshire beach sands as sea level rose and a shallow sea encroached across eastern New York. By late Cambrian time, the waves were lapping up against the edge of the Adirondacks, depositing beach sands. Layers in the Stockbridge formation range from gray sandy dolomite, interbedded green and maroon phyllite, blue quartz-pebble quartzite, and white to gray crossbedded marble.

Eastern margin of the Laurentian continent. Taconic thrust sheets from beyond the continental shelf were eroded, shedding muddy carbonate sediments and depositing them on top of the Stockbridge marble.

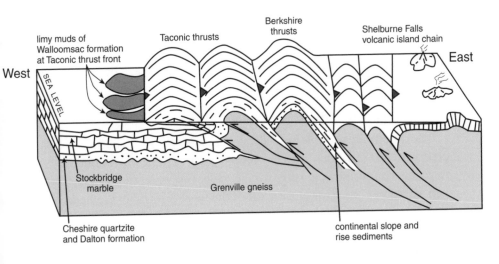

The Walloomsac formation of middle Ordovician age overlies the Stockbridge marble. As the Taconic collision shoved thrust sheets onto the continental shelf, they shed mud and silt in front of them. In sort of a precursory mélange, the mud and silt mixed with lime of the carbonate banks, forming the limy muds of the Walloomsac formation. The Walloomsac formation contains orangish-weathering marble, phyllite, and quartzite. Gray and white mottled lenses of former limestone contain bryozoan and coral fossils. Shaly limestone grades up into limy shale.

Grenville Thrust Sheets of the Berkshire Massif

Grenville rocks were originally part of the basement foundation of the eastern margin of Laurentia. You will have to go northwest to the Adirondack Mountains of New York or to the Canadian Shield to find Grenville basement rocks that have not been severely deformed. Grenville gneisses in the Berkshires are mainly metamorphosed volcanic and sedimentary rocks deformed in the Grenville mountain building event more than 1 billion years ago during the formation of the Rodinian supercontinent. Granites of Precambrian age intrude the Grenville gneisses. These hard granites and gneisses resist erosion and form the tops of many Berkshire mountains. A package of late Proterozoic sedimentary and volcanic rocks as well as some of Paleozoic age rests unconformably upon the older Grenville gneisses.

Washington Gneiss. The Washington gneiss is one of the most widespread Grenville sequences in the Berkshires. This gneiss forms the highest thrust sheet at the crest of the Berkshire Hills in Windsor, where it is associated with amphibolite, aluminous schist, quartzite, and greenish lime silicates—rocks consisting of calcium-bearing silicate minerals such as diopside. Before metamorphism, these rocks ranged from alumina-rich shales and gravels to lime-rich sediments and basaltic volcanic rocks. Metamorphism produced minerals rich in alumina—muscovite, biotite, sillimanite, kyanite, and garnet. The biotite and garnet also contain a good deal of iron. From this mineral assemblage, geologists deduced that the shaly and limy strata were derived from erosion of deeply weathered granitic rocks. These sediments were then deposited in an ocean basin more than 1.2 billion years ago.

The chemically variable lime silicates blossom with such beautifully colored minerals as pistachio green epidote, pink to red garnet, orangish red to yellow chondrodite, grass green diopside, and white to green tremolite and actinolite. These rocks are some of the most colorful in the Berkshires and contain collectable mineral specimens.

Tyringham Gneiss. The Tyringham gneiss, also widespread, is a very coarse-grained pink to gray granite gneiss. It often occurs as pegmatitic dikes containing feldspar crystals up to 4 inches long. It intruded the Washington gneiss and other rock formations of Precambrian age in the Berkshires about 1 billion years ago during the Grenville mountain building event. It characteristically contains hornblende and rods of quartz.

Stamford Gneiss. The Stamford granite gneiss, about 900 to 950 million years old, has coarse crystals of microcline feldspar in a groundmass of biotite, blue quartz, and plagioclase. The microcline crystals are commonly 2 to 3 inches long and zoned—they vary in composition from the core to the margin. Fine-grained epidote and muscovite formed in late phases of metamorphism as alteration products. The gneiss forms bulbous to elongate linear masses in the core of the Green Mountains in Vermont and within the thrust sheets of the Berkshire Massif.

Thrust Sheets from the Ocean Floor

Some thrust sheets of the Taconic mountain building event did not travel as far as the thrust sheets of the Taconic Mountains. Rock formations of the Berkshire Hills were thrust up from the continental slope, rise, and abyss. Deposits on the ocean floor were a mix of sediments shed from the continent to the west and from the Shelburne Falls volcanic islands to the east.

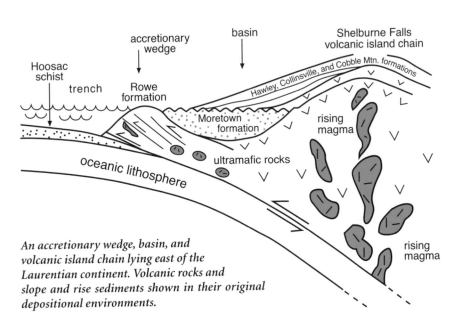

An accretionary wedge, basin, and volcanic island chain lying east of the Laurentian continent. Volcanic rocks and slope and rise sediments shown in their original depositional environments.

Hoosac Schist. The Hoosac schist of late Precambrian to Cambrian time was originally deposited on the continental slope or rise. It is mainly a medium- to coarse-grained schist with three micas—muscovite, chlorite, and biotite—and garnet and albite feldspar. The schist interbeds with marble and amphibolite. Characteristic and distinctive eyes of albite feldspar crystals up to 3/4 inch long stud the schist. The albite eyes range from colorless to white or black. The eyes and generally dark matrix distinguish the Hoosac schist from the overlying greenish Rowe schist.

The massive Hoosac Range of northwestern Massachusetts, including its steep western slope, is primarily Hoosac formation. The Hoosac Summit thrust fault, also called the Middlefield thrust, runs along the base of the Hoosac Range and carried this formation westward.

Rowe Schist. The Rowe schist of late Cambrian to Ordovician time is a distinctive, light green, fine-grained muscovite schist studded with tiny red garnets and black magnetite crystals. It contains white lenses of quartz that lie parallel to the prominent foliation, the planar texture of the rock. The formation contains white, gray, or black quartzite beds up to 1 foot thick and a dark green amphibolite with hornblende and plagioclase. The amphibolite, called the Chester amphibolite, sometimes contains abundant epidote and some garnet and sphene. The Rowe schist likely originated as basaltic volcanic ash deposits. It is younger than the Hoosac formation and older than the Moretown schist and the Hawley volcanic rocks.

North of the Massachusetts Turnpike, the upper, or younger, part of the Rowe schist contains abundant chlorite and is a deep green. South of the turnpike, big crystals of plagioclase, staurolite, kyanite, and biotite are common, and chlorite is less abundant, so the rock is gray rather than dominantly green.

Moretown Schist. The Moretown formation is a group of interbedded, light gray, sandy-textured granular rocks and schists with a distinctive pinstripe. The pinstripes—paper-thin layers of biotite alternating with thin, light layers of quartz and plagioclase feldspar—resemble those of tailored fabrics. This formation was originally a pile of sediments deposited in a basin above a subduction zone. Abundant amphibole-rich schist layers may have once been basaltic dikes that served as feeder pipes to the volcanic chain. Metamorphosed cherts form fine-grained pink quartz and garnet layers that preserve exquisite folds.

Ultramafic Rocks. Geologists have mapped some thirty-five bodies of Ordovician-age ultramafic rock—rock composed chiefly of iron- and magnesium-rich minerals—in western Massachusetts. These ultramafic rocks, generally lenses up to 2.5 miles long within the Rowe schist and in the basal

Folded Moretown pinstripe schist has undergone multiple episodes of deformation. In Florida Township south of the Deerfield River on Zoar Road.

part of the Moretown formation, are greenish gray serpentinite, some containing residual olivine, with talc-carbonate and soapstone. The rocks contain the minerals serpentine, olivine, talc, magnesite, tremolite, chlorite, chromite, and magnetite.

These ultramafic rocks were once pieces of oceanic crust that were scraped off as the ocean floor sank beneath the continental crust in a subduction zone. They merged with the wedge of sediments that accreted to the Laurentian continent to the west of the Shelburne Falls volcanic island chain. Many ultramafic bodies have been mined for talc and soapstone. Historically, soapstone was used in chemistry laboratories, kitchens, and laundry facilities, and for foot warmers in unheated bedrooms and horse-drawn sleighs.

Hawley Formation. The extremely variable Hawley formation consists mainly of green chlorite and hornblende schists, gray feldspar schists, and interbedded sulfidic black schists, all deposited in a volcanic environment. The 475-million-year-old Hawley formation, a middle Ordovician rock, rests unconformably on the top of an erosional surface of the Moretown

Quartz and mica schist of the Hawley formation containing fascicles of hornblende crystals. —Rolfe S. Stanley photo, U.S. Geological Survey

formation. It occurs only north of Blanford. South of Blanford, the Hawley formation interfingers with the Cobble Mountain formation, a micaceous schist and gneiss composed of sediments eroded from volcanic rocks.

A distinctive carbonaceous schist and quartzite unit contains a very fine-grained, pink quartz rock with garnets. These thin layers are commonly highly folded, preserving a photogenic record of deformation. Another distinctive schist unit contains clusters of hornblende crystals that resemble sheaves, or fascicles, similar to the cluster of arrows clutched in the left talon of the bald eagle on the one-dollar bill.

Shelburne Falls Volcanic Belt and Devonian Cover

A piece of the Iapetus Ocean floor separated Laurentia from a small microcontinent. As the two landmasses moved toward each other, the ocean floor dove beneath the microcontinent. This subduction of oceanic crust formed the Shelburne Falls volcanic island chain to the east of the Laurentian continental slope and rise.

A thick blanket of Devonian sediments covers rocks of the Shelburne Falls volcanic chain between the Berkshire Hills and the Connecticut Valley. The Devonian rocks are part of a gigantic slump, or landslide, that slid a relatively short distance west from mountains rising in central New England during the Acadian mountain building event. Geologists call the Devonian rocks a decollement, or slide fault. Informally called the Charlemont fault, this slide fault placed the Goshen and Waits River formations of Devonian time directly against the Ordovician Hawley formation to the west.

The underlying Ordovician sedimentary and volcanic strata of the Shelburne Falls volcanic chain are exposed in the cores of domes. Four domes—the Shelburne Falls, Goshen, Woronoco, and Granville—formed during the Acadian mountain building event of Devonian time, when pieces of the Gondwanan supercontinent landed against the Laurentian continent. Masses of rock escaping the pressure of the collision domed up. Over time, the tops of the domes eroded, forming windows through which we get a peek at the otherwise-hidden basement of Ordovician rocks.

The Acadian event also steepened strata along the eastern Berkshire Hills. The pronounced steepening in the dip of the rocks near Blandford and Windsor coincides with the western limit of metamorphism associated with the Acadian orogeny.

Geologists divide the overlying Devonian strata into limy and sandy mudstone formations that grade into one another: dark mica schist predominates in the Goshen formation, impure marble in the Waits River formation, and micaceous quartzite in the Gile Mountain formation. It is difficult to tell which formation you are looking at in any particular outcrop. The most prevalent rock in any given location determines which rock type geologists map.

The western part of the Goshen formation is well-bedded, micaceous quartzite and quartz schist grading upward into gray mica schist beds up to 10 inches thick. These beds are rich in aluminous minerals. The Goshen formation contains abundant sedimentary structures, such as graded beds, that allow geologists to determine the original orientation. Based on his "reading" of these structures, geologist Norman Hatch mapped the location of a large number of north-trending anticlinal and synclinal folds that formed in the Acadian collision. Some of these tightly compressed structures are 20 miles long.

GLACIAL GEOLOGY

The rounded hills of the western Massachusetts uplands are 300 to 400 feet above the valley bottoms. Rivers and streams flow in steep-sided valleys cut into the crystalline rocks of the uplands. This countryside does not immediately suggest that a continental ice sheet a few thousand feet thick once covered it, but a thin cover of ground moraine and polished and scratched bedrock in the higher elevations are evidence of glaciation.

The last Pleistocene ice sheet, the Wisconsinan, scoured the rugged uplands of the Berkshire Hills, scraping away the upper, weathered rind of the bedrock. As the ice moved over the freshly scraped bedrock, it scratched glacial grooves and striations into the ice-polished bedrock surface. These scratch marks, speaking a language that both professional geologists and amateurs understand, tell the story of a southeasterly icy journey toward the Connecticut Valley.

Hudson Valley Lobe

The Hudson Valley lobe was the westernmost glacial lobe of the Wisconsinan ice sheet in Massachusetts, covering the Berkshires to a depth of more than 1,000 feet. It generally traveled southward but spread out as it moved. The Hudson Valley lobe reached as far south as Long Island about 23,000 to 22,000 years ago. The orientation of forward- and retreat-direction indicators—such as glacial scratches, the orientation of elliptical drumlins, and the position of recessional meltwater features—tells us that the eastern part of the lobe spread out over the Berkshires in a southeasterly direction and melted back in the opposite direction. The ice overrode a few drumlins in the upland region, giving them a streamlined shape—the long axis parallel to the direction of ice movement. In general, the direction of ice movement was at right angles to the glacial front.

The major drainage systems of the eastern Berkshires—the Mill River, the Westfield River, and the West Branch of the Deerfield River—flow southeasterly, parallel to the main direction of glacial ice movement. Some smaller tributary streams flow in a more southerly direction, parallel to the trend of the foliation of the rock formations.

When the ice melted in the Berkshires, it laid down a blanket of unsorted rock fragments, ranging from clay to gigantic boulders—all part of the upper till layer. Here and there, big glacial boulders, or erratics, transported by the ice perch on hilltops. Distinctive boulder trains, such as the Middlefield granite train, have well-known bedrock sources that also tell us the direction of ice transport.

As the ice receded, it paused from time to time. During these standstills, the ice deposited glacial debris at the edge of the front. You can trace a series of sporadically connected morainal deposits of the same standstill from hill to hill and across valleys along the general northeast trend of the ice front. A study published in 1903 traced the winding pattern of as many as fifteen moraines across hills and valleys. One of these, the Tyringham-

Plainfield moraine, follows a zigzag path for 25 miles from the Connecticut line to northeast of Plainfield. The Lenoxdale Mountain moraine runs from southeast of Lenoxdale to northeast of Hinsdale. The Pittsfield-Dalton moraine formed when the ice stood still on the slopes of Mount Greylock.

The location of standstills is crucial in deciphering the complex glacial story. The standstill positions help geologists picture how the glacier receded, what the sequence of deposition was, and where glacial lakes, kames, and outwash plains were in relation to the melting ice.

Glacial meltwater picked up rock fragments, transported them to a variety of places, and deposited them as glacial outwash and deltas in ponds and temporary lakes. Some outwash materials were deposited over the top of blocks of ice beyond the melting front of the glacier. As the blocks melted, kettle ponds formed. Kames were deposited in contact with ice along the margins of valleys.

As a result of postglacial uplift, streams have eroded some of the previously deposited glacial deposits, notably, those in the bottom of valleys. Such erosion formed terraces at levels above that of the present river. These are especially noteworthy along the Westfield, the Farmington, and the Deerfield Rivers. Without benefit of an excavation pit along the valley wall, it is difficult, if not impossible, to distinguish between a kame deposit laid down on the melting tongue of residual glacial ice and a stratified river terrace. A kame may be a lumpy hillside deposit or a smooth-topped plain fairly high up along the valley wall. If undissected, stream terraces are continuous, planar deposits.

Glacial Lakes

Meltwater from the Hudson River lobe produced Glacial Lakes Great Falls, Housatonic, and Bascom in the Vermont-Stockbridge Valley. These curving finger lakes occupied marble-floored valleys between ridges capped by schists of the Taconic thrust sheets. Glacial Lake Great Falls straddled the Massachusetts-Connecticut line. Glacial Lake Housatonic occupied the valley and floodplain of the south-flowing Housatonic River from near Great Barrington to the present-day Pontoosuc Lake just north of Pittsfield. Glacial Lake Bascom occupied the lowland of the north-draining Hoosic River of the Hudson River drainage basin.

Glacial Lake Hitchcock, an extensive meltwater lake, occupied the Connecticut Valley. Shoreline features—wave-cut cliffs, deltas, and lake bottom sediments—are present along the margins of the Connecticut Valley lowland. Meltwater carried by the Westfield River deposited a 20-square-mile delta in Glacial Lake Hitchcock.

Some of the most important natural resources of the Berkshires consist of sand and gravel deposits transported by meltwater streams issuing from the front of the glacier. These deposits are nonrenewable until the next continental ice lobe covers the region! Sand and gravel deposits store valuable reservoirs of groundwater, form excellent recreational lands, and provide aggregate for construction and other human purposes.

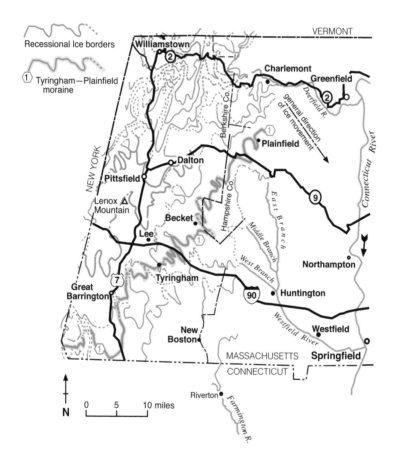

Moraines mark temporary, stationary positions of the receding glacial lobe.
—Modified from Taylor, 1903

ROADGUIDES TO THE BERKSHIRES

Interstate 90 (Mass Pike)
Westfield—West Stockbridge
41 miles

Wouldn't it be great to be able to take a journey to the center of the earth as Jules Verne did? An expedition along the Mass Pike across the Berkshires passes through rocks that formed 25 or 30 miles below the surface over 1 billion years ago—a trip deep within the earth *and* back in time. The trip across western Massachusetts is a great opportunity to examine rocks that form the Berkshire Hills on the east, the crest of the Berkshire Massif, the picturesque marble-floored Housatonic Valley, and the Taconic Mountains on the west. You'll cross belts of folds and thrusts trending more or less north-south, squeezed together like an accordion during the Taconic mountain building event of middle to possibly late Ordovician time.

If you want to examine extensive outcrops up close, consider driving U.S. 20, which parallels the Mass Pike. Access to U.S 20 is from the Westfield or Lee interchanges only. Mass Pike blasts a straight path up and over the Berkshire Hills to the Housatonic Valley. U.S. 20 follows the Westfield River to Chester, where it winds across the mountaintop along Jacob's Ladder. The roadguide for Sandisfield to Heath describes rocks along one segment of U.S. 20.

Westfield Delta

The glacial Westfield River deposited a large delta in the western arm of Glacial Lake Hitchcock. The volcanic ridge of East Mountain and Provin Mountain, which emerged as islands in the lake, separated the arm of the lake from the main body of water. Westfield sits on several terraces of the otherwise flat delta deposits of sand and gravel west of East Mountain. The 4 miles between the Westfield interchange and the Mass Pike bridge over the Westfield River cross gentle sand plains of the delta. The bridge is near the apex of the triangular delta, whose plain stands in sharp contrast to the rugged, tree-covered country of the crystalline Berkshire Hills looming up to the west.

A few outcrops of reddish sandstone and conglomerate stick up through the glacial sands just east of the Mass Pike bridge over the Westfield River. These beds of the New Haven arkose are part of the east-dipping Triassic red strata of the Connecticut Valley rift basin. Mass Pike crosses the normal

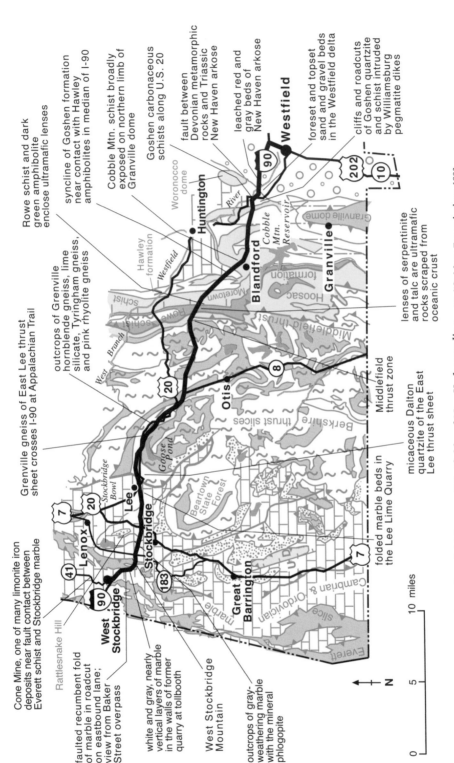

Cone Mine, one of many limonite iron deposits near fault contact between Everett schist and Stockbridge marble

Rowe schist and dark green amphibolite enclose ultramafic lenses

Grenville gneiss of East Lee thrust sheet crosses I-90 at Appalachian Trail

syncline of Goshen formation near contact with Hawley amphibolites in median of I-90

Cobble Mtn. schist broadly exposed on northern limb of Granville dome

Goshen carbonaceous schists along U.S. 20

fault between Devonian metamorphic rocks and Triassic New Haven arkose

leached red and gray beds of New Haven arkose

foreset and topset sand and gravel beds in the Westfield delta

cliffs and roadcuts of Goshen quartzite and schist intruded by Williamsburg pegmatite dikes

outcrops of Grenville hornblende gneiss, lime silicate, Tyringham gneiss, and pink rhyolite gneiss

lenses of serpentinite and talc are ultramafic rocks scraped from oceanic crust

micaceous Dalton quartzite of the East Lee thrust sheet

folded marble beds in the Lee Lime Quarry

faulted recumbent fold of marble in roadcut on eastbound lane; view from Baker Street overpass

white and gray, nearly vertical layers of marble in the walls of former quarry at tollbooth

West Stockbridge Mountain

outcrops of gray-weathering marble with the mineral phlogopite

Huntington

Blandford

Granville

Otis

Lee

Stockbridge

Lenox

West Stockbridge

Great Barrington

Westfield

Woronoco dome

Granville dome

Cobble Mtn. Reservoir

Hawley formation

Westfield River

West Branch

Rowe schist

Moretown formation

Hoosac formation

Berkshire thrust slices

Goose Pond

Beartown State Forest

Middlefield thrust zone

Middlefield thrust

Everett thrust slice

Cambrian & Ordovician marble

Rattlesnake Hill

Stockbridge Bowl

0   5   10 miles

N

Geology along I-90 between Westfield and New York state line. —Modified from Zen and others, 1983

Topset and foreset beds near the head of the Westfield glacial delta in western Westfield Township. —Carol Skehan photo

fault that bounds the western margin of the basin 300 feet east of the bridge. Because the fault dips to the east and trends southwest, some rift basin rocks are exposed on the mountain slope on the west side of the bridge. The Connecticut basin strata dropped down to the east relative to the rocks of the steep mountain front. Following uplift of the land, the ancient Westfield River cut a deep gorge through the crystalline rocks.

Goshen Schists

Just north of where I-90 crosses the Westfield River gorge, roadcut cliffs along I-90 and U.S. 20 and on the south face of Tekoa Mountain on the east side of the river sparkle with silvery schists and dark biotite- and quartz-rich strata. Large books of mica in white pegmatites flash in the sunlight. These schists and quartzites, the Goshen formation of Devonian age, cover older strata. The Devonian rocks slid like a giant landslide along a decollement, called the Charlemont fault, across rocks of the Shelburne Falls volcanic chain.

As I-90 heads upslope into the Berkshire Hills, the Goshen strata and pegmatites are extraordinarily well displayed for 1.1 miles in high roadcuts.

The Devonian Williamsburg granite intrudes folded, graded schist and amphibolite beds of the Devonian Goshen formation. On U.S. 20 west of Strathmore Park in Woronoco village.

Close-up view of Williamsburg pegmatitic granite in roadside cliff. Granite encloses two black fragments of biotite schist of the Goshen formation.

The abundant pegmatites are related to the Williamsburg granite, which intruded the Goshen formation in Devonian time during the late stages of the Acadian mountain building event.

To get your hands on an outcrop and examine the rocks, take U.S. 20 to Woronoco in Russell Township. Along Massachusetts 23, 600 feet west of its junction with U.S. 20, roadside outcrops of micaceous quartzite and quartz schist of the Goshen formation grade upward to the east into gray carbonaceous schists. These are interbedded with micaceous quartz and garnet schist, granular rocks full of green and black calcium silicate minerals, and some weathered calcareous schists. Primary structures, such as graded beds that indicate the top of the beds, are magnificently preserved in these rocks. The Goshen formation is also splendidly exposed along U.S. 20 for a distance of 0.9 mile north of the split from Massachusetts 23.

Granville Dome

The north end of the Granville dome, an elongate mass centered in Granville, extends north of the Mass Pike. The rocks of the inner core are Collinsville gneiss of early to middle Ordovician age, 485 to 475 million years old. Cobble Mountain schists of Ordovician age mantle the long northern end of the dome. At 1.1 miles west of Woronoco, Mass Pike leaves the Goshen schists, crosses over the Charlemont slide fault, and passes onto a 5-mile-wide expanse of the complexly folded and faulted Cobble Mountain schists. These light-colored schists, rich in quartz and feldspar, contain clasts of volcanic rocks. A smaller dome to the north, the Woronoco, also exposes the Cobble Mountain formation.

Farther north in Massachusetts when you are on the west side of the Charlemont fault, you are in the Hawley volcanic rocks of the Rowe-Hawley belt. The Hawley amphibolites extend from Vermont south all the way to the Blandford-Chester town line. South of this line, the Cobble Mountain formation appears. The dark volcanic rocks of the Hawley formation interfinger with the light sedimentary rocks of the Cobble Mountain formation between the town line and Blandford Center. A lateral variation in the depositional environment of the contemporaneous deposits created the interfingering.

Blandford Area

Scottish-Irish emigrants who settled in Blandford in 1735 called their village "Glasgow." In spite of an offer of a bell by the people of Glasgow, Scotland, if the name were retained, Provincial Governor Shirley, who had arrived from England on the *Blandford*, overruled the Glaswegians. These outcrop-studded, mountainous hills were soon covered with cow pastures, and a prosperous butter and cheese industry made this highlands village a wealthy, bucolic paradise.

The western margin of a synclinal fold of the Goshen formation surfaces in the median of the Mass Pike northwest of the town of Blandford. You can see the contact with the Hawley amphibolites 0.6 mile southeast of the North Street overpass over the Pike. The most southerly outcrops of Hawley volcanic rocks occur between North Blandford Road and Stage Road, 1 mile west of Blandford Center. Examine abundant outcrops of the Cobble Mountain formation in and around Blandford Center and along Russell Road, which passes under the Mass Pike.

The Moretown formation, a pinstriped schist with thin, alternating light quartz and dark biotite layers, lies below and west of the Cobble Mountain beds. Outcrops occur between 1.3 miles northwest of the North Street overpass in Blandford Township and 0.1 miles northwest of the turnpike overpass to the highway maintenance building near Old Chester Road. Some

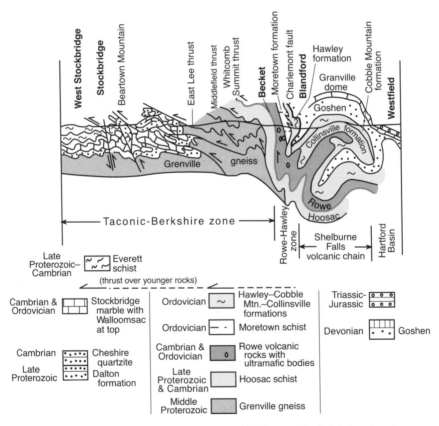

Schematic cross section from West Stockbridge to Westfield showing the interpreted structures of major tectonic zones. —Based on Zen and others, 1983

alternating layers contain garnet and staurolite, and north of the Mass Pike, also kyanite.

Just west of the Moretown schists and east of the Blandford-Becket town line, Mass Pike crosses thin bands of Rowe schist and Hoosac schist. The green Rowe schist encloses lenses of ultramafic rock, pieces of the ocean crust that were scraped off in the subduction zone associated with the Shelburne Falls volcanic chain. The Whitcomb Summit thrust separates the Rowe formation from the Hoosac formation. Two substantial outcrops of Hoosac schists surface along the westbound lane between 0.7 mile northwest of Old Chester Road overpass and 0.25 mile east of the Blandford-Otis town line. The Middlefield thrust fault forms the contact between the Hoosac formation and the Grenville gneiss.

Glaciated Crest of the Berkshire Hills

The high point on the Mass Pike is in Becket Township at 1,724 feet in elevation at mile marker 20.4. The Mass Pike briefly enters the Farmington River drainage before descending west into the Housatonic River basin. When the ice sheet covered the mountains here, meltwater flowed down the Farmington River and into Glacial Lake Hitchcock.

The ice sheet streamlined drumlins near U.S. 20 in Becket Township and deposited large boulders on the hilltops. Three large erratics of Grenville gneiss are visible from U.S. 20 near the south end of Greenwater Pond. The numerous ponds and mounds in Becket are a glacial landscape called knob-and-kettle topography.

Grenville Gneisses and Ordovician Granites

Between the Becket-Blandford town line and East Lee, a stack of thrust sheets of middle Proterozoic age pile on top of one another. These sheets are the Grenville gneisses of the Berkshire Massif. In Becket Township, the Mass Pike crosses at least five thrust sheets in succession. I-90 crosses onto the Berkshire Massif at the folded, east-dipping, Middlefield thrust fault near the town line between Blandford and Becket. The fault forms the base of the Hoosac schist. In the thrust zone, sheared Grenville gneiss enclosed slivers of the Hoosac formation.

Between West Becket and East Lee the Pike curves around to the west, mainly sticking to the East Lee thrust sheet, the lowest thrust in this part of the Berkshire Hills. The Pike stays just north of the Upper Goose Pond thrust that rides on top of the East Lee thrust. Grenville gneiss of the East Lee thrust sheet is exposed where I-90 crosses the Appalachian Trail. Here, you can see a black biotite and garnet schist rich in magnetite that interlayers with a lighter quartz schist. These schists share a contact with a massive greenish gneiss that contains diopside, a calcium- and magnesium-rich metamorphic mineral.

The western part of these Grenvillian thrust sheets includes a load of late Proterozoic Dalton quartzite and Cambrian and Ordovician sedimentary rocks—the Cheshire quartzite, Stockbridge marble, and Walloomsac formation. These sediments were deposited on the Laurentian continental shelf on top of the gneisses. The Stockbridge and Walloomsac formations are absent in the easternmost Grenville thrust sheets, indicating that these sheets originated east of the Ordovician continental shelf edge.

Granitic magma of late Ordovician age intruded along or melted from the thrust faults about 446 million years ago. These granites are light gray to white rocks that geologists call alaskites and trondhjemites—they contain essentially no dark minerals.

A view to the west in the Bowe Quarry shows a nearly horizontal, 200-foot-thick sheet of Ordovician bluish gray granite. Grenville gneiss and the granite are visible in the background. Bowe Quarry is 0.7 mile east of Algerie Road in Otis Township and 0.2 mile south of Mass Pike.

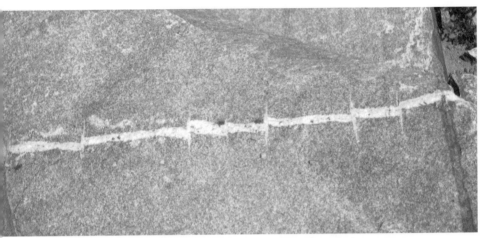

Close-up of an east-trending quartz vein cutting the Ordovician bluish gray granite. The quartz vein parallels one of three natural splitting directions of the granite. The vein is offset sinistrally along quartz-filled extension fractures that are parallel to another splitting direction.

Becket's early economy depended on its abundance of granite as well as hemlock bark used to tan hides. Ordovician granite was quarried in Becket Township from the Hudson and Chester Quarries 0.6 miles north of Mass Pike and in adjacent Otis Township from the Newall Quarry 0.3 mile south of the Pike. This granite was extensively used for monuments because it took a fair polish, and when hammered, became lighter in color.

Marble Valley

The steep west slope of the mountainous Berkshire Massif meets the Stockbridge marble valley at Lee. The East Lee thrust fault forms the break in slope where the western base of the Grenville gneisses meets the Stockbridge marble underlying the open fields in the fertile valley. This great fault carried the middle Proterozoic gneisses, along with their cover of Dalton formation, Cheshire quartzite, and Stockbridge marble westward across thick piles of these same Paleozoic formations in the Stockbridge Valley. The formations in the valley are part of an even deeper-seated, more westerly thrust sheet. Hills of faulted Cheshire and Dalton quartzites, including Rattlesnake Hill in the Stockbridge Bowl and The Pinnacle in Lee, interrupt the gentle profile of the Housatonic River valley.

You can see marble that contains the mineral phlogopite in a roadcut on the north side of the Mass Pike between U.S. 7 and Massachusetts 183 in

The south wall of the Lee Lime Quarry in Lee shows folded and faulted beds of Stockbridge marble. —T. N. Dale photo, U.S. Geological Survey

Stockbridge township. Phlogopite, a brown mica, is rich in magnesium. The gently dipping beds of marble here are weathering gray.

Glacial Lake Housatonic

After the ice sheet melted north of the Beartown and Monument Mountains, an ice dam continued to block the bend in the Housatonic River valley to the west of Monument Mountain. As the main body of ice receded, Glacial Lake Housatonic expanded to the drainage divide north of Pontoosuc Lake. Lake Mahkeenac in Stockbridge Bowl and Laurel Lake on the township line between Lee and Lenox are remnants of this glacial lake. The waves of the lake broke against Beartown Mountain.

On the west slope of the Berkshire Massif in Lee, I-90 cuts through spectacular deposits of sands and gravels of a large delta that formed at the edge of the lake. Foreset beds of the delta are visible in a sand and gravel pit near the mountain front just north of I-90 in Lee. The basic triangular form of the delta is still evident.

West Stockbridge Mountain

West of the scenic Lee and Stockbridge lowlands, the high and steep linear ridge of West Stockbridge Mountain rises abruptly from the marble

North wall of the Truesdell Quarry in West Stockbridge. Beds of Stockbridge calcite and dolomite marble are complexly folded and thrust faulted. —T. N. Dale photo, U.S. Geological Survey, circa 1923

valley. Two synclinal folds, overturned to the west, consist of dark green to gray Everett schist, a mudstone of early Cambrian time. The anticline between these synclines consists of biotite schist of the Walloomsac formation.

At the tollbooth near West Stockbridge, you can see white and gray, nearly vertical layers of marble in the walls of a former quarry. Just west of the tollbooth, a roadcut in the eastbound lane of the Mass Pike exposes a faulted recumbent fold in the marble. It is best viewed from the Baker Street overpass.

Richmond Furnace

Limonite, a hydrous form of iron, was mined near Richmond Furnace for high-tensile-strength iron for railroad car wheels. Limonitic iron ore is present along the thrust fault contact between the Everett schist and the underlying Stockbridge marble, either deposited by hot water circulating along the fault or deposited in swamps that were overrun by a Taconic thrust sheet.

The well-preserved Richmond Furnace still stands in the village of Richmond Furnace east of Massachusetts 41. Follow Furnace Road to Furnace Lane, at the end of which is the headquarters of the Richmond Iron Works. Obtain permission before following a path to the stone furnace, which has a characteristic stack face and casting arch. The furnace was one of the most important iron producers in the Salisbury district of southwestern New England before closing in 1923. Other iron furnaces operated in Van Deusenville, in Glendale, and at Shaker Mill Pond in West Stockbridge.

U.S. 7
Sheffield—Williamstown
49 miles

U.S. 7 follows the fertile, marble-floored valley of the Housatonic River basin between Sheffield and Lanesborough Townships, then climbs a drainage divide and drops into the Hoosic River basin at Williamstown. The Mohican Indian word *Housatonic* means "the place beyond the mountains." The rugged, heavily forested mountains of the Berkshire Massif thrust sheets rise to the east of Great Barrington, Stockbridge, and Pittsfield. Mountain ridges of Taconic thrust sheets, some of them forming linear ranges such as Lenox Mountain, and others forming high and massive features such as Mount Washington, rise mainly west of U.S. 7. Mount Greylock, an isolated Taconic thrust sheet lying east of U.S. 7, is the highest mountain in Massachusetts, at 3,491 feet.

Villages nestled between hills preserve much of the character and memory of the countryside from colonial times. The rock formations also have a collective memory, preserving geological events between middle Proterozoic and late Ordovician time, about 1,200 to 440 million years ago, and from the latest glacial epoch of the last few tens of thousands of years.

Housatonic River Valley

In Connecticut, the Housatonic River cut a deep canyon through middle Proterozoic rock—Grenville gneisses thrust west in the Taconic mountain building event. At the Massachusetts border, the valley of the south-flowing Housatonic River widens. Here, the Grenville rocks were not shoved as far west, leaving a broad expanse of marble that erodes more readily. The river meanders across its broad floodplain between Ashley Falls and Sheffield, creating numerous oxbow lakes.

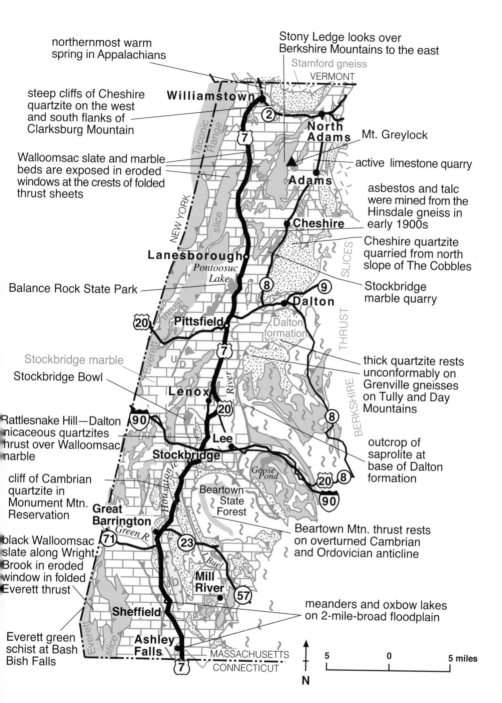

northernmost warm spring in Appalachians

Stony Ledge looks over Berkshire Mountains to the east

Stamford gneiss

VERMONT

steep cliffs of Cheshire quartzite on the west and south flanks of Clarksburg Mountain

Williamstown

North Adams — Mt. Greylock

active limestone quarry

Walloomsac slate and marble beds are exposed in eroded windows at the crests of folded thrust sheets

Adams

asbestos and talc were mined from the Hinsdale gneiss in early 1900s

Cheshire

Cheshire quartzite quarried from north slope of The Cobbles

Lanesborough

Pontoosuc Lake

Balance Rock State Park

Stockbridge marble quarry

Dalton

Pittsfield

Dalton formation

thick quartzite rests unconformably on Grenville gneisses on Tully and Day Mountains

Stockbridge marble

Stockbridge Bowl

Lenox

River

Rattlesnake Hill—Dalton micaceous quartzites thrust over Walloomsac marble

Lee

outcrop of saprolite at base of Dalton formation

cliff of Cambrian quartzite in Monument Mtn. Reservation

Stockbridge

Goose Pond

Beartown State Forest

Beartown Mtn. thrust rests on overturned Cambrian and Ordovician anticline

black Walloomsac slate along Wright Brook in eroded window in folded Everett thrust

Great Barrington

Green R.

Mill River

meanders and oxbow lakes on 2-mile-broad floodplain

Sheffield

Everett green schist at Bash Bish Falls

Ashley Falls

MASSACHUSETTS
CONNECTICUT

5   0   5 miles

N

Geology along U.S. 7 between Connecticut and Vermont.
—Modified from Zen and others, 1983

The Housatonic River basin is one of the most biologically diverse regions of Massachusetts, second only to Cape Cod for the large number of rare and endangered species. Marshes and oxbows are attractive habitat and nesting areas for Canadian geese, black and wood ducks, and other waterfowl. Deer, bear, and upland game and nongame wildlife abound. The bog turtle lives only in the river basin. Onota Lake in Pittsfield, at the headwaters of the Housatonic River, produces 40-inch northern pike and 3-foot tiger muskie.

From 540 to some 470 million years ago in Cambrian and middle Ordovician time, sandy beaches lined the coast, and carbonate banks grew on the Laurentian continental shelf. Now metamorphosed and contorted, they survive as quartzite and marble. Marble, or metamorphosed limestone, dissolves slowly in water, and caves and potholes often form. Unfortunately, caves in Stockbridge marble near Van Deusenville north of Great Barrington are now closed to the public.

Meltwater from the Hudson River lobe produced two curving finger lakes—Glacial Lake Great Falls and Glacial Lake Housatonic—in the Housatonic Valley. Glacial Lake Great Falls formed when a bedrock divide impounded meltwater in the southern Housatonic Valley of Massachusetts and northern Connecticut. The bedrock threshold, and thus the maximum level of the lake, was at an elevation of 630 feet just north of Falls Village, Connecticut. At maximum depth, the lake extended northward for 20 miles to the village of Housatonic in northern Great Barrington Township. Near Sheffield, fine-grained lake bottom sediments are more than 100 feet thick. The lake must have lasted for at least several hundred years for this volume of sediment to accumulate. The highest lake shoreline in Great Barrington is now at least 80 feet higher than the threshold to the south because of postglacial uplift and tilting to the northwest.

After the ice sheet melted north from the highlands of Beartown and Monument Mountains, it continued to block the lowlands to the west of Monument Mountain. From the bend in the Housatonic River south of Lee and East Lee, the river could not flow west beyond the Glendale section of Stockbridge. Thus, as the ice lobe receded, the second lake, Glacial Lake Housatonic, formed and expanded to the drainage divide north of Pontoosuc Lake. It also occupied the Monument Valley to Lake Buel in Monterey and may have extended about 3.5 miles up the Tyringham Valley along Hop Brook. Lake Mahkeenac in the Stockbridge Bowl and Laurel Lake are remnants of this glacial lake. A large lake delta formed at the edge of the lake in Lee. The lake drained when the ice dam near Monument Mountain finally melted from the valley.

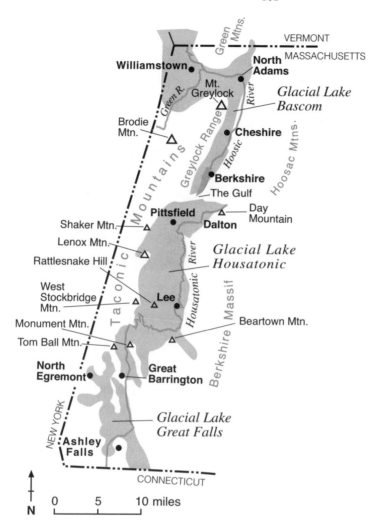

Maximum extent of Glacial Lakes Great Falls, Housatonic, and Bascom. —Modified from Stone and others, 1992; Dethier and Hamachek, 1998

Berkshire Massif

The Berkshire Massif, thrust sheets of Grenville gneisses and schists of middle Proterozoic age, form the highlands east of the Housatonic Valley. This basement rock of the Laurentian continent was thrust up from the depths in the Taconic mountain building event in Ordovician time. Numerous faults border the Berkshire Massif. Along the southern part of our route, a thrust fault at the base of Beartown Mountain carried middle Proterozoic gneisses onto an overturned anticline of Cambrian and Ordovician strata.

View from U.S. 7, 1.8 miles south of Great Barrington, looking east-southeast across sediments of Glacial Lake Great Falls to the June Mountain thrust sheet of Precambrian and Cambrian schists. The June Mountain thrust fault forms the sharp break in slope along the edge of the wooded area. —Heewon Taylor Khym photo

Monument Mountain Reservation

In Great Barrington Township on either side of U.S. 7, scenic mountains of Dalton and Cheshire quartzite of late Proterozoic age and early Cambrian age rise up abruptly from the valleys. Squaw Peak in Monument Mountain Reservation displays sheer cliffs of Cheshire and Dalton quartzites, part of a large fold overturned to the west. The quartzite contains original sedimentary features such as crossbedding, structures that formed as the sediments were deposited as beach sands. From these structures we can tell that beds on one limb of the fold are right side up and the beds on the other limb are upside down. The quartzite has been thrust over Stockbridge marble along the East Mountain thrust.

Mount Washington and the Everett Thrust Sheet

A Taconic thrust sheet, the Everett sheet, forms a dark gray to greenish cap of Everett schist across most of the higher mountaintops of western Massachusetts. The earliest thrust sheets in New York and westernmost Massachusetts were the first rocks shoved westward by the collision of the the Shelburne Falls volcanic island chain with Laurentia. These thrusts did not bite deep enough into the edge of the Laurentian continent to reach Grenville basement rocks, and so they do not contain Grenville gneiss.

Cambrian quartzite cliffs of Monument Mountain and Evergreen Hill, both in thrust sheets. View looking southwest from Ice Glen Road, southern Stockbridge Township.
—Heewon Taylor Khym photo

View to the southwest of Cambrian quartzite cliffs of Monument Mountain.
—Heewon Taylor Khym photo

Mount Washington, in the southwestern corner of Massachusetts, consists of a thrust sheet of Everett schist. You can see the Everett schist in the ruggedly scenic Bash Bish Falls State Park on the New York border.

An eroded window in the schist, in the middle of the mountain mass, exposes the underlying Walloomsac formation. The schist was thrust over the top of the Walloomsac limy muds. You can see these dark slates along Wright Brook and near the junction of Wright and Bash Bish Brooks.

Walloomsac Limy Muds at Laurel Lake

Look for the Walloomsac formation at Laurel Lake in Lee Township and as dark slates along U.S. 7 between Lenox and Pittsfield. One-tenth mile east of Laurel Lake, the small Pinnacle thrust sheet of Cheshire quartzite overlies the Walloomsac formation. The sheet hugs the east side of U.S. 20. Rattlesnake Hill, between Laurel Lake and Stockbridge Bowl, is made of Dalton micaceous quartzite thrust over Walloomsac marble.

Sheared Walloomsac marble at Laurel Lake. Here the Walloomsac formation is part of a refolded fold and consists dominantly of marble, although elsewhere it contains more shale. —Heewon Taylor Khym photo

Stockbridge Bowl

U.S. 7 passes through idyllic towns and villages near Stockbridge Bowl. Norman Rockwell lived and painted in Stockbridge. The setting in which Nathaniel Hawthorne wrote *The House of Seven Gables* is now the site of Tanglewood, the summer home of the Boston Symphony. With his poetic prose, Hawthorne memorialized my former home, the Jesuit novitiate called Shadowbrook.

Lake Mahkeenac, also called the Stockbridge Bowl, fills a glacially eroded depression in the marble beds. Here, gray-weathering Stockbridge marble is thrust over the top of the younger Walloomsac marble. The arcuate West Stockbridge Mountain of Everett schist to the west and north of Stockbridge Bowl accentuates the bowl's shape.

Balance Rock State Park

North of Pittsfield, an enormous 165-ton marble boulder sits on top of another marble boulder at Balance Rock State Park along Balance Rock Road west of Pontoosuc Lake. The melting glacier placed Balance Rock on top of a previously deposited glacial erratic.

View south from Shadowbrook on Richmond Road, looking over Lake Mahkeenac, the Stockbridge Bowl. —Heewon Taylor Khym Photo

Mount Greylock

Mount Greylock is an isolated thrust sheet that slid over the top of the continental shelf during the Taconic mountain building event. It is part of the Taconic thrust sheets, probably the oldest thrust of the event. Greylock schist, originally formed from sediments deposited on the continental slope of Laurentia, caps the mountain mass.

The summit of Mount Greylock is a great place to view the linear mountains and valleys of western Massachusetts and to see rocks close at hand. You can drive over the crest of the mountain on a paved road. If you're headed north, take Rockwell Road from Lanesborough to the summit. The first abundant outcrops of Greylock schist are at Rounds Rock, a prominent tree-covered hill about 3.5 miles north of the entrance to Mount Greylock State Reservation. About 1 mile farther north is Jones Nose, a bare cliff of schist shaped like a giant's nose that rises up to the north along the Appalachian Trail.

Aerial view of the east upper flank of Mount Greylock and the scar of a landslide that occurred in spring 1990. Note the 105-foot-tall war memorial on the summit. The linear feature to the south (left) of the landslide is an electric power line.
—Photo courtesy of North Adams Transcript

Western flank of the Hoosac Range, a series of thrust sheets, on the skyline. View from Greylock Glen near Gould Road looking southeast over the marble valley and Adams, in the middle distance. —Mary Havreluk photo

Stony Ledge is a rocky lookout that offers one of the most beautiful and spectacular sights in New England. Reach it by following 2-mile-long Sperry Road, which branches off 0.8 mile north of the junction of Greylock Road with Rockwell Road.

Memorial Tower, 105 feet high, marks the summit. It is constructed of Quincy granite and dedicated to those from Massachusetts who served in the military in major wars. From the summit you can glimpse a magnificent view of parts of five states. The Wisconsinan ice sheet passed over the top of Mount Greylock, leaving prominent glacial grooves oriented southeast to ornament the surface of Greylock schists.

Mining near Cheshire

The U.S. Gypsum Company mined Stockbridge marble along the western shore of Cheshire Reservoir. The marble was prepared as agriculture lime and lime whiting, which is used in paper manufacturing. Cheshire quartzite has been quarried from the north slope of The Cobbles, southeast of Cheshire, since 1813. The crushed quartzite is used for glass sand and abrasives, and for lining the bottom of drop forges.

In the early 1900s, the Berkshire Talc Manufacturing Company extracted asbestos and talc from ultramafic bodies within the Grenville gneiss. These limited bodies were mined 2 miles south of Cheshire on the west slope of North Mountain. They are in the same thrust block of the Berkshire Massif as similar deposits near Wauconah Falls Brook in Dalton Township.

Glacial Lake Bascom

The watershed divide at the village of Berkshire between the Hoosic and the Housatonic River basins began to impound the waters of Glacial Lake Bascom when the glacial lobe melted clear of it. The ice lobe to the north prevented the water from escaping in that direction. The divide is near Berkshire, at the present site of the Berkshire Mall. The spillway over the divide at Berkshire was 1,005 feet in elevation. The main body of the lake occupied the valley between Mount Greylock on the south and Clarksburg Mountain on the north.

From Williamstown, the lake extended north into the Pownal Valley of Vermont. The high-water stage of Lake Bascom, controlled by the spillway at Berkshire, lasted until the ice sheet melted north of Pownal, where an 895-foot spillway at Potter Hill, New York, lowered the lake level.

Brodie Mountain

North of Lanesborough Township, U.S. 7 passes between Mount Greylock to the east and Brodie Mountain to the west. The highway crosses the Housatonic-Hoosic River drainage divide near the Brodie Mountain Ski Resort. Brodie Mountain is part of the Everett sheet of the Taconic thrust sheets. Walloomsac slate and marble beds are exposed at the crest of folds in the thrust sheets. The crests have been eroded, forming a window through the bottom of the thrust sheet.

Sand Spring

Sand Spring in Williamstown is the only thermal spring in New England. The water is 72 to 76 degrees Fahrenheit at the surface and only moderately mineralized. Native Americans regularly camped at the spring, which is north of Williams College, just southeast of the junction of Massachusetts 7 and Massachusetts Avenue. Early settlers also partook of the waters, and a health spa was erected in 1880. Several mineral water bottling companies came and went, but a commercial spa still operates there.

The geologic setting of the spring is quite complex. It is near the western margin of the thrust boundary of the Green Mountain Massif and the eastern margin of the Giddings Brook thrust sheet of the Taconic Range. The Green Mountain thrust is probably the most deep-seated thrust of the

Laurentian continental shelf and provides a plausible deep-seated plumbing system for the spring. Lebanon Spring, some 17 miles to the southwest in New York, may be part of the same plumbing system. The spring water contains silica, probably from dissolved Cheshire and Dalton quartzites, and magnesia and carbon dioxide, probably from dissolved Stockbridge marble.

Massachusetts 2 (The Mohawk Trail)
Greenfield—Williamstown
43 miles

Massachusetts 2 follows the historic Mohawk Trail, a route blazed by Mohawk Indians traveling from New York to Massachusetts. The highway parallels the Deerfield River between Shelburne Falls and Charlemont, crosses over the steep Hoosac Range, and follows the Hoosic River between North Adams and Williamstown. Williamstown, incorporated in 1765, and Williams College, established in 1793, were named for Colonel Ephraim Williams, who, before leaving for the French and Indian War in 1755, left a bequest to establish a "free school in the township west of Fort Massachusetts called West Hoosac, provided [the town] be given the name of Williamstown."

Greenfield

A youthful crop of Liberty elms, which replaced the once-majestic elms that fell to Dutch elm disease, shades the wide streets of Greenfield, named for the fertile Connecticut Valley. The rocks of the valley were deposited in a rift basin about 200 million years ago. The red rocks on the west side of the valley are Sugarloaf arkose of late Triassic and early Jurassic age, an extensive alluvial fan deposit in the rift. Massachusetts 2 climbs west from the red rocks and lavas of the rift valley onto rocks that cover the Shelburne Falls volcanic island chain.

For a view of the valley, stop at Longview Tower west of Greenfield on Massachusetts 2. On a clear day, you can see into Vermont, New Hampshire, and Connecticut. Look east to see the basalt ridges in the rift valley and the domes in the Bronson Hill Upland. Goshen schist that contains the minerals garnet, staurolite, and actinolite is exposed in roadcuts near the tower. West of the tower, Massachusetts 2 curves around the north side of Greenfield Mountain, composed of gray Gile Mountain schists and amphibolites, before dropping back down to the Deerfield River valley.

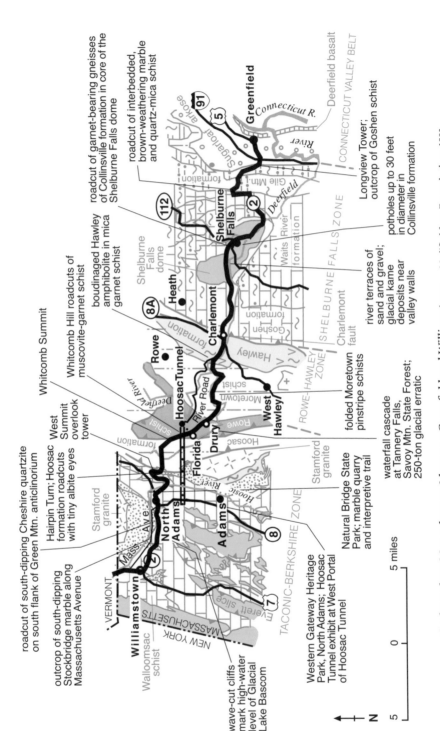

roadcut of south-dipping Cheshire quartzite on south flank of Green Mtn. anticlinorium

Whitcomb Summit

outcrop of south-dipping Stockbridge marble along Massachusetts Avenue

Hairpin Turn; Hoosac formation roadcuts with tiny albite eyes

Whitcomb Hill roadcuts of muscovite-garnet schist

boudinaged Hawley amphibolite in mica garnet schist

roadcut of garnet-bearing gneisses of Collinsville formation in core of the Shelburne Falls dome

roadcut of interbedded, brown-weathering marble and quartz-mica schist

Longview Tower; outcrop of Goshen schist

potholes up to 30 feet in diameter in Collinsville formation

West Summit overlook tower

river terraces of sand and gravel; glacial kame deposits near valley walls

folded Moretown pinstripe schists

waterfall cascade at Tannery Falls, Savoy Mtn. State Forest; 250-ton glacial erratic

Natural Bridge State Park; marble quarry and interpretive trail

Western Gateway Heritage Park, North Adams; Hoosac Tunnel exhibit at West Portal of Hoosac Tunnel

wave-cut cliffs mark high-water level of Glacial Lake Bascom

Greenfield

Connecticut R.

CONNECTICUT VALLEY BELT

Deerfield basalt

Gile Mtn.

Deerfield

Shelburne Falls

Waits River formation

SHELBURNE FALLS ZONE

Shelburne Falls dome

Heath

Goshen formation

Charlemont

Charlemont fault

Hawley

West Hawley

Moretown schist

ROWE-HAWLEY ZONE

Rowe

Rowe schist

Deerfield River

Hoosac Tunnel

River Road

Drury

Florida

Hoosac formation

Stamford granite

Stamford granite

Savoy schist

Hoosic River

North Adams

Adams

TACONIC-BERKSHIRE ZONE

Everett Slice

DEERFIELD ZONE

Mass. Ave.

Williamstown

Walloomsac schist

VERMONT

MASSACHUSETTS

NEW YORK

91

5

River

Sugarloaf arkose

112

2

8A

8

7

N

5   0   5 miles

Geology along Massachusetts 2 between Greenfield and Williamstown. —Modified from Zen and others, 1983

View from Longview Tower on Massachusetts 2 looking southeast at Greenfield and Massachusetts 2 (foreground), *Pocumtuck Range of basalt lavas* (dark, wooded ridge in middle foreground), *the Pelham dome rising up to middle skyline, the Holyoke Range on the extreme right, and the Warwick dome on the skyline at left.*

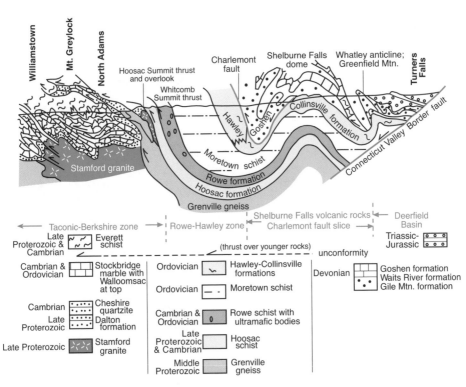

Geologic cross section along Massachusetts 2 between Greenfield and Williamstown.
—Modified from Zen and others, 1983

Shelburne Falls Volcanic Island Chain

Between Greenfield and Charlemont, the highway crosses over complexly folded Devonian schists. Beneath the blanket of resistant schists lies the Shelburne Falls volcanic island chain. The chain formed along the eastern margin of the Laurentian continent in Ordovician time. Erosion removed the tops of uplifted domes, exposing the older rocks in the dome cores. The Collinsville gneiss of Ordovician age forms the core of the Shelburne Falls dome. Glacial meltwater eroded potholes up to 30 feet in diameter in the gneiss in the bed of the Deerfield River. You can find them below Salmon Falls in the town of Shelburne Falls. Metamorphosed volcanic rocks and dikes of the core are exposed along the Deerfield River and in roadcuts along Massachusetts 2.

The Mohawk Trail

A lodge of Red Men of America erected the statue of a Mohawk Indian facing east with upraised arms. It stands near the Wishing Well on a terrace of the Deerfield River in Charlemont. This impressive monument, on what was once a warpath for the five Tribes of the Mohawk Nation, welcomes

Potholes in the Collinsville gneiss below Salmon Falls on the Deerfield River in the village of Shelburne Falls. —Christine C. Bronchuk photo

both the rising sun and travelers about to traverse the steep slopes of the Hoosac Range. The monument stands approximately on the Charlemont fault, which divides the Goshen formation of Devonian age to the east and the complex of thrust sheets along the Mohawk Trail to the west.

Rowe-Hawley Volcanic Belt

The mountainous area between Charlemont and Whitcomb Summit Overlook lies in the Rowe-Hawley zone. The well-exposed rocks of the Rowe-Hawley volcanic belt of Cambrian to Ordovician age lie west of the Shelburne Falls belt but were deposited and formed in the same volcanic chain. Devonian rocks, such as those between Greenfield and Charlemont, once covered these older rocks but have been eroded from this more elevated part of the Berkshire Hills.

This thick pile of volcanic strata was thrust from east to west along the Whitcomb Summit thrust over the top of the Hoosac thrust sheet. Between the Charlemont fault in Charlemont and the Whitcomb Summit thrust fault at the crest of the Hoosac Range, Massachusetts 2 passes through three intensely deformed formations. From east to west, they are the dark green Hawley volcanic rocks, the gray Moretown schists, and the emerald green Rowe schist.

Near-vertical foliation in Hawley volcanic rocks, west of Charlemont on Zoar Road on the north side of the Deerfield River, 1.2 miles southeast of Zoar.

You'll see abundant outcrops if you follow Zoar Road along the north side of the Deerfield River. This turns into River Road and takes you to the eastern portal of the Hoosac Tunnel, a 25,000-foot-long railroad tunnel through the middle of the Hoosac Range. Whitcomb Hill Road branches off from River Road near the east tunnel portal and connects with Massachusetts 2 near the east overlook at Whitcomb Summit.

Look for magnificent roadside cliffs of Rowe schist along Whitcomb Hill Road and River Road near Hoosac Tunnel. The distinctive, light green Rowe schist is studded with tiny red garnets and black magnetite crystals. White lenses of quartz parallel the prominent schistosity. A 400-foot-long and 10- to 20-foot-tall outcrop of orangish gray serpentine, an ultramafic lens from the earth's mantle, is exposed along Whitcomb Hill Road in Reed Brook Preserve near the town of Florida.

The Moretown schist is a thinly layered rock with a distinctive pinstripe—light layers of quartz and feldspar alternate with layers of biotite. The pinstripe character is poorly developed in the bottom, or westernmost, layers. These layers contain aligned, but irregularly shaped, clots of chlorite up to 3/4 inch long.

An extensive roadcut of folded Moretown pinstripe schist that has undergone multiple episodes of deformation. Along south side of the Deerfield River and Zoar Road in Florida Township.

View to the southeast from the 65-foot observation tower at 2,173-foot Whitcomb Summit, called Spirit Mountain by Native Americans. Beyond Clark Mountain (left), *the deeply incised Cold River* (center middle ground) *joins the Deerfield River. Rising above the relatively flat but incised skyline surface are Warwick dome, Pelham dome, and the Holyoke Range* (extreme right).

Hoosac Summit Thrust Sheet

The Hoosac formation of late Proterozoic to Cambrian age makes up a large part of the massive Hoosac Range, including its steep western slope. This formation has been carried westward on the Hoosac Summit thrust fault over the underlying Grenville gneisses of middle Proterozoic time. A short distance to the south in the Hoosac Range, Grenville gneisses and schists are well exposed, but along Massachusetts 2 the Hoosac thrust sheet completely covers the older rocks.

You can examine schists of the Hoosac formation, studded with albite crystals, at West Summit Overlook. Distinguish the Hoosac schist from the overlying green Rowe schists by color and by the presence of albite crystals in the Hoosac formation. Sometimes the Hoosac schist is notably green because of the mineral chlorite, but it still contains albite crystals. The greenish Hoosac formation surfaces at the Hairpin Turn Restaurant on Massachusetts 2.

West Summit Overlook

You can see three mountain ranges from the West Summit Overlook. Look southwest from the overlook, across the narrow Hoosic River valley to the imposing Mount Greylock. The base of Mount Greylock is Cambrian to Ordovician marble, and its top is Ordovician phyllite and schist thrust from the east onto the marble beds. It is part of the Greylock slice of the Taconic thrust sheets. If you look directly west down the Hoosic Valley from the overlook, you can see the overthrust sheets of the Taconic Mountains on the Massachusetts–New York border. To the northwest is a splendid view of the southern end of the Green Mountain Massif of Vermont.

The southern end of the Green Mountains, known locally as Clarksburg Mountain, is a gigantic, folded thrust block of middle Proterozoic gneisses wrapped by quartzite and marble strata. Massachusetts 2 and the Hoosic River skirt the foot of this imposing feature, which plunges to the southwest beneath the North Adams and Williamstown Valleys. The Cambrian Cheshire quartzite forms the southern slopes on Clarksburg Mountain. The core of the massif consists of Grenville gneiss intruded by the Stamford granite, part of the basement of Precambrian North America.

View from the observation platform, Natural Bridge, looking into the marble canyon of the deeply incised meanders of Hudson Brook. The brook plunges from pool to pool, each occupying a pothole eroded in the marble. —Mary Havreluk photo

Marble at Natural Bridge State Park

North Adams sits on the marble floor of the Vermont-Stockbridge Valley. The Stockbridge marble of Cambrian and Ordovician age, formerly a limestone bank on the shelf of Laurentia, lines the valley. The Taconic thrust sheets, the rocks of Mount Greylock, and the Hoosac thrust sheet have all been thrust over the top of the marble beds.

Natural Bridge State Park is in a very spacious marble quarry in the northeastern part of North Adams, west of Massachusetts 8 and 0.5 mile north of Massachusetts 2. The park was an active marble quarry from 1810 to 1947, a private tourist attraction from 1950 to 1983, and then a state park. As you enter the park, spectacular quarry walls of white to gray Stockbridge marble rise 80 to 100 feet above the road. A basalt dike intrudes the marble along the access road in the lower quarry level. In the upper level, a natural bridge of folded marble caps a steep-walled gorge. Hudson Brook, which is 50 to 60 feet below the natural bridge, carved this feature in part by dissolving the marble with acidic water and in part by erosion due to running water.

Quarries and mines in Stockbridge marble in western Vermont produced beautiful architectural and statuary marble comparable to that of Carrara, Italy. Today, limited quarrying of marble takes place in Massachusetts and Vermont, and a quarry on the east side of Mount Greylock is still active.

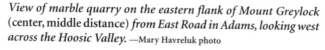

View of marble quarry on the eastern flank of Mount Greylock (center, middle distance) *from East Road in Adams, looking west across the Hoosic Valley.* —Mary Havreluk photo

Iron Ore and Lime Rock

Limonite, commonly referred to as bog-iron ore, is a hydrous form of iron. Limonitic iron ore has been mined and quarried along the contact of the Walloomsac schist and the underlying Stockbridge marble, either deposited by hot water circulating along the contact or in bogs on the top of the Stockbridge marble.

The New England Lime Company quarry, north of Renfrew in Adams Township, is one of only a few active quarries producing lime in Massachusetts. Abandoned quarries in Adams and North Adams produced quicklime but ceased operations in the late 1800s when Portland cement came into general use. Specific layers within the Stockbridge marble produce high-calcium and high-magnesium lime.

Glacial Lake Bascom

Glacial Lake Bascom filled the valley of the Hoosic River as the last ice sheet melted. The Hoosic River originates in southernmost Lanesborough Township. The drainage divide that separates the north-flowing Hoosic River from the south-flowing Housatonic River is near the Berkshire Mall in Berkshire north of Pittsfield. A bedrock spillway at 1,005 feet in elevation along this divide controlled the water level during the early stages of the lake. In time, glacial debris accumulated above the bedrock and increased the spillway's elevation. The bedrock spillway itself may have been higher originally, cut down to its present level by flowing water. Kame deltas and terraces are common in the Hoosic Valley from Cheshire Reservoir north to North Adams.

A break in the slope on the northern side of Mount Williams, at the northern edge of the Mount Greylock range, is a wave-cut cliff that formed during the highest lake level. The break in slope is about 1,050 to 1,060 feet in elevation. The accumulation of debris on the bedrock spillway, as well as postglacial uplift of the land, account for the apparent discrepancy between the elevation of the bedrock dam at Berkshire and the high-water mark in North Adams Township. The maximum depth of Glacial Lake Bascom was between 450 and 500 feet.

U.S. Geological Survey geologist and Williams College professor T. Nelson Dale named Glacial Lake Bascom for John Bascom, a wealthy landowner who donated Mount Greylock to the Commonwealth of Massachusetts. He was the father of Florence Bascom, the second woman to earn a doctorate in geology in the United States. She had a distinguished career in teaching at Bryn Mawr College and in research with the United States Geological Survey.

Stages in the retreat of ice and the history of Glacial Lake Bascom.
—Modified from Bierman, 1986; Bierman and Dethier, 1986

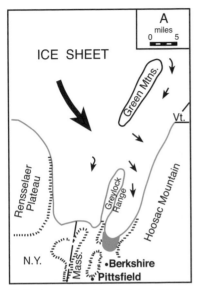

Meltwater accumulates between the glacier on the north and the drainage divide at 1,005 feet in elevation at Berkshire, north of Pittsfield.

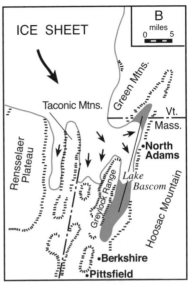

Lake Bascom forms initially by northward melting near Cheshire and southward melting near Stamford, Vermont. Shrinkage of the ice along its eastern margin allows the water near Stamford to drain southward past North Adams into the main body of Lake Bascom.

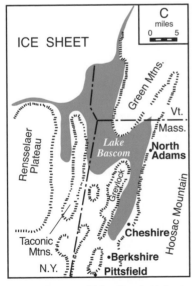

Lake enlarges to fill the Hoosic Valley to an elevation of 1,005 feet.

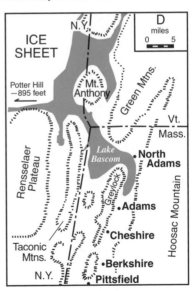

When ice melts clear of the spillway at Potter Hill at 895 feet in elevation, the lake drains to that level.

Massachusetts 9
Northampton—Pittsfield
42 miles

Before the construction of the Massachusetts Turnpike, two main roads crossed from Boston to the Pittsfield area, U.S. 20 and Massachusetts 9. U.S 20 ascends steeply through Chester, climbing 600 feet in 4 miles. The designers of Massachusetts 9, north of U.S. 20, found a much easier path across the rough terrain of the Berkshire Hills. The Grenville gneisses at the crest are extremely hard rock—difficult to blast through—but Massachusetts 9 crosses an exceptionally narrow band of these gneisses.

Massachusetts 9 follows the valley of the West Branch of the Mill River from Northampton upstream to Goshen, where the road then crosses a small divide into the Westfield River drainage and follows the river to West Cummington. The highway then crosses over the gneisses at the Berkshire Summit in Windsor and drops down through the Dalton Valley to Pittsfield.

Shorelines of Glacial Lake Hitchcock

An extensive meltwater lake, Glacial Lake Hitchcock, occupied the Connecticut Valley about 15,000 to 14,000 years ago. The elevation and position of the shorelines of the lake changed as the lake grew over its 3,000-year life span. Northampton, built on the gentle terrain of lake bottom clays, is near the western shoreline of Glacial Lake Hitchcock. Smith College campus was about 140 to 160 feet below the maximum surface elevation of the lake. In 1946, an excavation for a tunnel between two campus buildings exposed varved clays, seasonal deposits laid down in Glacial Lake Hitchcock.

Prominent wave-cut cliffs and wave-formed platforms mark the high-water shoreline. Most of the drumlins and other hills that rise above 300 feet in elevation in Northampton Township have well-developed wave-cut cliffs on their eastern slopes. Waves cut a cliff at the north end of the Northampton Country Club in Leeds Village and along the lower slopes of the drumlin on which the Veterans Administration Hospital is situated along Massachusetts 9 northwest of the city. Erosion along the western shore of Glacial Lake Hitchcock exposed a vast number of bedrock outcrops in the area north and south of West Brook and west of Chestnut Plain Road. Round Hill, a drumlin north of the Smith College campus, and Baker Hill, a drumlin between Elm Street and Mill River, stood just below and just above the level of Lake Hitchcock, respectively.

West of I-91 in Hatfield, shoreline features are present at an elevation of 300 feet along the mountain front in northwestern Hatfield and 330 feet in northern Whatley and Deerfield—an average rise of 4 to 5 feet per mile.

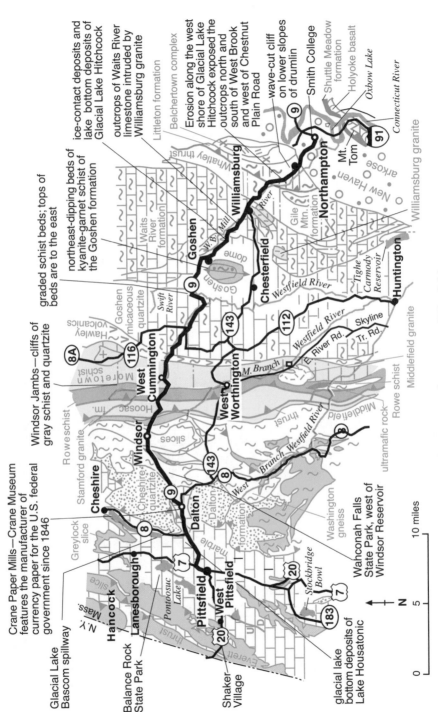

Crane Paper Mills—Crane Museum features the manufacturer of currency paper for the U.S. federal government since 1846

Glacial Lake Bascom spillway

Balance Rock State Park

Windsor Jambs—cliffs of gray schist and quartzite

ice-contact deposits and lake bottom deposits of Glacial Lake Hitchcock

northeast-dipping beds of kyanite-garnet schist of the Goshen formation

graded schist beds; tops of beds are to the east

outcrops of Waits River limestone intruded by Williamsburg granite

Littleton formation

Belchertown complex

Erosion along the west shore of Glacial Lake Hitchcock exposed the outcrops north and south of West Brook and west of Chestnut Plain Road

Shuttle Meadow formation

Holyoke basalt

Oxbow Lake

wave-cut cliff on lower slopes of drumlin

Smith College

Connecticut River

New Haven arkose

Williamsburg granite

Geology along Massachusetts 9 between Northampton and Pittsfield.—Modified from Zen and others, 1983

As the region rebounded, the shoreline tilted up to the north. This uplift, the result of releasing the earth's crust from the weight of the continental glacier, occurred perhaps as long as 4,000 years after the glacier had wasted away.

A Meeting of Ice Lobes

The Connecticut Valley ice lobe and the Hudson River ice lobe met near the western edge of the Connecticut Valley. The Connecticut Valley lobe moved south down the valley, but near its western boundary, it moved in a west-southwesterly direction. The Hudson Valley lobe moved generally in a southeasterly direction across the Berkshire Hills and receded essentially along the same lines.

Drumlins north and south of Massachusetts 9 in the boundary area trend in a variety of directions, reflecting irregular ice movement near the lobe boundaries. The long axes of many drumlins in the city of Northampton trend south-southwest, probably streamlined by the Connecticut Valley lobe.

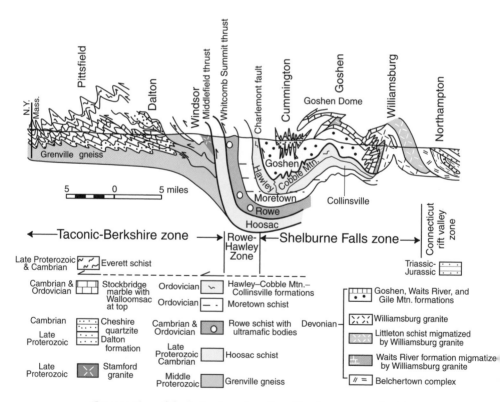

Cross section of the bedrock geology from Northampton to the New York line. —Interpreted from Zen and others, 1983

In Hatfield Township and in the eastern half of Williamsburg Township, the long axes of drumlins are oriented south-southwest to south. In the western part of Williamsburg, long axes of drumlins, such as O'Neil Hill and Nash Hill, are oriented southeast to south-southeast, but another drumlin, also on Nash Street, trends southwest.

Rift Basin Boundary
The Connecticut Valley is an ancient rift basin that formed in Mesozoic time as the Atlantic Ocean rifted open. Northampton sits at the western boundary of the basin. Sedimentary rocks of the basin contact the crystalline rocks of the western uplands along a border fault. Just west of the big loop in the Connecticut River in Northampton, I-91 and U.S. 5 barely have enough room to pass between the river and the crystalline rocks to the west. Here, the Connecticut River has meandered up against the western edge of the rift basin. On Massachusetts 9, you cross the faulted western margin of the basin at Florence, where the foothills rise from the sandy, terraced lowlands.

Belchertown Complex and Whately Thrust
North of Florence, west of I-91, and south of Whately, a block of rock is isolated from a complex that mainly occurs east of the Connecticut Valley at Belchertown. The Belchertown complex, an igneous pluton, intruded the Bronson Hill domes about 380 million years ago in middle Devonian time. The pluton is zoned, with an inner granitic core and hornblende gneiss rind. The main body bulges west toward the Connecticut Valley east of Amherst. The continental rift of Mesozoic time must have split the complex, separating the small fault block north of Florence from the main body.

Massachusetts 9 passes to the south of the fault block, bounded on the west by the east-dipping Whately thrust. The thrust pushed the fault block to the west over a mass of Littleton schists of Devonian age during the Acadian mountain building event. The schists are impregnated with the Belchertown granite.

Lake Deposits in Mill River Valley
Massachusetts 9 follows the Mill River between Northampton and Williamsburg, then continues northwest to Goshen along the West Branch of the Mill River. Lakebeds and ice-contact deposits occur at high elevations in Goshen Township. The largest deposit, a little more than 1 mile long and one-half mile broad, is east of the village of Goshen along East Wing Hill Road and Rogers Brook. The thickest lake bottom deposits lie between 1,150 and 1,260 feet in elevation. The temporary lake probably

ponded on or near the receding glacier, 600 to 700 feet above the level of Glacial Lake Hitchcock.

Devonian Landslide

Between the Whately thrust and Cummington, Massachusetts 9 crosses a wide belt of Devonian sedimentary rocks. These strata include the dark mica schists of the Goshen formation, impure marble of the Waits River formation, and micaceous quartzite of the Gile Mountain formation. These shales, sandstones, and limy mudstones slid west a relatively short distance from rising mountains in central New England to where they now rest on top of the Shelburne Falls volcanic chain. This giant slide, along a decollement called the Charlemont fault, placed the Devonian rocks directly against the Ordovician Hawley formation to the west.

The Acadian mountain building event folded and refolded the blanket of Devonian rocks. Luckily for geologists, the Goshen formation contains abundant sedimentary structures such as graded beds that allow us to determine the original orientation of the sedimentary beds, despite tight, overturned folds. In graded beds, coarse particles settle out of the water first, followed by smaller and smaller particles. A very fine-grained layer of

Graded, thinly bedded aluminous schist and micaceous quartzite of the Goshen formation. Though the tops of these graded beds are toward the left, they dip to the right, indicating that they are overturned. Along Massachusetts 9, 1 mile west of Lithia. —Norman Hatch photo, U.S. Geological Survey

clay tops the sequence. When metamorphosed, the fine-grained shale layer becomes a coarse-grained schist due to the growth of garnet and other chunky minerals.

Williamsburg Granite

To the north and south of Massachusetts 9 in Williamsburg Township and west to the village of Swift River, the Devonian sedimentary strata have soaked up the intrusive magma of the Williamsburg granite. You can still recognize the metasedimentary rocks, but the granite impregnated them pervasively about 373 million years ago in a late stage of the Acadian mountain building event in Devonian time. Whether the granite intruded them before or after the block of rock slid onto the Shelburne Falls chain is as yet undetermined. Though the main body of granite is near Williamsburg, dikes and sills intrude as much as 4 miles north and 10 miles south of Massachusetts 9. The heat from granitic intrusions into the siliceous, aluminous carbonate rocks of the Waits River formation produced spectacular and beautiful metamorphic minerals such as kyanite, andalusite, sillimanite, staurolite, garnet, diopside, and tremolite.

Goshen Dome

Domes formed during the Acadian mountain building event. The Devonian cover rocks later eroded from the crests of the domes, exposing the otherwise hidden Ordovician sedimentary and volcanic strata of the Shelburne Falls volcanic chain. Ordovician volcanic rocks are exposed in the core of the Goshen dome southwest of Goshen.

Besides creating a window to the past, the Goshen dome also shaped the Westfield River drainage. At its junction with the Swift River, the Westfield River deviates from the standard southeast-trending direction of major rivers in the Berkshire Hills. The Goshen dome likely formed an insurmountable obstacle in Cretaceous time when an ancient river began shaping the landscape. The Westfield River heads straight south to Huntington before resuming a southeasterly trend.

Chesterfield Gorge

For a great side trip from Swift River village to the Chesterfield Gorge, head south on Fairgrounds Road in Cummington. In Chesterfield, the road's name changes to Cummington. Proceed south through West Chesterfield to River Road, which leads to the gorge. About one-quarter mile north of West Chesterfield, originally called New Hingham, are the remains of a former dam that was almost totally washed away in a flood. From 0.6 mile north of West Chesterfield to the gorge, you'll pass through a wild and splendid valley with several rapids and two waterfalls. The Chesterfield Gorge is a magnificent box canyon carved into north-trending Devonian schists of the Goshen formation. At the upper end of the gorge, near the public view spot, are the remnants of a high bridge for an old stage road that operated from 1769 to 1875.

Continental Slope Rocks

Just east of West Cummington, Massachusetts 9 crosses onto the Rowe-Hawley belt of Ordovician strata. The Rowe-Hawley belt was thrust over the Hoosac formation along the Whitcomb Summit thrust fault. The breadth of outcrop of these formations is only about 5 miles because the thrust sequences here are nearly vertical.

The Hawley, Moretown, Rowe, and Hoosac formations were originally deposited on the continental slope and rise of the Laurentian craton and in, and adjacent to, the Shelburne Falls volcanic island chain. The volcanoes produced the volcanic rocks of the Hawley formation, which were deposited

Folded, alternating light and dark layers of the Moretown pinstripe schist west of Plainfield. The white, folded vein of quartz in the upper part of the photo has been folded in harmony with the folded Moretown rocks, whereas the white quartz vein below intruded later.
—Rolfe S. Stanley photo

on the eroded surface of the Moretown schist. Near Massachusetts 9, the Hawley formation consists of metavolcanic rocks and sulfidic black slates.

West Cummington is the type locality of cummingtonite, a brown amphibole that contains iron, magnesium, calcium, and manganese. This mineral is found in schists associated with the Hawley volcanic rocks that at one time supported mining of manganiferous ore near Charlemont.

The Windsor Jambs, in Windsor State Forest to the north of Massachusetts 9 along the Westfield River, are 70- to 80-foot cliffs of gray schist and quartzite of the Rowe formation. The Westfield River flows swiftly at the Jambs, dropping 1,000 feet in its first 14 miles. Like its two other branches, the main branch of the Westfield River originates in steep, forested hills and wooded valleys.

The Hoosac formation of the Hoosac thrust sheet has characteristic crystals of albite feldspar that range from barely visible spots to half-inch crystals. The Hoosac formation was pushed up along the Middlefield thrust fault and over thrust sheets of the Grenville gneisses of the Berkshire Massif.

Berkshire Massif

The Grenville gneisses, the oldest rocks of the eastern margin of Laurentia, cap the western crest of the Berkshire Hills. Three kinds of middle Proterozoic gneiss make up this 3-mile-wide band of thrust sheets in the Berkshire Massif. The 950-million-year-old Stamford granite gneiss forms a thin band just west of the Middlefield thrust fault. This gneiss has coarse crystals of microcline in a groundmass of biotite, quartz, and plagioclase. West of the Stamford unit are bands of the Washington gneiss and a gray, well-layered biotite gneiss. The Washington gneiss is a rusty-weathering rock with such metamorphic minerals as muscovite, biotite, sillimanite or kyanite, and garnet. The gray, layered gneiss contains beds of amphibolite, quartzite, and lime silicate.

Dalton Valley

The Dalton Valley is nestled between Weston Mountain on the north and Day Mountain on the south, both composed of late Proterozoic Dalton formation, Cambrian Cheshire quartzite, and some fault sheets of Grenville gneisses. These rocks have been thrust over the Stockbridge marble beds of the Housatonic Valley, which are late Cambrian through middle Ordovician in age.

The northern part of Glacial Lake Housatonic occupied the Dalton Valley. Massachusetts 9 cuts through spectacular delta sands and gravels on the west slope of the Berkshire Hills. Topset and foreset beds of the delta were laid down during the high-water stage at 1,040 feet in elevation. The Crane Museum, which exhibits the complete history of the Crane Paper Mills established in Dalton in 1801, is situated on delta sands of Glacial Lake Housatonic.

The drainage divide between the Hoosic and Housatonic River basins, near the present site of the Berkshire Mall in Lanesborough, separated Glacial Lake Housatonic to the south from Glacial Lake Bascom to the north. Ice-contact deposits occur at higher elevations northeast of Gulf Road and also southeast of the divide near the Berkshire Mall.

Wahconah Falls State Park is 2 miles east of Dalton at the headwaters of Wahconah Falls Brook at the Windsor Reservoir spillway. The brook cascades over greenish gray schists of talc and serpentinite, metamorphosed ultramafic rocks that intruded the Grenville gneisses before the Taconic mountain building event. Mine shafts at the top of the falls accessed the talc bodies.

Side Trip to the Taconic Range

Follow U.S. 20 west to the Taconic Range, which lies between Shaker Village on its east flank and Lebanon Springs in easternmost New York on its west flank. The Everett thrust sheet forms the range here, thrust over the marble beds of Stockbridge Valley in the Taconic collision during Ordovician time. The Everett sheet was subsequently folded at least once. The rocks of the Everett sheet consist of late Proterozoic to early Cambrian Everett schist and Nassau formation, a greenish phyllite with beds of quartzite. The Mettawee slate member of the Nassau formation is a notably lustrous green and purple rock that was quarried extensively for distinctive paving and roofing slate. These sediments are similar in many respects to the higher-grade Hoosac and Rowe sediments farther east, probably all deposited on the continental slope and rise of Laurentia and then thrust onto the continental shelf.

A Circuitous Route through the Berkshire Hills
Sandisfield—Heath
70 miles

If you're headed north or south across western Massachusetts, you could take U.S. 7 or I-91, well-behaved routes that follow valleys filled with glacial sands and gravel. But if you're interested in seeing the mountains—not

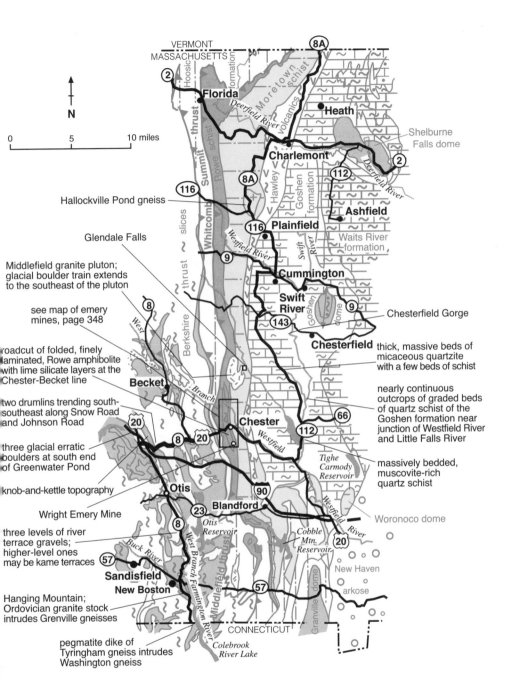

N

0 5 10 miles

VERMONT
MASSACHUSETTS

Florida

Heath

Hoosac Formation

Deerfield River

Moretown schist

Charlemont

Shelburne
Falls dome

Deerfield River

Summit thrust

Rowe schist

8A

116

Hawley Volcanics

Goshen formation

112

Ashfield

Whitcomb thrust slices

Hallockville Pond gneiss

Plainfield

116

Waits River
formation

Glendale Falls

Westfield River

Swift River

9

Middlefield granite pluton;
glacial boulder train extends
to the southeast of the pluton

Cummington

Swift
River

143

Goshen dome

9

Chesterfield Gorge

see map of emery
mines, page 348

Chesterfield

thick, massive beds of
micaceous quartzite
with a few beds of schist

roadcut of folded, finely
laminated, Rowe amphibolite
with lime silicate layers at the
Chester-Becket line

Berkshire thrust

West Branch

nearly continuous
outcrops of graded beds
of quartz schist of the
Goshen formation near
junction of Westfield River
and Little Falls River

Becket

two drumlins trending south-
southeast along Snow Road
and Johnson Road

Chester

Westfield

66

112

massively bedded,
muscovite-rich
quartz schist

three glacial erratic
boulders at south end
of Greenwater Pond

20

8 20

Tighe
Carmody
Reservoir

knob-and-kettle topography

Otis

90

Wright Emery Mine

Blandford

Woronoco dome

three levels of river
terrace gravels;
higher-level ones
may be kame terraces

23

8

Otis
Reservoir

Cobble
Mtn.
Reservoir

Westfield River

20

New Haven

57

Buck River

West Branch Farmington River

Sandisfield

New Boston

57

Granville dome

arkose

Hanging Mountain;
Ordovician granite stock
intrudes Grenville gneisses

Middlefield thrust

CONNECTICUT

pegmatite dike of
Tyringham gneiss intrudes
Washington gneiss

Colebrook
River Lake

Geology of the Berkshire Hills between Sandisfield and Heath.
—Modified from Zen and others, 1983

from the lowlands but up close and personal—consider this circuitous route. It winds through the heart of the rugged, crystalline Berkshire Hills, where tilted old rocks that once made up the eastern margin of Laurentia now form the backbone of the north-south trending mountains. This route accesses some unusual geologic features, including the emery mines in Chester and the Middlefield granite.

Because of the dominant southeast trend of the rivers and the northeast trend of the rock foliation, roads in the Berkshire Hills do not head straight north through the mountains. Rather, they follow one stream for awhile, then cross over a divide and follow another stream, zigzagging through the region. We'll follow Massachusetts 8 along the Farmington River between the Connecticut border and Becket, U.S. 20 over the mountains between Becket and Huntington, Massachusetts 112 up tributaries of the Westfield River between Huntington and Cummington, Plainfield Road to Plainfield, Massachusetts 116 over the mountains between Plainfield and Massachusetts 8A, and Massachusetts 8A along tributaries of the Deerfield River to Heath and the Vermont line.

Glacial Farmington River

Three terrace levels accent the Farmington River valley. The postglacial river probably cut the lowest terrace. The higher terraces may be kames, deposited along the edge of the melting ice lobe of the last ice age. Without benefit of an excavation pit, it is sometimes difficult to distinguish between a kame deposit laid down on the melting tongue of ice and a river terrace deposit, especially if a river cuts a terrace in a kame!

When the ice sheet covered the mountains near Becket, meltwater flowed down the Farmington River and into the southern end of Glacial Lake Hitchcock in Connecticut. Enough meltwater passed through this valley to deposit a large delta in the lake at Farmington.

Grenville Gneisses

The main rocks of the Berkshire Massif, the Washington and Tyringham gneisses that you can see in mountainous canyons along Massachusetts 8, were metamorphosed during the Grenville episode of mountain building about 1.1 to 1.2 billion years ago. These gneisses formed the basement rock of the eastern Laurentian continent. During Ordovician time, about 470 million years ago, these old rocks were thrust up from the depths in the Taconic mountain building event. They became a series of stacked thrust sheets piled up like shingle blocks on the edge of the North American continental shelf.

View looking to the northwest along the upper end of Colebrook River Lake on the Massachusetts-Connecticut state line. A high kame terrace (heavily wooded area on the right) *is composed of coarse gravel, cobbles, and boulders. Tree stumps mark a lower river terrace. Hanging Mountain, a stock of Ordovician granite that cuts Grenville gneiss, stands to the left of the valley. Chestnut Hill is to the right.*

Grenville rocks underlie our route between the Connecticut state line and Bonny Rigg Corners in Becket Township. A very coarse-grained pink pegmatite dike of Tyringham gneiss intrudes the Washington gneiss at the state border. Both the Washington and Tyringham gneisses are well exposed in clifflike roadcuts along Massachusetts 8. A younger granite dike of Ordovician age intrudes the dike of the Tyringham gneiss.

Ordovician Granites

An Ordovician pluton forms Hanging Mountain, a nearly vertical 300-foot cliff, 1.5 miles south of New Boston. Magma of the Ordovician granites in the Berkshires probably formed toward the end of the Taconic mountain building event and intruded the Grenville thrust sheets after thrusting was finished.

Ordovician granites also intrude the gneisses near Otis and Chester. These unnamed granites are pale bluish gray to white, with plenty of muscovite, plagioclase, microcline, and quartz. They contain almost no dark minerals. Quarrying of this attractive granite was big business in Becket, Otis, and Chester Townships several decades ago and continues on a limited scale in Chester. Builders worldwide continue to seek this bluish gray stone.

Close-up of Washington gneiss of middle Proterozoic age. Stretched, boudinaged lenses of quartz are parallel to the foliated layers.

View looking northwest at gray dike of Tyringham granite gneiss of middle Proterozoic age, which crosscuts black, massive amphibolite of the Washington gneiss.

Ice Sheet in Becket

The southeast-flowing ice sheet streamlined drumlins near U.S. 20 in Becket. Note the drumlin visible from Snow Road and the one from Johnson Road. The ice sheet also deposited large boulders—three erratics of Grenville gneiss are visible along U.S. 20 near the south end of Greenwater Pond. The numerous ponds and mounds in Becket are a glacial landscape called knob-and-kettle topography.

Thrust Sheets of the Continental Slope

U.S. 20 near Bonny Rigg Corners has long been known as Jacob's Ladder—a biblical reference suggesting vistas of beautiful mountain country opening up to heaven after a steep ascent on the highway. The foot of the ladder is 0.6 mile east of Bonny Rigg Corners, an old stagecoach crossroads. The ladder crosses thrust sheets that are younger than the Grenville thrusts exposed in the Berkshire Massif farther south and west. Massachusetts 8, heading north to Becket Center, follows the Middlefield thrust fault, the boundary between the Grenville rocks to the west and the Hoosac-Moretown schists to the east. The Hoosac sheet was thrust over the top of the Grenville gneisses.

The dominant rock of the Hoosac schist is a fine-grained, micaceous garnet schist that typically has albite feldspar eyes up to one-half inch long that peek out at you. Rusty-weathering schists are common, as are the minerals sillimanite, kyanite, and staurolite. The Hoosac sediments were deposited in fault-block basins on the Grenville basement along the margin of the Laurentian continent. Basaltic lavas, though not abundant, are commonly associated with Hoosac sediments; the magma oozed up from below the Grenville crust along rift basin faults.

Between Bonny Rigg Corners and about 3 miles southeast of Chester, where Sanderson Brook empties into the West Branch of the Westfield River, late Proterozoic through Ordovician strata are well exposed in roadcuts. This package of strata consists of late Proterozoic to Cambrian age Hoosac beds, Cambrian to early Ordovician Rowe basaltic volcanic rocks and green shaly sediments, pinstriped Moretown schists, and middle Ordovician volcanic rocks and shaly rocks of the Hawley formation. The Rowe-Hawley rocks form a distinctive group of fault-bounded strata—a band of rocks that extends not only across Massachusetts but also through and beyond adjacent states.

The Rowe schist consists of light green to black schists topped by the Chester amphibolite. The schist probably originated as deep oceanic sediments scraped off in the accretionary wedge at the edge of the Shelburne Falls volcanic island chain. Outcrops of finely laminated and folded amphibolite schist interlayered with lime silicates surface on the south side of U.S. 20 at the Chester-Becket town line.

Emery in the Chester Amphibolite

The town of Chester lies about in the center of a 3-mile-wide distinctive sequence of Cambrian and Ordovician micaceous schists and amphibolites that enclose scattered ultramafic bodies. Iron in the form of magnetite was

Highly folded Chester amphibolite schist of the Rowe formation. View looking northeast from U.S. 20 in Chester to the east of Blandford Road.

Close-up of tightly folded Chester amphibolite.

discovered in Chester in the 1830s, and in 1854, Herman S. Lucas began searching for ore bodies of iron within the ultramafic rocks. Drill bits wore out unusually fast, so Charles T. Jackson, a famous New England geologist, came to investigate, and in 1864 he discovered emery. Emery is a gray to black, impure variety of corundum, one of the hardest minerals around. It measures 9 on the Mohs hardness scale. By comparison, diamond measures 10 and quartz 7.

The emery deposits occur as lenses and layers within 100 feet of the top of the Chester amphibolite. These lenses, as much as 400 feet long and 20 feet wide, consist of a mixture of magnetite and corundum with a number of other minor minerals such as ilmenite, hematite, and rutile. The Chester emery deposits may have once been a highly aluminous and iron-rich soil that formed on the weathering surface of the Chester amphibolite and converted to emery lenses by metamorphism and deformation. Hot water circulated through the highly permeable lenses, introducing calcium, sodium, potassium, boron, and silica, which resulted in a variety of minerals and concentrations of tourmaline. A pegmatite in southern Chester Township yielded a substantial amount of mica. The Hamilton Memorial Library in Chester houses a collection of 1,500 mineral specimens, including some from Chester, donated by Hilda Burdick, a coworker of Herman Lucas at the emery mines.

The Hamilton Emery and Corundum Company, one of the oldest manufacturers of emery in the country, erected a mill in 1901. The Wright Emery Mine began operation as an open pit mine in the early 1900s. Later, miners sank an 87-foot vertical shaft, and a horizontal shaft penetrated about 900 feet of rock. Adits of several old mines are on a high hill called Big Rock west of the town hall in Chester.

Talc is mined from a large ultramafic body north of Chester along Middlefield Road on the Chester-Middlefield town line. The talc body within the ultramafic body ranges from 20 to 60 feet wide, is more than 800 feet long, and has been worked to a depth of 30 feet.

Middlefield Granite

About 4 miles north of Chester on the Chester-Middlefield line, and only 0.25 mile northeast of the ultramafic body with the talc mine, sits an elliptical pluton of Middlefield granite. It intruded the Moretown and Rowe schists 447 million years ago in late Ordovician time. Thrusting of the Taconic mountain building event has not offset this granite, so it must have been complete by the time of the intrusion. The heat of the thrusting collision probably generated the magma. Middlefield granite is a distinctive, gray or buff rock with 3-inch-long microcline feldspar crystals often lining

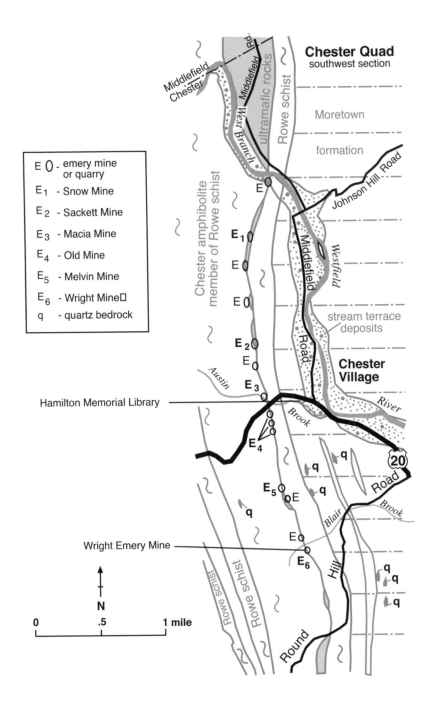

Chester quadrangle emery mines. —Modified from Hatch and others, 1970

up and imparting a foliation. The granite contains distinctive crystals of biotite, muscovite, epidote, and garnet.

A trail of boulders of Middlefield granite fans out to the southeast of the pluton. The base of the ice sheet tore large blocks of granite from the bedrock, transporting them to the southeast.

Glendale Brook cascades 150 feet over bedrock steps of Middlefield granite and Moretown schist. To access Glendale Falls State Park, follow Skyline Trail (Chester Hill Road) at Huntington for about 3 miles. Then turn right on East River Road, which follows the Middle Branch of the Westfield River upstream to Glendale Brook, a tributary.

Devonian Rocks

The Charlemont fault, 2 miles southeast of Chester on U.S. 20, separates the Cambrian and Ordovician strata west of it from the broad expanse of Devonian strata to the east. The Devonian strata, a thick blanket of sediments overlying the rocks of the Shelburne Falls volcanic island chain, slid an undetermined but probably relatively short distance west along the Charlemont fault from the rising Acadian mountains in central New England. Both the Taconic and the later Acadian mountain building events deformed the rocks of the volcanic chain; only the Acadian event of middle Devonian time affected the overlying Devonian strata.

The Acadian mountain building event, the final welding of the Gondwanan microcontinents to Laurentia, squeezed the rocks together, forming large folds and domes in the Shelburne Falls belt and overlying Devonian cover rocks. Four domes lie east of our route. From south to north, they are the Granville, Woronoco, Goshen, and Shelburne Falls domes.

Along U.S. 20 between Chester and the Westfield River at Huntington and north toward Plainfield, a thick sequence of early Devonian strata is exposed— the well-bedded quartz-rich shaly rocks of the Goshen formation. The western part of the Goshen formation consists of micaceous quartzite and quartz schist grading upward into dark gray schist rich in aluminous minerals in beds up to 10 inches thick. Outcrops surface near the junction of the Westfield and Little Falls Rivers. Depending on your highway map, the Knightville Reservoir may appear to inundate the confluence of these rivers. The dam at Knightville does not impound water, and you should see ample outcrops, not water, along Massachusetts 112.

Farther upstream along the Little Falls River near South Worthington, 20-foot-thick, right-side-up, massive beds of micaceous quartzite of the Goshen formation dip to the west. Fine-grained tops of the beds contain garnets and more mica than the lower, coarse-grained beds. Lime silicate and limy beds weathering to brown interlayer with the quartzite. The limy beds are similar to the dominant rock type in the Waits River formation.

Chesterfield Gorge

Take Massachusetts 143 east from Massachusetts 112 to West Chesterfield to see the Chesterfield Gorge, where the Westfield River flows swiftly through a steep canyon of Goshen schist. The meandering shape of river canyons in the Berkshire Hills indicates the land surface at one time was essentially flat, and the rivers migrated across the gentle landscape. But now, the rivers flow through steep valleys cut 1,200 to 1,400 feet deep into the crystalline rocks of the uplands.

Hallockville Pond Gneiss

Two miles north of Plainfield near Hallockville Pond on the west side of our route, Massachusetts 8A crosses a strongly foliated, white to gray gneiss with large crystals of microcline. The 2-mile-long body of Hallockville Pond gneiss, formerly a granodiorite pluton, intruded the Moretown formation about 479 million years ago in Ordovician time. The pluton was probably associated with the subduction zone at the edge of the Laurentian continent.

Mineral Prospects near Charlemont

The Charlemont fault that divides the Hawley formation from the Goshen formation cuts through Charlemont. Massachusetts 8A runs west of the fault in the Hawley formation between Charlemont and Vermont. The Hawley formation here consists of interbedded amphibolite and schist with some distinctive pink garnet layers and equally distinctive fascicles of hornblende or actinolite crystals. The easternmost part of the Hawley formation, adjacent to the Charlemont fault, is amphibolite with coarse crystals of plagioclase.

A number of abandoned mines or prospects exist in the northern part of the Rowe-Hawley belt near Charlemont. These include a body of talc and magnesite rock; a copper, iron, and zinc ore body; a few garnet prospects; and two manganese prospects. Cummingtonite, a brownish amphibole mineral that typically contains calcium and manganese, is present in schists of the the Hawley formation. It was named for the town of Cummington, where it was first described.

The now-abandoned village of Davis, north of Charlemont, was a beehive of mining activity. From 1882 to 1910, an iron and copper mine operated near the Davis Mine Brook, a tributary to Mill Brook.

GLOSSARY

accretionary wedge. A triangular mass of sediment on the landward side of an oceanic trench composed of debris scraped off the upper surface of the subducting, or down-going, plate.

actinolite. Greenish amphibole mineral that grows as needle or fibrous crystals in metamorphic schists and altered mafic and ultramafic igneous rocks.

alaskite. A very pale granite nearly without dark minerals.

albite. A sodium-rich plagioclase feldspar mineral that is colorless to white.

alkalic. An igneous rock that contains more sodium and/or potassium than needed to form feldspar with the limited amount of silica.

alluvial fan. A relatively flat to gently dipping, fan-shaped wedge of loose rock material deposited by a stream, especially in a semi-arid region, where it flows out of a narrow valley onto a broad valley or plain.

amphibole. A group of dark, iron- and magnesium-rich minerals that are commonly present in igneous and metamorphic rocks. Hornblende is the most common amphibole and crystallizes into black needles.

amphibolite. A medium- to coarse-grained metamorphic rock composed of hornblende and plagioclase.

amphibolite schist. A strongly foliated amphibolite that can be readily split.

andalusite. A metamorphic silicate mineral that occurs in alumina-rich schist and gneiss. It has the same chemical composition as kyanite and sillimanite.

andesite. A grayish volcanic rock that consists mostly of plagioclase and one or more of the dark minerals, biotite, hornblende, or pyroxene. It is intermediate in composition between basalt and rhyolite.

anticline. In layered rocks, a folded arch with the oldest rocks in the center.

anticlinorium. A bulge, or large regional fold—often consisting of a series of smaller folds—in rock layers.

arkose. A sandstone composed mainly of quartz and orthoclase feldspar derived from weathering granite.

ash. Small shreds of lava that escape in the air during a volcanic eruption. Ash consolidates into tuff.

asthenosphere. A zone of partially melted rock within the earth's mantle that lies beneath the solid lithosphere.

barrier island. A long, narrow, coastal sandy island parallel to the shore.

basalt. A black or very dark gray volcanic rock that consists mainly of microscopic crystals of plagioclase feldspar, pyroxene, and perhaps olivine.

basement. The fundamental rocks of the continental crust, mainly granite, schist, and gneiss.

batholith. A mass of coarsely granular igneous rock, generally granite, that is exposed over an area greater than about 40 square miles.

bedrock. Solid rock exposed in place or that underlies unconsolidated superficial sediments.

biotite. Dark mica, a platy mineral, is rich in potassium, iron, and magnesium. It is a minor but common mineral in igneous and metamorphic rocks.

boudinage. A structure in strongly deformed metamorphic rocks in which a layer, vein, or sill is stretched, thinned, and broken into bodies resembling boudins, or sausages.

boulder train. A linear or fan-shaped distribution of boulders and other rock fragments from the same bedrock source, transported by a glacier.

brachiopod. A double-valved marine invertebrate with a calcareous shell. Also called a lampshell.

breccia. A rock broken into angular to rounded fragments held together by a fine-grained matrix. Produced by any one of a variety of processes.

calc-alkaline. An igneous rock containing plagioclase feldspar with silica between 56 and 61 percent. Commonly found in continental crust, island arcs, and continental margins.

calcareous. Rich in calcium carbonate. Said of a rock consisting of more than 50 percent calcium carbonate.

caldera. A large, collapsed volcanic crater.

carbonaceous. Rich in carbon, such as coal.

chert. A sedimentary rock composed mainly of microscopic crystals of quartz, usually occurring as concretionary segregations in limestone.

chlorite. A green, platy, micaceous mineral characteristic of low-grade metamorphic mafic rocks—rocks with iron and magnesium.

clay. A family of mud-forming minerals commonly resulting from weathering of granite.

conglomerate. A coarse-grained sedimentary rock composed of pebbles, cobbles, or boulders set in a fine-grained matrix of silt or sand. Often called a puddingstone in Massachusetts.

continent. A large landmass that is, or was, comparable in size to a modern continent. Laurentia, a Precambrian continent, evolved into the larger North American continent.

continental rise. The gently sloping edge of the continent that is adjacent to the abyssal region of the ocean.

continental shelf. The gently dipping part of the continental landmass between the shoreline and the more steeply dipping continental slope.

continental slope. The most steeply sloped part of the continental margin, between the shelf and the rise.

crossbed/crossbedding. A sedimentary bed, usually in sand or silt, that is at an angle to the main bedding.

crust. The upper surface of the lithosphere. Continental crust consists mainly of granite, gneiss, and schist; oceanic crust consists of basalt.

dacite. A fine-grained volcanic rock intermediate in composition between the felsic volcanic rock rhyolite and the mafic rock andesite. It is the approximate chemical equivalent of the intrusive igneous rock granodiorite.

decollement. A large-scale slide associated with deformation. Overlying rocks detach from and slide along a fault surface over the underlying rocks.

delta. A nearly flat accumulation of clay, sand, and gravel deposited in a lake or ocean at the mouth of a river.

dike. A sheet of igneous rock that forms when molten magma fills a crosscutting fracture in layered rock.

diopside. A white to green metamorphic mineral of the clinopyroxene group, commonly found in metamorphosed limestone.

diorite. A plutonic igneous rock intermediate in composition between granite and gabbro.

dip. The sloping angle of a planar surface in rocks such as a sedimentary bed or metamorphic foliation.

dome. A circular or ellipical uplift of rock, sometimes in the form of an anticline or a mass of igneous rock.

drumlin. A streamlined deposit of glacial till elongated in the direction of ice movement and having an elliptical profile.

epidote. A typically pistachio green mineral formed in low-grade metamorphic rocks derived from alumina- and iron-bearing limy sediments. Also, an alteration mineral in mafic igneous rocks.

erratic, glacial. A block of rock transported by glacial ice and deposited at a distance from the bedrock outcrop from which it was derived.

esker. A long, narrow, commonly sinuous ridge deposited by a stream flowing in a tunnel beneath the ice sheet.

estuary. The lower end of a river where ocean tides mix seawater with fresh water.

fault. A fracture or zone of fractures in the earth's crust along which blocks of rock on either side have shifted.

feldspar. The most abundant rock-forming mineral group, making up 60 percent of the earth's crust and including calcium, sodium, or potassium with aluminum silicate. Includes plagioclase feldspars (albite and anorthite) and alkali feldspars (orthoclase and microcline).

felsic. An adjective used to describe an igneous rock composed of light-colored minerals, such as quartz, feldspar, and muscovite.

felsite. A general term for felsic igneous rocks whose minerals are too fine grained to be identified.

flow banding. Compositionally distinctive layers or lenses of magmatic rocks that indicate the flow direction of the magma.

fluvial. Refers to rivers or streams.

foliation. A textural term referring to planar arrangement of minerals or structures in any kind of rock.

foreset beds. Inclined sedimentary layers of sand or silt deposited on the margin of a delta.

formation. A body of sedimentary, igneous, or metamorphic rock that can be recognized over a large area. It is the basic stratigraphic unit in geologic mapping.

gabbro. A dark igneous rock consisting mainly of plagioclase and pyroxene in crystals large enough to see with a simple magnifier. Gabbro has the same composition as basalt but contains much larger mineral grains.

garnet. A family of silicate minerals with widely varying chemical compositions. Garnets occur in metamorphic and igneous rocks and are usually reddish.

gneiss. A coarse-grained metamorphic rock with a streaky grain due to parallel alignment of mineral grains, usually in bands of light- and dark-colored minerals.

Gondwana. A supercontinent of Precambrian time.

graded bed. A sedimentary bed in which particle size progressively changes, usually from coarse at the base to fine at the top.

granite. An igneous rock composed mostly of orthoclase feldspar and quartz, in grains large enough to see without using a magnifier. Most granites also contain mica or amphibole.

granodiorite. A group of coarse-grained plutonic rocks intermediate in composition between granite and diorite.

Grenville(ian). Late Precambrian gneiss and schist that metamorphosed about 1.1 to 1.2 billion years ago and formed the basement rock of the supercontinent Laurentia.

headland. A steep cliff jutting out from the coast into the ocean.

hornblende. An iron and calcium silicate mineral, the most common of the amphibole group. It commonly crystalizes into blackish needles in igneous and metamorphic rocks.

Iapetus Ocean. The ocean that existed in the general position of the Atlantic Ocean before the assembly of the continental masses that made up the Pangaean supercontinent at the end of Paleozoic time.

igneous rock. Rock that solidified from the cooling of molten magma.

kame. A variety of stratified landforms deposited by meltwater streams in contact with the ice of a glacier.

kettle. A bowl-shaped depression or hole in glacial drift, such as Walden Pond, formed by the inclusion of a block of ice within the drift. A depression forms when the ice melts.

kyanite. A blue or light green metamorphic mineral having the same chemical formula as andalusite and sillimanite, but formed at medium temperatures and high pressures.

Laurentia. The largest continental nucleus in the Rodinian supercontinent. It broke free of the supercontinent between 750 and 550 million years ago and eventually became the North American continent.

lava. Molten rock erupted on the surface of the earth.

lime silicate. A rock consisting chiefly of calcium- or magnesium-bearing silicates such as diopside formed by the metamorphism of impure limestone or dolomite. Also called calc-silicate.

limestone. A sedimentary rock composed of calcium carbonate.

limy mud. A sediment rich in calcium, iron, and alumina, the metamorphism of which produces lime silicate.

lobe. The rounded, terminal edge of a continental glacier.

longshore drift. The movement of sand and other fine-grained material along the shore by an ocean current.

mafic. An adjective used to describe an igneous rock composed of dark minerals such as hornblende, biotite, and pyroxene.

magma. Molten rock within the earth.

mantle. The part of the earth between the interior core and the outer crust.

marble. Metamorphosed limestone.

metamorphic rock. Rock derived from preexisting rock that changes mineralogically and texturally in response to changes in temperature and/or pressure, usually deep within the earth.

metamorphism. Recrystallization of an existing rock. Metamorphism typically occurs as temperature and pressure increases.

metasedimentary rock. A sedimentary rock that has been metamorphosed.

metavolcanic rock. A volcanic rock that has been metamorphosed.

mica. A family of silicate minerals, including biotite and muscovite, that crystallize into thin flakes. Micas are common in many kinds of igneous and metamorphic rocks.

micaceous. Containing micas such as muscovite, biotite, and/or chlorite.

microcline. A potassium-rich alkali feldspar, a common rock-forming mineral.

microcontinent. A small, isolated fragment of continental crust.

migmatite. A composite of metamorphic rock, commonly gneissic, mixed with igneous rock crystallized from magma melted out of or injected into the gneiss.

moraine. A landform made of glacial till, typically a ridge deposited at the edge of a glacier.

mountain building event. An event in which rocks are folded, thrust faulted, metamorphosed and/or uplifted. Intrusive and extrusive igneous activity often accompanies it.

muscovite. A common, colorless to light brown mineral of the mica group. It is present in many igneous, metamorphic, and sedimentary rocks.

mylonite. A metamorphic rock produced by shearing of rock masses past each other, resulting in brittle fracturing of certain minerals such as feldspar and micas and recrystallization of ductilely deformed quartz.

normal fault. A fault in which rocks on one side move down relative to rocks on the other side in response to extensional forces.

olivine. An iron and magnesium silicate mineral that typically forms glassy green crystals. A common mineral in gabbro, basalt, and peridotite.

orthoclase. A potassium-rich alkali feldspar, a common rock-forming mineral. It forms at higher temperatures than microcline.

ostracode. A small bivalved crustacean, often preserved as a fossil.

outwash. Sand and gravel deposited by meltwater from the receding glacier.

Pangaea. A supercontinent that assembled about 300 million years ago. It broke into the modern continents beginning about 200 million years ago.

pegmatite. An igneous rock, generally granitic, composed of extremely large crystals.

peridotite. A coarse-grained ultramafic igneous rock consisting mainly of olivine. The earth's mantle consists mainly of peridotite.

phyllite. A metamorphic rock intermediate in grade (and grain size) between a slate and a schist. Very fine-grained mica typically imparts a lustrous sheen.

plagioclase. A feldspar mineral rich in sodium and calcium. One of the most common rock-forming minerals in igneous and metamorphic rocks.

pluton. A large intrusion of igneous rock.

porphyry. An igneous rock in which larger crystals exist in a fine-grained but completely crystalline groundmass.

puddingstone. A popular term for conglomerate, especially applicable to well-rounded, brightly colored pebbles in a finer matrix, reminiscent of plums in a pudding.

pyroxene. A family of silicate minerals that occur mostly in dark, mafic igneous and metamorphic rocks.

quartz. A mineral form of silica. Quartz is one of the most abundant and widely distributed minerals in rocks. It comes in a wide variety of forms, including clear crystals, sand grains, and chert.

quartzite. A metamorphic rock, composed of mainly quartz, formed by the metamorphism of sandstone.

quartz porphyry. Term applied to rhyolites that have coarse-grained quartz crystals.

rhyolite. A felsic volcanic rock, the extrusive equivalent of granite. It contains quartz and feldspar in a very fine-grained groundmass.

rift, continental. A long, narrow trough that marks a rupture in the earth's crust.

Rodinia. Early supercontinent consisting of a cluster of all major landmasses that existed by about 750 million years ago.

sand. Weathered mineral grains, most commonly quartz.

sandstone. A sedimentary rock made primarily of sand.

saprolite. A soft, earthy, clay-rich decomposed rock formed by chemical weathering.

schist. A metamorphic rock that is strongly foliated due to an abundance of platy minerals.

serpentine. A group of rock-forming minerals derived from the alteration of iron- and magnesium-rich silicate minerals and having a greasy or silky luster and soapy feel. A rock composed of serpentine is called *serpentinite.*

shale. A deposit of clay, silt, or mud solidified into more or less solid rock.

sill. An igneous intrusion that parallels the planar structure of the surrounding rock.

sillimanite. A needle-shaped alumina silicate mineral with the same chemical composition as andalusite and kyanite. Sillimanite forms at very high temperatures and pressures of regional metamorphism and near the intrusive or contact border with igneous rocks.

silt. Weathered mineral grains larger than clay but smaller than sand.

siltstone. A sedimentary rock made primarily of silt.

slate. Slightly metamorphosed shale or mudstone that breaks easily along parallel cleavage surfaces.

sphene. A wedge-shaped accessory mineral in granitic rocks and calcium-rich metamorphic rocks.

strata. Layers of rocks deposited sequentially.

subduction zone. A long, narrow zone where a plate, usually an oceanic plate, descends into the mantle below a continental plate at a collision boundary.

submarine fan. Mass of loose rock fragments deposited at the mouth of canyons on the ocean floor.

supercontinent. A clustering of many or all continental masses on earth into one major landmass, such as has occurred at least three times in geologic history.

syenite. An igneous rock containing feldspar, plagioclase, one or more mafic minerals, and very little quartz.

syncline. In layered rocks, a folded trough with the youngest rocks in the center.

talc. A soft, platy, hydrous magnesium silicate mineral. It is commonly associated with mafic and ultramafic rocks derived from the oceanic crust.

talus. An accumulation of rock fragments derived from and resting at the base of a cliff or rocky slope.

tectonic. The forces involved in large-scale continental plate movements that create mountain building events.

terrane. A large-scale assemblage of rocks that share a more or less common origin and history.

thrust fault. A low-angle fault in which rocks above the fault moved up and over rocks below the fault.

thrust sheet/thrust slice. A body of rock above a thrust fault.

till. Unsorted and unstratified sediment deposited directly from glacial ice. Likely to contain rock fragments of all sizes.

tombolo. A sand or gravel bar that connects an island with the mainland or with another island.

topset beds. Horizontal sedimentary beds on the top surface of a delta. They usually cover foreset beds.

tremolite. A white to dark gray amphibole mineral. A constituent of commercial talc, often occurring in limy metamorphic rocks.

trench, oceanic. A narrow, elongate depression that develops where the ocean floor begins its descent into a subduction zone at a collisional plate boundary.

trilobite. A three-lobed marine arthropod that lived from Cambrian to Permian time.

trondhjemite. A very light-colored granitic rock consisting mainly of sodic plagioclase and quartz.

tuff. Volcanic ash more or less consolidated into solid rock.

twin crystals. The symmetrical intergrowth of two crystals, often reflected around a common plane.

ultramafic rocks. Black to dark green rocks that are more mafic than basalt, consisting mainly of iron- and magnesium-rich minerals such as hypersthene, pyroxene, or olivine. They make up the oceanic crust and mantle.

unconformity. A break or gap in the geologic record where one rock unit is overlain by another that is not next in the stratigraphic succession.

varves. From the Swedish *varv* meaning "layer," referring to alternating layers of fine-grained sediment deposited annually in a standing body of glacial meltwater. The lighter, silt-size particles represent the summer layer and the darker, clay-size particles represent the winter layer.

vein. A deposit of minerals that fills a fracture in rock.

ventifact. A sandblasted, wind-sculpted pebble.

vesicular. The spongelike texture of a lava containing rounded gas holes, or vesicles, produced by the expansion and loss of gases from near the top of a lava flow.

weather. To soften, crumble, or discolor because of exposure to atmospheric agents such as water.

zircon. A silicate mineral that contains uranium, an essential element in radiometric age dating. A common but minor mineral in many rocks.

SELECTED READING

Alden, W. C. 1925. The Physical Features of Central Massachusetts. *U.S. Geological Survey Bulletin* 760:13–105.

Anstey, Robert L. Jr. 1979. Stratigraphy and Depositional Environments of the Early Cambrian Hoppin Slate of Southeastern New England. *Northeastern Geology* 1:9–17.

Ashwal, L. D., W. L. Gerhard, P. Robinson, R. E. Zartman, and D. J. Hall. 1979. The Belchertown Quartz Monzodiorite Pluton, West-Central Massachusetts: A Tectonic Acadian Intrusion. *American Journal of Science* 279(8)936–69.

Bierman, P. 1986. *Lake Bascom and the Deglaciation of Northwestern Massachusetts Field Trip for NAGT.* Unpublished.

Bierman, P., and D. P. Dethier. 1986. Lake Bascom and the Deglaciation of Northwestern Massachusetts. *Northeastern Geology* 8(1/2):32–43.

Billings, M. P. 1979. Boston Basin, Massachusetts. In *The Mississippian and Pennsylvanian Carboniferous Systems in the United States—Massachusetts, Rhode Island, and Maine.* Ed. J. W. Skehan, D. P. Murray, J. C. Hepburn, M. P. Billings, P. C. Lyons, and R. G. Doyle, A1-A30. U.S. Geological Survey Professional Paper 1110-A.

Bird, J. M., and J. F. Dewey. 1970. Lithosphere Plate-Continental Margin Tectonics and the Evolution of the Appalachian Orogen. *Geological Society of America Bulletin* 81:1031–60.

Bowring, S. A., J. P. Grotzinger, C. E. Isachsen, A. H. Knoll, S. M. Pelechaty, and P. Kolosov. 1993. Calibrating rates of early Cambrian evolution. *Science* 261:1293–98.

Brigham-Grette, J. 1991. Glacial and Deglacial Landforms of the Connecticut Valley, North-Central Massachusetts. In *Geology of Western New England, NAGT Field Trip Guidebook and Proceedings.* Ed. L. Matson, 32–51. Greenfield, Mass.: Greenfield Community College.

Chute, N. E. 1959. *Glacial Geology of the Mystic-Lakes Fresh Pond Area, Massachusetts.* U.S. Geological Survey Bulletin 1061-F:186–216.

————. 1966. *Geology of the Norwood Quadrangle Norfolk and Suffolk Counties, Massachusetts.* U.S. Geological Survey Bulletin 1163-B.

Crosby, Irving. 1928. *Boston through the Ages: the Geological Story of Greater Boston.* Boston: Marshall Jones Company.

Crosby, I. B. 1939. Ground water in the Buried Valleys of Massachusetts. *Journal of the New England Water Works Association* 53(3):372–83.

Crosby, W. O., and A. W. Grabeau. 1900. Glacial Map of Lake Bouvé. Boston Society of Natural History Occasional Papers, IV:25.

Dale, T. N. 1923. *The Commercial Granites of New England.* U.S. Geological Survey Bulletin 738.

Dalziel, I. W. D. 1997. Neoproterozoic-Paleozoic Geography and Tectonics: Review, Hypothesis, Environmental Speculation. *Geological Society of America Bulletin* 109(1):16–42.

da Silva, Manuel Luciano. 1971. *Portuguese Pilgrims and Dighton Rock: The First Chapter in American History.* Bristol, R.I.: Nelson D. Martins.

Dethier, D. P., and M. Hamachek. 1998. Sedimentology of Fine Lacustrine Deposits from Glacial Lake Bascom. *Northeastern Geology and Environmental Sciences* 20(3):192–99.

Finch, R. 1996. *The Smithsonian Guides to Natural America: Southern New England, MA, CT, RI.* Washington, D.C.: Smithsonian Books; New York: Random House.

Goldstein, A. G. 1982. Lake Char Fault in the Webster, Massachusetts Area: Evidence for West-Down Motion. In *NEIGC Guidebook for Field Trips in Connecticut and South Central Massachusetts.* Ed. R. Joesten and S. S. Quarrier, 375–94. Hartford, Conn.: The State Geological and Natural History Survey of Connecticut, The Natural Resources Center.

Hartshorn, J. H. 1967. *Geology of the Taunton Quadrangle, Bristol and Plymouth Counties, Massachusetts.* U.S Geological Survey Bulletin 1163-D.

Hatch, N. L. Jr., S. A. Norton, and R. G. Clark Jr. 1970. *Geologic Map of the Chester Quadrangle, Hampden and Hampshire Counties, Massachusetts.* U.S. Geological Survey, GQ Map-858. Scale 1:24,000.

Hatch, N. L. Jr., and others, ed. 1991a. *The Bedrock Geology of Massachusetts.* U.S. Geological Survey Professional Paper 1366-A-D.

————. 1991b. *The Bedrock Geology of Massachusetts.* U.S. Geological Survey Professional Paper 1366-E-J.

Hepburn, J. C., R. Hon, G. R. Dunning, R. H. Bailey, and K. Galli. 1993. The Avalon and Nashoba Terranes (Eastern margin of the Appalachian Orogen in southeastern New England). In *Fieldtrip Guidebook for the Northeastern United States: 1993 Geological Society of America Annual Meeting and 85th Annual New England Intercollegiate Geological Conference, Boston, Massachusetts.* Ed. J. J. Cheney and J. C. Hepburn, X1-X31. Amherst: University of Massachusetts.

Hepburn, J. C., G. R. Dunning, and R. Hon. 1995. Geochronology and Regional Tectonic Implications of Silurian Deformation in the Nashoba Terrane, Southeastern New England. In *New Perspectives in the Appalachian-Caledonian Orogen.* Ed. J. Hibbard, C. R. van Staal, and P. A. Cawood, 349–65. Geological Association of Canada Special Paper 41.

Hermes, O. D., and R. E. Zartman. 1992. Late Proterozoic and Silurian Alkaline Plutons within the Southeastern New England Avalon Zone. *Journal of Geology* 100:477–86.

Hitchcock, Edward. 1841. *Final Report on the Geology of Massachusetts. Vol. I & Vol. II*. Amherst, Mass.: J. S. and C. Adams.

Hoffman, P. F. 1991. Did the Breakout of Laurentia turn Gondwanaland Inside-Out? *Science* 252:1409–12.

International Union of Geological Sciences. 1989. Global Stratigraphic Chart. *Supplement to Episodes* 12(2).

Johnson, D. 1925. *The New England–Acadian Shoreline*. New York: John Wiley & Sons.

Kales, E., and D. Kales. 1976. *All about the Boston Harbor Islands*. Boston: Herman Publishing Inc.

Karabinos, P., S. D. Samson, J. C. Hepburn, and H. Stoll. 1998. Taconian Orogeny in the New England Appalachians: Collision Between Laurentia and the Shelburne Falls arc. *Geology* 26(3):215–18.

Karabinos, P., and B. F. Williamson. 1994. Constraints on the Timing of Taconian and Acadian Deformation Western Massachusetts. *Northeastern Geology* 16(1):1–8.

Kaye, C. A. 1964a. Illinoian and Early Wisconsin Moraines of Martha's Vineyard, Massachusetts. In *Geological Survey Research*, U.S. Geological Survey Professional Paper 501-C:C140–43.

————. 1964b. Outline of Pleistocene Geology of Martha's Vineyard, Massachusetts. In *Geological Survey Research*, U.S. Geological Survey Professional Paper 501-C:C134–39.

————. 1976. Beacon Hill end moraine, Boston: New explanation of an important urban feature. *Geological Society of America Special Paper* 174:7–20.

Kelley, J. T., D. F. Belknap, and D. M. FitzGerald. 1993. Sea-Level Change, Coastal Processes, and Shoreline Development in Northern New England. In *Fieldtrip Guidebook for the Northeastern United States: 1993 Geological Society of America Annual Meeting and 85th Annual New England Intercollegiate Geological Conference, Boston, Massachusetts*. Ed. J. J. Cheney and J. C. Hepburn, G-1–G-30. Amherst: University of Massachusetts.

Koteff, C. 1964. *Geology of the Assawompset Pond Quadrangle, Massachusetts*. U.S. Geological Survey, GQ Map-265. Scale 1:24,000.

————. 1974. The Morphologic Sequence Concept and Deglaciation of Southern New England. *Glacial Geomorphology*. Ed. D. R. Coates. Binghampton, N.Y.: Publications in Geomorphology, State University of New York.

————. 1980. Deglacial History of Glacial Lake Nashua, East-Central Massachusetts. In *Late Wisconsinan Glaciation of New England*. Ed. G. J. Larson and B. D. Stone, 129–43. Dubuque, Iowa: Kendall Hunt Publishing Co.

Koteff, C., and F. Pessl Jr. 1981. *Systematic Ice Retreat in New England.* U.S. Geological Survey Professional Paper 1179.

LaForge, Lawence. 1932. *Geology of the Boston Area, Massachusetts.* U.S. Geological Survey Bulletin 839.

Larsen, F. D., and C. Koteff. 1988. Deglaciation of the Connecticut Valley: Vernon, Vermont, to Westmoreland, New Hampshire. In *NEIGC Guidebook for Field Trips in Southwestern New Hampshire and Southeastern Vermont and North-Central Massachusetts.* Ed. W. A. Bothner, 103–25. Durham, N.H.: University of New Hampshire.

Larson, G. J. 1982. Nonsynchronous Retreat of Ice Lobes from Southeastern Massachusetts. In *Late Wisconsinan Glaciation of New England.* Ed. G. J. Larson and B. D. Stone, 101–14. Dubuque, Iowa: Kendall Hunt Publishing Co.

Leatherman, S. P. 1988. *Cape Cod Field Trips from Yesterday's Glaciers to Today's Beaches.* College Park, Md.: Coastal Publication Series, Labratory for Coastal Research, University of Maryland.

Little, Richard D. 1986. *Dinosaurs, Dunes and Drifting Continents: The Geohistory of the Connecticut Valley.* Greenfield, Mass.: Alley Geology Publications.

Lougee, R.J. 1957. Hanover in the Ice Age. *Dartmouth Alumni Magazine* 50:24–29.

Lyons, P. C., and H. B. Chase Jr. 1976. Coal flora and stratigraphy of the northwestern Narragansett Basin. *NEIGC Guidebook for Field Trips to the Boston Area and Vicinity.* Ed. B. Cameron, 405–27. Princeton, N.J.: Science Press.

Massachusetts Department of Environmental Management, and Metropolitan District Commission. 1998. Brochure on Boston Harbor Islands State Park.

Mather, K. F. 1952. Glacial Geology in the Buzzards Bay Region and Western Cape Cod. In *Geological Society of America Guidebook for Fieldtrips in New England,* Field Trip no. 4, 119–42. Boulder, Colo.: Geological Society of America.

McDonald, N. G. 1982. Paleontology of the Mesozoic Rocks of the Connecticut Valley. In *NEIGC Guidebook for Field Trips in Connecticut and South Central Massachusetts.* Ed. R. Joesten and S. S. Quarrier, 143–72. Hartford, Conn.: The State Geological and Natural History Survey of Connecticut, The Natural Resources Center.

McIntyre, W. G., and J. P. Morgan. 1962. Recent Geomorphic History of Plum Island, Massachusetts, and Adjacent Coasts: Atlantic Coastal Studies. In *A Study of the Marine Resources of the Parker River–Plum Island Sound Estuary.* Ed. W. C. Jerome Jr., A. P. Chesmore, and C. O. Anderson Jr. Massachusetts Department of Natural Resources, Division of Marine Fisheries. Monograph Series no. 6.

Meissner, R., P. Sadowiak, S. A. Thomas, and BABEL Working Group. 1994. East Avalonia, the third partner in the Caledonian collisions: Evidence from deep seismic reflection data. *Geologische Rundschau* 83:186–96.

Murray, D. P. 1988. *Rhode Island: The Last Billion Years.* Kingston: University of Rhode Island.

Oldale, R. N. 1992. *Cape Cod and The Islands, the Geologic Story.* East Orleans, Mass.: Parnassus Imprints.

Oldale, R. N., and R. A. Barlow. 1986. *Geologic Map of Cape Cod and the Islands, Massachusetts.* U.S. Geological Survey Miscellaneous Investigation Series Map I-1763. Scale 1:100,000.

Oldale, R. N., and C. J. O'Hara. 1984. Glaciotectonic origin of the Massachusetts coastal end moraines and a fluctuating late Wisconsinan ice margin. *Geological Society of America Bulletin* 95:61–74.

Olsen, P. E., N. G. McDonald, P. Huber, and B. Cornet. 1992. Stratigraphy and Paleoecology of the Deerfield Rift Basin (Triassic-Jurassic, Newark Supergroup), Massachusetts. In *NEIGC Guidebook for Field Trips in the Connecticut Valley Region of Massachusetts and Adjacent States, Vol. 2.* Ed. P. Robinson and J. B. Brady, 488–535. Amherst, Mass.: University of Massachusetts.

Palmer, A. 1983. Decade of North American Geology: Geologic Time Chart. *Geology* 11:504.

Press, F., and R. Siever, eds. 1994. *Understanding Earth.* New York: W. H. Freeman and Co.

Rast, N., J. W. Skehan. 1993. Changing Tectonic Environments of the Avalon Superterrane and the Nashoba Terrane in Massachusetts. *Journal of Geodynamics* 17:1–20.

Rast, N., J. W. Skehan, and S. Grimes. 1993. Highlights of Proterozoic Geology of Boston. In *Fieldtrip Guidebook for the Northeastern United States: 1993 Boston GSA, Vol. 2.* Ed. J. J. Cheney and J. C. Hepburn, S1-S16. Amherst, Mass.: University of Massachusetts.

Raymo, C., and M. E. Raymo. 1989. *Written in Stone: A Geological History of the Northeastern United States.* Old Saybrook, Conn.: The Globe Pequot Press.

Robinson, P. 1979. Bronson Hill Anticlinorium and Merrimack synclinorium in Central Massachusetts. In *The Caledonides in the U.S.A.: Geological Excursions in the Northeast Appalachians.* Ed. J. W. Skehan and P. H. Osberg, 126–74. Weston, Mass.: Weston Observatory.

Robinson, P., and R. Goldsmith. 1991. Stratigraphy of the Merrimack Belt, Central Massachusettts. In *The Bedrock Geology of Massachusetts.* Ed. N. L. Hatch Jr., and others, G1-G37. U.S. Geological Survey Professional Paper 1366-E-J.

Rodgers, J., compiler. 1985. *Bedrock Map of Connecticut.* Connecticut Geological and Natural History Survey in cooperation with the U.S. Geological Survey.

Ross, M. E. 1994. *A Geologic Tour of the Headlands Rockport, Massachusetts.* Rockport, Mass.: Geoprof.

————. 1995a. *A Geologic Tour of Halibut Point State Park, Rockport, Massachusetts.* Rockport, Mass.: Geoprof.

————. 1995b. *A Geologic Tour of Stage Fort Park, Gloucester, Massachusetts.* Rockport, Mass.: Geoprof.

Sabin, S., and M. Whatley. 1999. *A Visitor's Guide to the Cape Cod National Seashore.* Orleans, Mass.: Eastern National, Thompson's Printing.

Skehan, J. W. 1979. *Puddingstone, Drumlins, and Ancient Volcanoes: A Geologic Field Guide along Historic Trails of Greater Boston.* Dedham, Mass.: WesStone Press.

————. 1983. Geological Profiles Through the Avalonian Terrane of Southeastern Massachusetts, Rhode Island and Eastern Connecticut, U.S.A. In *Profiles of Orogenic Belts.* Ed. N. Rast and F. Delaney, 275–300. American Geophysical Union, Geodynamics Series.

————. 1993. Walden Pond: Its geological setting and the Africa connection. In *Thoreau's World and Ours: A Natural Legacy.* Ed. E. A. Schofield and R. C. Baron. Golden, Colo.: North American Press.

————. 1997. Assembly and Dispersal of Supercontinents; The View from Avalon. *Journal of Geodynamics* 23(3/4):237–62.

Skehan, J. W., and N. Rast. 1995. Late Proterozoic to Cambrian Evolution of the Boston Avalon Terrane. In *New Perspectives in the Appalachian-Caledonian Orogen.* Ed. J. P. Hibbard, C. R. van Staal, and P. A. Cawood, 207–25. Geological Association of Canada Special Paper 41.

Skehan, J. W., N. Rast, and S. Mosher. 1986. Paleoenvironmental and Tectonic Controls of Sedimentation in Coal Forming Basins of Southeastern New England. In *Paleoenvironmental and Tectonic Controls in Coal Forming Basins of the United States.* Ed. P. C. Lyons and C. L. Rice, 9–30. Geological Society of America Special Paper 210.

Skehan, J. W., and J. W. Ring. 1982. A Field Guide to the Geology of Franklin Park. *Franklin Park Coalition Bulletin.*

Stanley, S. M., ed. 1993. *Exploring Earth and Life Through Time.* New York: W. H. Freeman and Co.

Stone, B. D., W. L. Lapham, and F. D. Larsen. 1992. Glaciation of the Worcester Plateau, Ware-Barre Area, and Evidence for the Succeeding Late Woodfordian Periglacial Climate. In *NEIGC Guidebook for Field Trips in the Connecticut Valley Region of Massachusetts and Adjacent States, Vol. 2.* Ed. P. Robinson and J. B. Brady, 467–87. Amherst, Mass.: University of Massachusetts.

Stone, B. D., and J. Peper. 1982. Topographic Control of the Deglaciation of Eastern Massachusetts: Ice Lobation and the Marine Incursion. In *Late Wisconsinian Glaciation of New England.* Ed. G. J. Larson and B. D. Stone, 145–66. Dubuque, Iowa: Kendall Hunt Publishing Co.

Strahler, A. N. 1988. *A Geologists View of Cape Cod.* East Orleans, Mass.: Parnassus Imprints.

Taylor, F. B. 1903. The Correlation and Reconstruction of Recessional Ice Borders in Berkshire County, Massachusetts. *Journal of Geology* 11:323–63.

Thompson, M. D., O. D. Hermes, S. A. Bowering, C. E. Isachsen, J. R. Besancon, and K. L. Kelly. 1996. Tectonic implications of Late Proterozoic U-Pb Zircon ages in the Avalon of southeastern New England. *Geological Society of America Special Paper* 304:179–91.

Torsvik, T. H., M. A. Smethurst, R. van der Voo, A. Trench, N. Abrahamsen, and E. Halvorsen. 1992. Baltica: A synopsis of Vendian-Permian paleomagnetic data and their paleotectonic implications. *Earth Science Review* 33:133–52.

Tougias, M., and R. Laubach. 1996. *Nature Walks in Central Massachusetts (Worcester County through the Connecticut River Valley)*. Boston, Mass.: Appalachian Mountain Club Books.

Tucker, R. D., and W. S. McKerrow. 1995. Early Paleozoic Chronology: A review in light of new U-Pb zircon ages from Newfoundland and Britain. *Canadian Journal of Earth Sciences* 32:368–79.

Wise, D. U., and E. C. Belt. 1991. Geologic cross section, northern Connecticut Valley. In *Field Trip Guidebook and Proceedings, Eastern Section of the National Association of Geology Teachers*, 1–26. Greenfield, Mass.: Greenfield Community College.

Wise, D. U., J. F. Hubert, and E. C. Belt. 1992. Mohawk Trail Cross Section of the Mesozoic Deerfield Basin: Structure Stratigraphy and Sedimentology. In *NEIGC Guidebook for Field Trips in the Connecticut Valley Region of Massachusetts and Adjacent States, Vol. 1*. Ed. P. Robinson and J. B. Brady, 170–98. Amherst, Mass.: University of Massachusetts.

Zen, E-an, ed. 1983. *Bedrock Geologic Map of Massachusetts*. Compiled by R. Goldsmith, N. M. Ratcliffe, P. Robinson, and R. S. Stanley. Reston, Va.: U.S. Geological Survey. Scale 1:250,000, 3 sheets.

INDEX

Acadian mountain building event, 11–13, 35, 60, 219–21
Acton, 176
Acushnet diorite, 134, 138
Acushnet River, 133
Agassiz, Louis, 18–19
Agawam River, 135, 165
alaskite, 134, 139, 152, 159
Alden, William C., 21, 199, 268, 280
Alleghanian mountain building event, 11–15, 38, 52, 54, 56, 60, 141, 188
Amherst, 18, 247, 249, 251, 266, 271–72
Ammonoosuc volcanic rocks, 221, 262–63, 270–71, 277–78
amphibolite, 41, 346–47
Andover, 110, 146
Andover granite, 34, 40, 42, 110, 122, 143, 146, 154–56, 172, 179
Antevs, Ernst, 31, 230, 233
Apple Island, 67
aquifers, 193, 253–54
Ararat, Mount, 89, 96
Arlington, 174–75, 180
Ashley Falls, 310
Assabet, Glacial Lake, 23, 63
Assabet diorite, 40, 42
Assabet River fault, 152
Assawompsett Pond, 165, 169
Assonet fault, 16, 34, 55, 141, 168–69, 208–9
Athol, 37, 144, 219, 262–65, 281
Athol normal fault, 262–63
Atlantic Hill, 72, 189
Attleboro, 113, 116–20
Auburn, 107, 234–35
Avalonian mountain building event, 12
Avalonian volcanic chain, 43, 46, 48–49

Avalon terrane, 10–12, 14–16, 35, 37–38, 42–53, 55, 60, 112, 115, 122–23, 129, 130, 139, 193, 196; geologic events in, 45
Ayer, 182–83, 201–2
Ayer granite, 36, 38, 110, 172, 181, 202, 234, 236, 267

Bailey, Richard, 72–73
Balance Rock State Park, 311, 317
Barnstable, 83, 85, 87–88
Barre, 232, 280
basalt, 7, 11, 73, 126, 225–26, 246–47
Bascom, Glacial Lake, 23, 297, 313, 320, 322, 330–31, 333
Bash Bish Falls State Park, 311, 316
beaches, barrier, 83, 121, 125, 142, 212–13
Beachmont, 70, 75
Beacon Hill, 1, 70, 96, 100–102, 104
Beartown Mountain, 308, 311–13
Becket, 305, 307, 342, 345
Belchertown, 271
Belchertown complex, 81, 218, 226, 240, 266, 270–71, 335
Bellingham, 158–59
Belmont, 174–75
Berkshire (town), 320
Berkshire Hills, 6, 8, 12–13, 27, 283–84. See also Berkshires
Berkshire Massif, 286, 291, 307–8, 313, 339, 283–84
Berkshires, 282–98
Bernardston, 226, 251, 257
Bernardston nappe, 218, 257
Berwick formation, 36, 38, 108–10, 144–46, 181
Beverly, 216–17

367

Billingsgate Island, 89
Bird Island, 67
Blackstone River, 35, 107, 193–98, 200, 267
Blandford, 294–95, 303–5
Block Island (R.I.), 21, 25–26, 62
Bloody Bluff fault, 16, 37, 42–45, 107–8, 122–23, 151, 156–57, 174, 176. *See also* Lake Char fault
Blue Hills, 49, 65, 133, 183–85, 188
Blue Hills quartz porphyry, 52, 59, 126, 131–33
Blue Hills volcano, 52–53, 131, 183–85, 204–6
Bonny Rigg Corner, 345
Boston, 17–18, 46, 49, 51, 61, 71, 77, 96–104, 173–74, 188. *See also* Boston Basin; Boston Harbor; drumlins
Boston Basin, 34, 45–46, 48–51, 54, 71–74, 112, 130, 133, 171, 188
Boston Harbor, 66–75, 150, 188; drumlins, 74–75; Inner Islands, 67–74; Outer Islands, 67, 74–75
Boston Harbor Islands National Recreation Area, 1, 66–67
Bourne, 83, 164–65, 169–70, 193
Bouvé, Glacial Lake, 23, 63–65, 189–90
Bouvé, Thomas T., 65, 189
Bowring, Samuel, 51
Boylston, 183, 200–201
Braintree, 189, 126
Braintree slate, 45, 51, 133, 188–89
Brewster, 87, 89
Brewster Islands, 66, 73–74
brick industry, 108, 150, 168, 174
Bridgewater, 168, 192
Brighton volcanic rocks, 45, 49–50, 71, 126
Brimfield, 239–40
Brockton, 63, 168, 208
Brodie Mountain, 320
Bronson Hill domes, 221–22, 240–41, 262–64, 270. *See also specific dome names*
Bronson Hill Upland, 27, 218–22, 231–32
Bronson Hill volcanic belt, 13–14, 35, 219, 221–22, 240–41, 262–64, 285. *See also* Laurentian terranes
Brookfield, 268–69; drumlins, 27
Buckland, William, 19
Bunker Hill, 70, 133, 187

Burlington mylonite, 44–46, 50, 106, 110–11, 126–28, 157–58, 175
Buzzards Bay, 82, 133, 135–37, 163, 169
Buzzards Bay moraine, 23, 26, 62–63, 83, 85–86, 134–36, 170
Buzzards Bay ice lobe. *See* Narragansett Bay–Buzzards Bay lobe

Calf Island, 74–75
Cambrian time, 1–2, 14–15; rocks of 49, 51–52, 56–57, 112–13, 118–19, 133, 162. *See also* fossils
Cambridge, 49, 111, 150, 171–74
Cambridge slate, 45, 49, 51, 71, 73, 75, 104, 112, 126, 171, 174
Campbell Hill fault, 38, 204
Canton, 112–15
Cape Ann, 65, 123–25, 210–17
Cape Ann granite, 33–34, 45, 53, 211, 215
Cape Cod, 29–30, 32, 35, 47, 60, 62, 81–96, 193
Cape Cod, Glacial Lake, 23, 61, 63–64, 82–83, 88
Cape Cod Bay ice lobe, 21–23, 26, 61–65, 81, 84–85, 88, 90, 123, 135, 144, 192, 208–9
Cape Cod Canal, 83, 85, 88, 169, 193
Cape Cod National Seashore, 84
Castle Island, 69–70
Castle Neck River, 213
Castle Rock, 211
Char, Lake, 195
Charlemont, 324–25, 339, 350
Charlemont fault, 295, 325, 350
Charles, Glacial Lake, 23, 63, 103, 116, 127, 130–31
Charles River, 35, 103–4
Charlestown, 133, 187
Charlton, 236
Charlton City, 237
Chatham Light, 89, 94
Chelmsford, 36, 112, 142–43, 146, 150–52
Chelmsford granite, 33, 104, 143, 145, 148–50, 191
chert, 42, 152
Cheshire quartzite, 288–89, 300, 306–7, 314, 316, 319–20, 322, 328, 339
Chester, 343, 345–49
Chester amphibolite, 292, 346–48
Chesterfield Gorge, 337–38, 350

Chicopee Brook, 273-76
Chicopee delta, 230, 242, 244–45
Chicopee River, 219, 241–42, 244, 272
Clapp, Charles H., 216
Clarksburg Mountain, 311, 328
clay, 165, 172–74
Claypit Pond, 172, 174
Clinton, 36–37, 146, 181, 183, 193, 202, 225
Clinton–Newbury fault, 6, 16, 35–38, 40, 107, 110, 121–22, 142, 145–46, 150–53, 180–81, 193, 196, 203–4, 222, 225, 235, 267
Clough quartzite, 221, 263, 277–78
coal, 1, 55, 59, 116, 163
coal basins, 54–57. *See also* Narragansett Basin; Norfolk Basin
coastal plain, ancient, 77–78
Cobble Mountain formation, 294, 300, 303–4
Cochituate, Lake, 103
Coes, Loring J. Jr., 268
Cohasset, 96, 190
Collinsville formation, 322, 324
Concord, 103, 176–77, 179
Concord, Glacial Lake, 23, 63, 127, 176–77
concrete, 33, 35, 148, 151
Connecticut River, 244, 248–50
Connecticut River valley, 6–7, 12–13, 18, 22, 30, 32, 60, 111, 220–21, 225, 227, 232, 242–57, 264–65, 335. *See also* Deerfield Basin; Hartford Basin; Northfield outlier
Connecticut Valley Border fault, 226, 241, 257, 264–65
Connecticut Valley ice lobe, 21, 23, 228, 274, 334
Cosgrove Tunnel, 37–38, 156, 194, 203–4
Coskata-Coatue Wildlife Refuge, 82
Coys Hill granite, 218, 220, 223, 234, 240, 261, 266, 269–70, 279
Crane Wildlife Refuge, 211, 213
Cretaceous rocks, 18, 76–78
Crosby, Irving, 150
Cummington, 336–39

Dale, T. Nelson, 330
Dalton formation, 288–89, 300, 311, 339–40

Dalton Valley, 320, 339–40
Danvers, 121–22, 212, 217
Dartmouth granite, 134, 139
Davis, 351
DeCordova Museum and Sculpture Park, 178–79
Dedham, 103, 129
Dedham batholith, 34, 158, 161–62
Dedham granite, 5, 44–48, 53, 71, 104, 110, 126, 129–30, 172, 174–75, 188, 190–91, 206–8
Deerfield, 231, 332
Deerfield basalt, 251–52, 256
Deerfield Basin, 14, 218, 225–27, 251–57
Deerfield River, 297, 324; West Branch of the, 296–97
Deer Island, 66–67, 70, 73, 75
Deitz, Bob, 287
deltas, glacial, 22, 30, 231, 242, 244, 253–54, 299–301. *See also* Glacial Lake Hitchcock
Dennis, 83, 85, 88–89
Devens, 182–83, 201
Devenscrest, 202
Dever State School, 170
Devonian time, 2, 10, 12, 17, 35; rocks of, 13, 223, 237, 294–95, 324, 336–37, 349–50
Dighton, 208
Dighton conglomerate, 45, 55, 58–59, 134, 140–42, 205, 208–10
Dighton Rock State Park, 208–9
dinosaur footprints, 18, 227, 247–48
Dinosaur Footprint State Reservation, 248
diorite, 7–8, 38, 206
domes. *See* Bronson Hill domes; Shelburne Falls domes
Dorchester, 185
Douglas Woods anticline, 194, 196–98
Dracut, 143, 147–48
Dracut diorite, 36, 38, 143, 146–47
Drumlin Farm Wildlife Sanctuary, 176, 181
drumlins, 22, 27–28, 124, 173, 180–81, 190, 213, 232, 268, 296, 305, 334–35, 345
Dry Hill gneiss, 220, 262–64
Dudley, 196
dunes. *See* sand dunes

Dunstable, 202
Duxbury, 184, 190

East Berlin formation, 244–47
Eastham, 85
Eastham outwash plain, 90–91
Easthampton, 246, 248
East Harbor, 96
East Lee, 312
East Lee thrust, 305
East Longmeadow, 244
East Millbury, 107
East Mountain, 244, 246, 299
Easton, 168
East Springfield, 242
East Walpole, 116
Eliot formation, 36, 143–45, 223
Elizabeth Islands, 85
Ellisville moraine, 26, 62, 134–36, 184, 193
Emerson, Benjamin K., 216
erratics. See glacial erratics
Erving, 263
eskers, 22, 25–26, 97, 103, 189, 213, 266, 268, 273–74
Essex County, 108, 216
Everett schist, 287, 300, 311, 314–16, 340
Everett thrust, 311, 314–16

Fairhaven, 134, 139–40
Fall River, 64, 133–34, 205, 208
Fall River granite batholith, 34, 45, 48–49, 53, 58–59, 134, 137–40, 161, 165, 169, 205, 209–10
Fall River spillway, 135, 166, 209
Falmouth, 85
Farley, 264
Farmington River Valley, 297, 305, 342–43
faults and thrusts, 5–6, 37–38, 49, 188, 196, 217, 235, 249, 262–63, 287, 289–91, 305–6, 314–16, 318–19, 327, 345–46. See also Bloody Bluff fault; Clinton–Newbury fault
Fish Brook gneiss, 40–42, 108, 110, 122, 143, 178
Fiske Hill, 172, 176
Fitchburg, 258–60
Fitchburg granite, 33, 38, 203–4, 218, 220, 225, 258–61
Fitch formation, 221–22, 263

florence, 335
florida, 326
Forest Lake, 278
Fort Phoenix State Reservation, 134, 139
fossils, 1–2, 17–18, 49, 57, 59–60, 81, 112–13, 118–19, 121, 123, 162, 203, 255–56, 267. See also trilobites
Foxborough, 113, 115–16, 165
Framingham, 47, 105–6, 158
Franklin, 159, 161–62
Franklin pluton, 45, 53, 161
French River, 193–95
Fresh Pond, 109, 111, 150, 171–74

gabbros, 7–8, 48, 53, 104–5, 110, 115, 122, 127–28, 174, 225
Gardner, 261
Gardner belt, 218, 222, 224, 239, 261, 269, 278–79
Gay Head, 76–78, 81
Gay Head moraine, 76
geologic time, xii, 1–2
George's Island, 66–67, 70
Georgetown, 41, 122
Giant conglomerate, 52, 59, 114, 126, 131–32, 206
Gilbertville, 278–79
glacial deposits, 23, 166, 335–36. See also lakebeds; outwash plains; till; varves
glacial erratics, 33, 89–92, 104–5, 190, 193, 209, 214, 260, 280–81, 296, 305, 317, 345
glacial lakes, 23, 30–31, 63–65, 297–98. See also specific lake names
glaciation, 18–32, 61–65, 228–33, 265, 274–76, 295–98, 305, 332–36, 342–45. See also glacial lakes; ice lobes
Glastonbury dome, 13, 218, 226–27, 240–41, 271, 277
Gleasondale, 153, 26
Glendale, 310
Gloucester, 33, 210–15
gneiss, 41, 154, 179, 196, 221, 262–64. See also specific formation names
Gondwana, 8–16, 35, 37, 43, 46, 72, 128, 219–22, 290–91
Gondwanan terranes, 14–16. See also Avalon terrane; Meguma terrane; Merrimack terrane; Nashoba terrane
Goshen dome, 337

Goshen formation, 295, 300–304, 322, 333, 336–38, 341, 349–51
Governor's Island, 67
Grafton, 105–7, 267
Granby tuff, 246–47
granite, 7, 33, 38, 42, 45, 46–50, 128, 139–40. *See also* rift granite; quarrying
Granite Railway, 185–88
Granville dome, 218, 295, 300, 303
Graves, The, 66, 74
Great Barrington, 288, 297, 312, 314
Great Falls, Glacial Lake, 23, 297, 312–13
Greenfield, 252–53, 321–24
Greenfield Community College Rock Park, 251, 256–57
Green Island, 66, 74
Green Mountain anticlinorium, 282, 291, 322, 328
greensand, 78
Grenville gneiss, 1, 13, 286, 288, 290, 300, 305–7, 310, 313, 320, 327–28, 339, 342, 345
Grenville mountain building event, 8, 12, 287, 290–91, 342
Greylock, Mount, 1, 287, 297, 318–19, 328
Greylock schist, 287, 318
Greylock thrust, 282

Hadley township, 248–49
Halibut Point State Park, 211, 216–17
Halifax, 208
Hallockville Pond gneiss, 350
Hamilton, 213
Hampden, 226
Hampden basalt, 244, 246–47
Hanging Mountain, 343
Hanover, 191
Hardwick fault, 266
Hardwick granite, 33, 218, 220, 223, 259, 261, 269–70, 278–79, 281
Hartford (Conn.), 228, 246
Hartford Basin, 11, 14, 218, 225–27, 245–50, 272
Hartshorn, Joseph, 29
Harvard, 154, 180–81
Harvard conglomerate, 35, 152
Harwich, 83, 85, 87, 89
Harwich outwash plain, 23, 83–86, 90
Hatch, Norman, 295

Hatfield, 248, 332
Hawley formation, 293–94, 300, 303–4, 322, 325, 339, 345, 350
Heath, 341
Hemlock Gorge, 97, 104
Hindale gneiss, 311
Hingham, 71, 188–90
Hingham granite, 184, 190–91
Hitchcock, Glacial Lake, 22–23, 31, 228–33, 242–45, 251, 272, 297, 332–35; deltas of, 244, 253–54, 299, 301
Hitchcock, Rev. Edward, 19–20, 228, 247–48, 272, 284
Hitchcock volcanic rocks, 246–47
Hixville fault, 134, 139
Hockomock Swamp, 135, 165–66, 168, 208
Hoffman, Paul, 51
Hog Island, 213
Hog Rock moraine, 135–36
Holbrook, 189
Holden, 226
Holyoke, 225
Holyoke basalt, 244, 246–478
Holyoke Range, 226, 228, 272, 326
Hoosac formation, 292, 322, 327, 339, 345
Hoosac Range, 282, 285–86, 292
Hoosac Tunnel, 322, 326
Hoosic River, 328, 330
Hope Valley alaskite gneiss, 106, 152, 158–59, 196–98
Hopkinton, 103
Hoppin Hill, 113, 118–20, 162
Houghs Neck, 71
Housatonic, Glacial Lake, 23, 297, 308, 312–13, 330, 333
Housatonic River, 307–8, 310–12
House Rock, 184
Hudson, 41, 151, 155
Hudson Valley ice lobe, 21, 23, 26, 296–97, 312, 334
Hull, 67, 188, 190
Huntington, 349
Hutton, James, 3
Hyannis, 83

Iapetus Ocean, 10, 13, 220
ice ages, 20–32
ice lobes, 61–62, 120. *See also specific lobe names*

igneous rocks, 3, 7–8, 38, 41, 47, 91, 115, 222, 338–39. *See also* gabbros; granites
Illinoian ice sheet, 20–21, 27, 61, 78, 80, 232
Indian Head Hill diorite, 40, 42
Indian Head Hill granite, 40, 42, 60, 156, 176
Ipswich, 123, 212–13
Ipswich River, 122
iron mining, 1, 148, 284, 309–10, 330
Islands, The. *See* Martha's Vineyard; Nantucket

Jahns, Richard, 199
Johnson, Douglas, 135
Jones River, 63–65, 135, 168, 192, 209
Jurassic time, 11, 60

kames, 22–23, 29–30, 33, 62–63, 85–86, 108, 143, 166–69, 177, 183, 192, 201, 266, 269, 275–76, 278–79, 297, 330
Kempfield dome, 262–63
kettles, 22, 29, 81, 84–87, 90, 103, 111–12, 135, 169, 171–74, 242, 254
Kingston, 63, 192
Kittery quartzite, 36, 120–21
knob-and-kettle topography, 85–86, 305, 345. *See also* kames; kettles
Koteff, Carl, 26, 199

lakebeds, 22–23, 31
Lake Char fault, 194, 196–98. *See also* Bloody Bluff fault
Lakeville, 208
Lancaster, 181, 202
Landing, Edward, 189
Lanesborough, 320, 340
Lanesville, 215
Laurel Lake, 308, 312, 316
Laurentia, 8, 10–13, 35, 46, 219–22
Laurentian terranes, 14–16, 285–95
Lawrence, 142, 144–46, 150–51
Leadmine Hill, 238–39
Lebanon Springs, 340
Lee, 307–8, 312
Lenoxdale Mountain moraine, 297
Leominster, 171–72, 180, 182–83, 201–2, 258
Lexington, 43, 174
limonite, 300, 309, 330

Lincoln, 178, 181
lime silicates, 172, 290
Lithia, 336
Little Brewster Island, 66, 74
Little Calf Island, 66, 74
Little Quittacas Pond, 165, 169
Littleton, 179, 181
Littleton schists, 203, 223–24, 234, 238, 258–63, 335
Long Island (Boston Harbor), 66–67, 69–70, 73
Long Island (N.Y.), 20–21, 84–85
Long Island Sound, 43, 198
longshore currents, 83, 91–94
Longmeadow, 243–45
Lovell Island, 66–67
Lowell, 110, 143, 146, 151
Lowell National Historic Park, 145
Ludlow, 240–42
Lunenburg, 203
Lyell, Sir Charles, 19, 78
Lynnfield ultramafic deposit, 217
Lynn volcanic rocks, 45, 49–50, 109–10, 211, 217

Magnolia, 213
Mahkeenac, Lake, 308, 312, 317
Manchester-by-the-Sea, 212–13, 217
manganese, 339, 350
Manomet River. *See* Monument River
Mansfield, 55, 113, 116, 159, 163
marble, 300, 307, 312. *See also* Stockbridge marble; Walloomsac formation
Marblehead, 32, 211, 217
marine sediments, 23, 123, 144
Marion, 134
Marlboro formation, 42, 44, 97, 107–8, 152, 155–56, 178–79
Marlborough township, 157
marshes, 82, 96, 138, 165
Marshfield, 190
Martha's Vineyard, 18, 20, 22, 61, 76–82, 84, 91
Martha's Vineyard moraine, 23, 25–26, 62, 76, 78, 81, 134
Mashpee pitted plain, 23, 29, 63–64, 83–87, 136
Massabesic complex, 13, 34, 38–39, 203–4, 218
Masslite Quarry, 57, 159, 163

mastodon teeth, 32
Mattapoisett, 134
Mattapan volcanic rocks, 45–46, 49–50, 129
Medford, 109–10
Medford dike, 60, 109, 111
Medway, 103
Meguma terrane, 10–11, 13–16, 35, 43, 56, 60, 128
Melrose, 112
Menotomy Rocks Park, 172, 175
Merrimack, Glacial Lake, 23, 202
Merrimack River, 35, 108, 111–12, 124–25, 142, 145–46, 150–51, 173
Merrimack terrane, 10, 14–16, 35–36, 38, 41, 60, 108–10, 120–21, 144–45, 181–82, 193, 196, 202–3, 222–25. See also specific belt names
Merrimacport, 145, 150
Mesozoic time, 2, 11, 18, 225–27
metamorphic rocks, 4, 37, 48
Methuen, 108–9, 143–44, 150
Methuen, Glacial Lake, 23, 108, 143, 150
Middleboro, 163–66, 168
Middleborough Township, 165, 169, 208
Middleborough moraine, 26, 63, 135–36
Middlefield granite, 292, 296, 341, 347, 349
Middlefield thrust, 16, 286, 292, 300, 305, 339, 345
Middlesex County, 108
Middlesex Fells Reservation, 111
Middleton, 122, 217
Middleton Basin, 34, 45, 60, 121, 123
migmatites, 4–5, 12, 38
Milford, 103, 160
Milford granite, 33, 45–48, 53, 97, 105–6, 158–61, 195–96, 198
Milford granite (N.H.), 204
Millers Falls, 264
Millers River, 35, 219, 265, 272, 281
Millis, 161
Mill River, 296, 335–36
Millstone Hill granite, 194, 202, 266
Milton, 188
Minechoag Mountain, 241
mineralogy, 4, 7–8, 42, 259, 260–61, 290, 347, 350
mining: asbestos, 311, 320; copper, 350–51; emery, 347–49; graphite,

238–39; limonite, 330; manganese, 339, 350; nickel, 148; talc, 311, 320. See also iron
Minute Man National Historical Park, 43, 172, 176
Mohawk Trail, 324–25
Monks Hill moraine, 23, 26, 136, 193
Monomoy Island, 89, 93–94
Monomoy Island National Wildlife Refuge, 84
Monponsett Pond, 168
Monson, 274–76
Monson dome, 218, 240, 262–63, 270–71, 277
Monson gneiss, 221–23, 240, 262–63, 270, 277–78
Monument Mountain, 308, 312, 314
Monument River, 82, 88
Moon Island, 66, 69–70
moraines, 21–25, 134–35, 193, 296–98. See also Sandwich moraine; Buzzards Bay moraine; Martha's Vineyard moraine
Moretown formation, 292–93, 304, 322, 325–26, 345
mountain building events, 12–13
Mount Hope Bay, 47, 133, 139, 141, 165–66, 208
Mount Hope fault, 49, 188
Mount Toby formation, 252–53
mud balls, 257
Muddy Brook, 278
Muehlberger, William, 199–200
Myles Standish State Forest, 169, 193
mylonitic rocks, 4–5, 37–38, 40, 44, 47, 152, 155, 157–58, 175, 196–98
Mystic Lakes, 109, 111, 150, 173
Mystic River, 103

Nahant, 49, 66
Nantasket Beach, 66–67, 71–72, 190
Nantucket, 18, 20, 23, 25, 61–62, 77–82, 84, 91
Nantucket moraine, 23, 26, 79
nappes, 220–21, 224
Narragansett Bay, 133, 135
Narragansett Bay–Buzzards Bay ice lobe, 21, 23, 26, 61–63, 65, 81, 112, 116, 120, 130, 134–35, 190, 192
Narragansett Basin, 17, 46–47, 54–59,

113, 134–42, 140–42, 162–65, 191–92, 207–8
Narragansett Pier granite, 13, 56, 139–40
Nashoba belt, 181–82
Nashoba formation, 41–42, 97, 107, 146, 154–55, 172, 179, 235
Nashoba terrane, 10, 14–16, 35, 37–44, 60, 107–8, 110, 122, 142, 146, 151–53, 176, 178–81, 193, 196
Nashua, Glacial Lake, 23, 63, 65, 182–83, 198–202
Nashua Basin, 198
Nashua belt, 35–36, 38, 202–4, 222, 225, 235–36
Nashua River valley, 35, 182–83, 193, 198–202, 268
Natick, 47, 103, 105
Native Americans, 50, 67, 113, 131, 147, 174, 195, 238, 240, 320, 324–25
Natural Bridge State Park, 322, 328–29
Nauset Beach, 84, 86, 89–91, 94, 96
Nauset fault, 16, 34, 60
Nebo, Mount, 158, 161
Needham, 99, 129
Neponset, Glacial Lake, 23, 63, 116, 127
Neponset fault, 49, 188
Neponset River, 35, 116, 187
New Bedford, 134, 139
New Boston, 343
Newbury, 121–22, 146
Newburyport, 121, 124–25, 150–51
Newburyport batholith, 34, 120–21, 143–45
Newbury volcanic rocks, 38, 40, 121, 123
New Haven arkose, 244–47, 299–300
Newton, 103, 130
Newton Corner, 104
Newton Highlands, 103–4
Newton Lower Falls Park, 104
Newtonville, 97, 102
Nickerson State Park, 87
Nippenicket, Lake, 168
Nissitissit River, 198, 202
Nobscot Hill, 97, 106–7, 158
Nobscot mylonite, 47, 106, 157–58
Nomans Land, 76, 80
Norfolk Basin, 17–18, 46, 54–55, 58–59, 112–15, 131, 133, 162, 188, 204. *See also* coal
North Adams, 288, 329

Northampton, 226, 243, 245, 248–51, 333–35
North Andover, 142, 146, 150
North Attleboro, 118–19, 162
Northborough, 156
North Chelmsford, 173
North Dighton, 208
North Eastham, 90
North Easton, 207
Northern Border fault, 16, 34, 49, 110, 172, 174–75
Northfield outlier, 257
North Harwich, 88
North Nashua River, 198
North Oxford, 193
North Randolph, 206
North Scituate Basin (R.I.), 54–55
North Spencer, 226
North Truro, 91
North Weymouth, 190
North Wilbraham, 242
Norton, 55, 159, 162–66, 208
Norwood, 113, 116, 129
Nut Island Tunnel, 51, 73–74

Oakdale formation, 35–36, 38, 172, 180–82, 202, 204, 223, 235–36, 268
Ordovician time, 12–15, 285; granites of, 305, 341, 343–44
Orleans, 86, 92
Oshburn Bog, 169
Otis, 306–7, 343
outwash plains, 22, 27, 62–63, 88, 90, 108, 169. *See also* kames; kettles; sand and gravel
Oxbow Lake, 249–50
Oxbow National Wildlife Refuge, 182
oxbows, 248–50, 311–12
Oxford, 196, 198, 235, 237

Paleozoic time, 2, 10, 13, 37
Palmer, 233, 240, 242, 275–76, 278
Pangaea, 8, 10–11, 13, 111, 225
Parker River fault, 143
Parker River National Wildlife Refuge, 121, 213
Partridge formation, 221, 239, 262–63, 270, 278
Paskamanset River, 133
Paxton, 237

Paxton schists, 182, 202–3, 220, 223–25, 234, 236–37, 239, 258–62, 268; Bigelow Brook member, 234, 237–38
Peabody, 120–21, 123, 126, 210–11, 214–15
Peabody granite, 34, 45, 53, 127, 212, 215
Pearl, Lake, 162
peat, 56, 81, 108, 116, 135, 137, 141, 168, 238
Peddocks Island, 66–67, 70–71
Pelham dome, 218, 220, 222, 226, 241, 262–64, 270–72, 323
Pemberton Hill, 100, 102
Pennsylvanian coal basins, 35, 38, 58–59, 112, 134, 267
Pennsylvanian time, 13, 17, 35
Peper, John, 233
Permian time, 11, 13
Petersham, 280–81
Phillipston, 261
Pilgrim Heights, 92, 96
Pilgrim Lake, 89, 94–96
Pine Hills moraine, 184, 193
Pine Swamp, 166–67
Pin Hill, 35, 180, 201
Pittsfield, 297, 333–34
Plainfield, 349
Plainfield quartzite, 196–98
Plainville, 57
plate tectonics, 6–7, 52
Pleistocene Ice Age, 2, 18, 33. See also glaciation
Plum Island, 121–25, 213
Plymouth, 169, 192–93
Plymouth Rock, 33, 184, 190
Plympton, 192
Pondville conglomerate, 55, 58–59, 114, 119, 126, 162, 206
Ponkapoag fault, 16, 34, 48–49, 188–89
Poplar Mountain gneiss, 262, 264
Portland formation, 241–42, 245–46, 248
potholes, 195
Pownal, 320
Pratt Museum at Amherst College, 18, 247, 251, 272
Precambrian time, 1–2, 8, 12–14, 43, 46, 72
Princeton, 258, 260
Province Lands, 87, 93–95

Provincetown, 88–89, 94, 96
Provin Mountain, 244, 246, 299
puddingstone. See Roxbury conglomerate
Purgatory Chasm, 194, 198

Quabbin Reservoir, 270–71, 279–80
Quaboag, Glacial Lake, 23, 273–75
Quaboag River, 35, 240, 242
quarrying, 33, 36, 133, 147–48, 160, 163, 183, 185–88, 202, 215–16, 258, 329–30, 343
Quidnet outwash, 79, 81
Quinapoxet River, 198, 200
Quincy, 1, 67, 71, 183–89
Quincy granite, 1, 33–34, 45, 52–53, 68, 104, 126, 131, 133, 183–85
Quinebaug formation, 197
Quinsigamond, Lake, 200, 203, 268
Quittacas moraine, 134, 136

Race Point, 89, 93–94
radiometric age dating, 2, 13, 36, 41, 230
Rainsford Island, 66, 69
Ragged Hill gneiss, 273, 279
Randolph, 131, 183, 189, 205–6
Rattlesnake Hill granite, 115, 204
Raynham, 27
Raynham Center, 166
Reading, 215
redbeds, 112–15. See also Wamsutta redbeds
Reed Brook Preserve, 326
Rehoboth 208
Renfrew, 330
Rhode Island, 47, 54–56
Rhode Island batholith. See Milford granite
Rhode Island formation, 45, 55, 58–59, 116–18, 140–41, 165
rhyolite, 7, 121
Richard T. Crane Jr. Memorial Reservation, 213
Richmond Furnace, 1, 309–10
rift basins, 6, 54–55, 225–27, 272, 321, 335. See also Boston Basin; Pennsylvanian coal basins
rift granite, 52–53, 215
Rings Island, 145
Riverside, 256
Rochester township, 135

Rock House Reservation, 266, 269–70
Rockingham belt, 36–38, 108, 144–46, 151, 181, 202, 222, 225
Rocking Stone Public Park, 273, 281
Rockland, 189
Rockport, 212–17
Rodinia, 8, 10, 12
Ronkonkoma moraine, 21, 25–26
Rowe-Hawley volcanic belt, 14, 16, 282, 303, 305, 325–26, 338
Rowe schists, 292, 300, 325–26, 341, 345–46
Rowley, 123
Roxbury conglomerate, 45, 49–51, 71, 97, 104, 126, 130; Dorchester member, 51, 70; Squantum member, 51, 70
Roxbury flats, 99
Royalston, 273, 281
Rum Rock, 273, 280
Russell, 303

Sabbatia, Lake, 168, 170
Sachuest conglomerate, 45, 58–59, 140–41
Sagamore Bridge, 184, 193
Salem gabbro, 211
Salem Neck, 216–17
Salisbury, 37, 120–21, 142–45
Salisbury Beach, 142. See also Plum Island
Salt Pond, 89–92
sand, 31–32, 63, 93, 170, 244–45
sand and gravel, 22, 30, 33, 62, 131, 182–83, 192–93, 298. See also deltas; outwash plains
sand dunes, 87–89, 94–96, 242–45
Sandisfield, 341
Sand Spring, 320–21
Sandwich moraine, 23, 26, 62–63, 84–85, 88–89, 134–35, 170, 193
Sandy Neck, 83, 93
Sangamon interglacial stage, 78–80
Sankaty Head, 61, 78, 80
Sayles, S. A., 72
Scarborough, 103
Schenk, Paul, 60
schists, 41, 154, 235, 260–61
Scituate, 185, 190–91
Scorton moraine, 184, 193

Scotland Road fault, 121
Scusset River, 82
Seabrook, 121
sea level changes, 65, 81, 87, 91–94, 102
Sears, John Henry, 216–17
Seekonk, 134, 140–42
Shaker Village, 340
Sharon, 115
Sharon Upland, 34, 54, 115, 188, 204
Sharpners Pond diorite, 34, 40, 42, 110, 122
Shawmut, 100
Shawsheen gneiss, 40–41
Shawsheen-Merrimack, Glacial Lake, 23, 63, 108–9, 150
Shawsheen River, 108, 150
Sheffield, 310–11
Shelburne Falls, 324
Shelburne Falls domes, 282, 295, 303, 337, 349
Shelburne Falls volcanic belt, 13–14, 282, 285, 291, 294–95, 305, 321–22, 324, 336–38
Shirley, 182–83
Shrewsbury, 146, 155, 268
Shuttle Meadow formation, 227, 245–46
Siasconset outwash deposit, 81
Silurian time, 15, 17
Silver Lake, 168, 192
Sippican River, 133
Smith College, 229, 332
Snipatuit moraine, 136
soapstone, 293
Somerset, 208
Somerville, 112, 171
South Barre, 279
Southborough, 37, 47, 156, 203
Southbridge belt, 222–25, 236–38
Southbridge syncline, 258, 268
South Channel lobe, 21, 23, 61–62, 81, 84, 88, 90
South Chatham, 87
South Dartmouth, 139
South Deerfield, 253
South Hadley, 227
South Hingham, 65
South Monson, 233, 273, 275
South Norwood, 116
South Wellfleet, 90
South Worthington, 349–50

Spencer, 268
spillways, 30–31, 92, 135, 166, 209–10
Springfield, 232, 234, 242, 244
Spy Pond, 150, 172–74
Squannacook River, 198
Squantum Head, 66, 71–73
Squibnocket moraine, 80
Stage Fort Park, 211
Stamford gneiss, 291, 339
Stillwater River, 198, 200
Stockbridge, 288, 308, 312
Stockbridge Bowl, 317
Stockbridge marble, 289–90, 300, 306–8,
    319–20, 322, 329, 339
Stony Brook fault, 115, 126
Straw Hollow diorite, 40, 42, 60, 156
Sturbridge, 147, 237–39
Sudbury, 42
Sudbury, Glacial Lake, 23, 63, 97, 103,
    127, 176–77
Sugarloaf formation, 251–53, 321
Sugarloaf Mountain State Reservation,
    251, 253
Sunderland delta, 231, 251, 253–54
supercontinents, 8–11
Swansea, 141–42
Swift River, 242, 280, 337
syenite, 115, 216

Taconic mountain building event, 12–
    13, 285–87, 291–95
Taconic Range, 6, 287–88, 340
Taconic thrust sheet, 14, 16, 282
Tadmuck Brook schist, 40–41, 146, 153–
    54, 178–80
talc, 293, 311, 320, 347
Tatnic Hill formation, 196–98
Taunton, 27, 29, 168, 208
Taunton, Glacial Lake, 23, 63–65, 135,
    165–68, 192, 208–9
Taunton River Basin, 35, 135, 140–41,
    165, 208
Templeton, 259–62
terrane. See Avalon terrane; Meguma
    terrane; Merrimack terrane; Nashoba
    terrane
thermal springs, 320–21
Three Rivers, 242
thrusts. See faults and thrusts

till, glacial, 25, 69–70, 89–91, 165, 173,
    213–15, 231–32, 281. See also
    drumlins
Toby, Mount, 251, 254
Tom, Mount, 245, 247
tombolos, 68–69, 74, 190–91
Topsfield, 121, 123
Topsfield granodiorite, 121–22
Toulmin, Priestly, III, 216
Townsend, 194, 204
Tremount, 99, 100
trilobites, 14, 17, 51–52, 133, 189, 289
Tucker, Robert, 51
Tuckernuck Island, 79
Tully dome, 218, 262, 281
Tully River, 281
Turners Falls, 252, 255–56, 259, 262
Turners Falls sandstone, 252, 255–57
Tyngsboro, 226
Tyringham gneiss, 291, 300, 341–44
Tyringham-Plainfield moraine, 296–97

ultramafic rocks, 292–93, 320, 347
uniformitarianism, 3–4

Van Deusenville, 310, 312
varves, glacial, 229–30
Vermont-Stockbridge Valley, 283–84
Vernon, Mount, 100
volcanic rocks, 7–8. See also igneous
    rocks; Blue Hills volcano; Avalon
    volcanic chain

Wachusett Lake, 260
Wachusett-Marlboro Tunnel, 37–38,
    156, 203–4
Wachusett Mountain, 21, 258, 260
Wachusett Mountain belt, 13, 181–82,
    199, 202, 204, 222–224, 236, 258–60,
    268
Wachusett Reservoir, 37, 97, 198–99, 200
Wahconah Falls State Park, 333, 340
Waits River formation, 271, 295, 333,
    336–37
Wakefield, 127
Walden Pond, 29, 177–78
Walden fault, 110, 126
Waldo Lake, 206
Walloomsac formation, 286, 290, 306,
    309, 311, 316–17

Walpole, 116
Waltham, 103, 128–29, 174
Wamsutta redbeds, 54–55, 58–59, 113–14, 119, 126, 159, 162, 205–6
Ware, 219, 222–23, 270, 279
Ware belt, 239, 261, 269, 278–79
Wareham, 133–35, 137
Wareham pitted outwash plain, 63–64, 135–36, 169
Wareham River, 169
Ware River, 242
Warner, Mount, 225, 249
Warren, 239–240, 279
Warwick dome, 218, 262
Washington, Henry S., 216
Washington, Mount, 314
Washington gneiss, 290, 339, 341–44
Wayland, 5, 103, 128
Webb State Park, 71
Webster, 193–98, 202
Webster Lake. See Char, Lake
Weir River, 190
Wekepeke fault, 16, 38, 172, 182, 204, 267
Wellesley, 103
Wellfleet outwash plain, 89–90
Wells State Park, 238
Wendall Depot, 263
Wenham, 215
Wequaquet Lake, 83, 87
West Acton, 179
West Berlin, 204
Westboro quartzite, 45–47, 50, 97, 105–6, 110, 128, 152, 158–59, 160–61
Westborough, 151–52, 158–61
West Bridgewater, 168
West Brookfield, 224
West Cummington, 338–39
Westerly granite, 13
Westfield, 299–301, 304
Westfield delta, 299–301
Westfield River, 231, 244, 296–97, 339, 345
Westford, 202
Westminster, 258
West Newbury, 122
Weston, 104–5, 127, 129–30
Weston Observatory, 17, 57, 128, 149, 229
West Quincy, 65, 188

West Springfield, 242
West Stockbridge, 300, 304, 309–10
West Stockbridge Mountain, 308–9
West Summit Overlook, 327–28
West Wareham, 169
West Warren complex, 240, 279
Westwood, 129
Westwood granite, 45–46, 49–50, 110, 126, 129, 162, 206
Weweantic River, 169
Weymouth, 49, 189–90
Weymouth formation, 45, 49, 51, 189
Weymouth granite, 104
Whatley, 332
Whatley thrust, 16, 282, 335
Whipples, Glacial Lake, 23, 273, 278
Whitcomb Summit, 325–27
Whitcomb Summit thrust, 16
White's Pond, 178
Wilbraham, 241–42
Williamsburg granite, 300, 303, 333, 337
Williamstown, 311, 320, 322–23
Wilmington, 111
Winchester, 151
Windsor, 290, 295
Windsor State Forest, 333, 339
Winimussett, Glacial Lake, 23, 273, 279
Winnetuxet River, 165, 208
Winthrop, 66–67, 70, 75
Winthrop Peninsula, 75
Wisconsinan ice age, 1, 20–21, 27, 61, 78, 80–82, 231–32. See also glaciation
Woburn, 110–11, 150–51
Wolfpen lens, 152, 157
Woonsocket Basin, 46, 54–55, 159
Worcester, 27, 35, 38, 182, 195, 198–99, 202–3, 225, 236, 266–68
Worcester formation, 35, 38, 172, 181, 202, 235
Worlds End Reservation, 66, 71, 189
Woronoco, 303
Wrentham, 113, 162

Yarmouth, 85
Yarmouth Port, 88

Zoar, 325

ABOUT THE AUTHOR

Longtime teacher and researcher, Jim Skehan is professor emeritus in the Department of Geology and Geophysics at Boston College and director emeritus of Weston Observatory. The National Association of Geology Teachers named Skehan Teacher of the Year in 1976. Skehan holds a doctorate in geology from Harvard University as well as a master's in theology from Weston College. As a Jesuit priest and geologist, he actively promotes dialogue between scientists and theologians. While studying Iceland's active geology in 1970, he officiated the first mass ever celebrated on Surtsey Volcano, a newly formed island in the North Atlantic.

We encourage you to patronize your local bookstore. Most stores will order any title they do not stock. You may also order directly from Mountain Press, using the order form provided below or by calling our toll-free, 24-hour number and using your VISA, MasterCard, Discover or American Express.

Some geology titles of interest:

____ROADSIDE GEOLOGY OF ALASKA	18.00
____ROADSIDE GEOLOGY OF ARIZONA	18.00
____ROADSIDE GEOLOGY OF S. BRITISH COLUMBIA Can. $25.00 / U.S.	20.00
____ROADSIDE GEOLOGY OF COLORADO, 2nd Edition	20.00
____ROADSIDE GEOLOGY OF HAWAII	20.00
____ROADSIDE GEOLOGY OF IDAHO	20.00
____ROADSIDE GEOLOGY OF INDIANA	18.00
____ROADSIDE GEOLOGY OF MAINE	18.00
____ROADSIDE GEOLOGY OF MASSACHUSETTS	20.00
____ROADSIDE GEOLOGY OF MONTANA	20.00
____ROADSIDE GEOLOGY OF NEBRASKA	18.00
____ROADSIDE GEOLOGY OF NEW MEXICO	18.00
____ROADSIDE GEOLOGY OF NEW YORK	20.00
____ROADSIDE GEOLOGY OF NORTHERN and CENTRAL CALIFORNIA	20.00
____ROADSIDE GEOLOGY OF OHIO	24.00
____ROADSIDE GEOLOGY OF OREGON	16.00
____ROADSIDE GEOLOGY OF PENNSYLVANIA	20.00
____ROADSIDE GEOLOGY OF SOUTH DAKOTA	20.00
____ROADSIDE GEOLOGY OF TEXAS	20.00
____ROADSIDE GEOLOGY OF UTAH	20.00
____ROADSIDE GEOLOGY OF VERMONT & NEW HAMPSHIRE	14.00
____ROADSIDE GEOLOGY OF VIRGINIA	16.00
____ROADSIDE GEOLOGY OF WASHINGTON	18.00
____ROADSIDE GEOLOGY OF WISCONSIN	20.00
____ROADSIDE GEOLOGY OF WYOMING	18.00
____ROADSIDE GEOLOGY OF THE YELLOWSTONE COUNTRY	12.00
____EVIDENCE FROM THE EARTH: Forensic Geology and Criminal Investigation	20.00
____FINDING FAULT IN SOUTHERN CALIFORNIA: An Earthquake Tourist Guide	18.00
____GEOLOGY OF THE LAKE SUPERIOR REGION	22.00
____GEOLOGY OF THE LEWIS & CLARK TRAIL IN NORTH DAKOTA	18.00
____GEOLOGY UNDERFOOT IN CENTRAL NEVADA	16.00
____GEOLOGY UNDERFOOT IN DEATH VALLEY AND OWENS VALLEY	16.00
____GEOLOGY UNDERFOOT IN ILLINOIS	15.00
____GEOLOGY UNDERFOOT IN SOUTHERN CALIFORNIA	14.00
____GEOLOGY UNDERFOOT IN SOUTHERN UTAH	18.00
____GLACIAL LAKE MISSOULA AND ITS HUMONGOUS FLOODS	15.00

Please include $4.00 per order to cover postage and handling.

Send the books marked above. I enclose $_____

Name_____

Address _____

City/State/Zip _____

☐ Payment enclosed (check or money order in U.S. funds)

Bill my: ☐ VISA ☐ MasterCard ☐ Discover ☐ American Express

Card No. _____ Expiration Date:_____

Signature _____

MOUNTAIN PRESS PUBLISHING COMPANY
P.O. Box 2399 • Missoula, MT 59806 • Order Toll-Free 1-800-234-5308
E-mail: info@mtnpress.com • Web: www.mountain-press.com